Fundamentals of Freeze-Drying

Fundamentals of Freeze-Drying

J. D. Mellor

Commonwealth Scientific and Industrial Research Organization
Division of Food Research
North Ryde, New South Wales
Australia

1978

ACADEMIC PRESS
LONDON NEW YORK SAN FRANCISCO
A Subsidiary of Harcourt Brace Jovanovich, Publishers

Academic Press Inc. (London) Ltd
24–28 Oval Road
London NW1

US edition published by
Academic Press Inc.
111 Fifth Avenue,
New York, New York 10003

Library of Congress Catalog Card Number: 77-94297
ISBN: 0-12-490050-X

Printed in Great Britain by
J. W. ARROWSMITH LIMITED
Winterstoke Road
Bristol BS3 2NT

Foreword

If we investigate into the past history of mankind, we find the basic need of preserving alterable natural products which are used essentially as food. Refrigeration dates back to pre-Roman times and was routine practice under Nero, and stabilization of perishable goods by drying has been resorted to by primitive people for millenniums. However, the combination of both is a rather recent achievement in technology and, in fact, it has been in wide usage only in the last two or three decades. By freezing unstable aqueous material and subsequently drying it by sublimation from the frozen state, one can prepare stable, dehydrated products which can be stored at room temperature over extended periods, keeping the integrity of their biological, nutritional and organoleptic properties provided they are protected from oxygen, moisture and light.

This elegant technique, often referred to as freeze-drying or lyophilization, was born, originally, in the biomedical field where it played a major role for the preparation and preservation of blood plasma, with E. W. Flosdorf, R. I. N. Greaves and F. Henaff, and of antibiotics, with E. B. Chain, during the last world war. Since then, it has known a continuous success in many new areas ranging from vitamins to tissue extracts, steroids to vaccines, and recently pure chemicals to foods. Curiously enough, the food applications which look the most promising have been also the most difficult to mature despite the fact that, historically, they might have been the very first. It is claimed, indeed, that in pre-Columbian times, the Incas were securing their food into large depots, the tambos, after stabilization of the products by sun drying in the heights of the Andes; low temperature, high altitude, progressive dehydration by sublimation of ice by radiant heat under reduced pressure—here are exactly the conditions of freeze-drying.

Lyophilization has been practised for a long time as a routine operation based on daily experience. However, the ever growing complexity of the treated materials, the increasing demand for more standardized, high quality, efficiently finished products and the constant evolution of the freeze-drying plants themselves have made it essential to approach the whole process in a precise, systematic, scientific way. This is why, starting in 1958 with the first International Courses of Lyophilization, a sustained effort has been made to understand the physicochemical basis of the process and to monitor its industrial applications so that they fit, exactly, their specific goals.

Basically an engineer, but with a deep knowledge of biology and food science, J. D. Mellor has been able, because of his solid scientific background, to present an excellent book on the "Fundamentals of Freeze-Drying". This was a difficult, painstaking and time-consuming task, requiring a thorough analysis of the published works followed by a clever synthesis in order to present a clear comprehensive view of a technique which requires a very wide multidisciplinary approach.

Mellor's book has met this challenge and I am sure that it will prove most useful to all those dealing with freeze-drying in the laboratories, in the design and development offices of equipment makers and in the processing plants. It will also, undoubtedly, be of very great help for university students and engineers who are keen to remain in this challenging area.

July 1978

LOUIS REY
Nestlé S.A.,
Vevey,
Switzerland.

Preface

Until recently few people would have thought it possible that living things could be artificially desiccated and later restored to their original state. However, the idea is not a new one—it was first put forward as long ago as 1862 by the romantic novelist About, and later by P. Becquerel (1936) as part of his theory concerning the continuance of life in the universe.

Drying by means of wind, the sun's rays, or latterly, artificial heating, gives a product which is easier to store and distribute than the undried substance. But for many *biological* materials the advantages of drying are as yet unproven.

The process of lyophilization, i.e. drying by freezing biological materials in a vacuum, followed by benign heating, as in "freeze-drying", was for long regarded as no more than a scientific curiosity. It was only when research had led to a proper understanding of the denaturing, histological and organoleptic properties of freeze-dried labile products that efficient techniques for freeze-drying were developed. Today the future of freeze-drying, including its application to living things, is assured.

The present book provides a comprehensive background to the fundamental principles of freeze-drying, for the use of both research and technical workers, and also makes a critical examination of the basic research literature.

The book is divided into three parts: Part 1, General Principles; Part 2, Cyclic-Pressure Operation; Part 3, Applications.

Part 1, starting with an historical survey, deals with the classical theory of ice sublimation of a freely exposed ice surface. It shows the limitations of this naïve approach with respect to the theory of freeze-drying where the resistance to heat and vapour flow in the dried shell is involved. Next, the problem is analysed in terms of heat and vapour transfer controlling the rate of drying. This analysis employs quasi-steady-state or moving-boundary models for the ice interface between the dry and the frozen portions of the material being dried.

The important physical parameters upon which the analysis depends—for instance, permeability and thermal conductivity coefficients—are discussed and methods of measuring them are described in detail.

Ultimately a truly reversible drying process might be evolved for complex living things, based on analyses of the freezable water in biological systems, residual water levels, and safe drying temperatures. These important considerations are fully discussed.

It is stressed that theoretical research is moving towards a unifying concept of freeze-drying and towards the possibility of securing rapid modes of heat transfer and precise process control. The mathematical equations essential to the development of these ideas are given in appendices.

Part 2 deals with cyclic-pressure operation, a freeze-drying process newly developed by the author, which gives up to 50 per cent reduction in the long drying runs normally encountered in freeze-drying.

Besides chapters describing the physical principle and the equipment, another with an appendix concerns theoretical difficulties imposed by cycling. These are analysed in several ways: first by iterative procedures and then by analytical methods to obtain the pressure variation and temperature wave in the product.

Part 3, dealing with applications, lays emphasis on design theory of freeze-drying equipment and laboratory apparatus for foods, vaccines and microorganisms, histological specimens and for novel uses. Other chapters deal with vacuum cooling, freezing and temperature of incipient melting, rehydration and specific aspects of the preparation and storage of freeze-dried products.

In conclusion I am grateful to the Commonwealth Scientific and Industrial Research Organization for leave to write this book and to Mr M. V. Tracey for his unfailing support and encouragement.

I should like to express my grateful appreciation to my colleagues: Mr J. Middlehurst and Dr N. S. Parker for technical editing and for assistance with the revision of Chapters 3 and 7 and Dr J. R. Vickery, for many useful suggestions and for editing Chapters 1, 9, 11, 12 and 14.

In the preparation of line drawings, and basic editorial work, Mr A. H. Seppings has been of inestimable help, especially in eliminating many of my fastidious corrections. My debt of gratitude cannot be adequately expressed in words.

I especially want to thank Dr Marina Gheorghiu for assistance with the bibliography and topics of medical interest.

I am grateful to Mrs G. A. Moore and Miss R. Johns for producing a clear typescript for the publisher.

Paris
July 1978 J. D. MELLOR

Contents

Nomenclature

The mathematical symbols appearing in the text are the ones commonly used in the literature of freeze-drying, exceptions being the constants employed. The text contains also shorter lists of symbols where appropriate.

The properties of water vapour are denoted by symbols without subscripts, ice or interface by subscript i and sublimation by s.

The dimensions conform to SI units.

a	constant
	drying coefficient
	length [m]
	molar flux factor
	radius [m]
a'	intercept
	constant
$4a$	latus rectum (parabola)
a_w	water activity
a'_w	water activity: freezing point
$A = \lambda_d(T_0 - T_s)/\lambda_f(T_0 - T_L)$	
	area [m^2]
	constant
	postulated substance
A'	internal surface area: pores [m^2]
$[A]$	concentration: A at t [$kmol\ m^{-3}$]
dA	element of cross-section of tube [m^2]
dA'	element of wall area [m^2]
$[A]_0$	concentration: A at $t = 0$ [$kmol\ m^{-3}$]
b	constant
	correction term (depending on λ and K)
	length [m]
	radius [m]

b'	slope
b_1, b_2	dimensionless coefficients
b_3, b_4	constants defined by equations (5.44) and (5.45)
$b\lambda_e$	mean free path: non-equilibrium vapour
$B =$	$\lambda_d \Delta H / \lambda_f c_g (T_L - T_0)$
	constant
c	concentration [kmol m^{-3}]
	constant
	length [m]
	specific heat [kJ kg^{-1} K^{-1}]
c_1	constant (Dyer and Sunderland's notation)
c_1, c_2	surface concentration [kmol m^{-3}]
c_f	concentration at boundary f [kmol m^{-3}]
	specific heat: frozen solid [kJ kg^{-1} K^{-1}]
c_G	concentration: glycerol [kmol m^{-3}]
c_j	outer concentration: aroma component j [kmol m^{-3}]
$c_{j,i}$	inner concentration: aroma component j [kmol m^{-3}]
$c_{j,0}$	concentration: aroma component j, initially [kmol m^{-3}]
$c_{j,z}$	concentration: aroma component j in incremental distance z [kmol m^{-3}]
c_l	concentration at boundary 1 [kmol m^{-3}]
c_L	specific heat: liquid [kJ kg^{-1} K^{-1}]
c_N	concentration: sodium chloride [kmol m^{-3}]
c_0	concentration at boundary 0 [kmol m^{-3}]
c_p	molar specific heat at constant pressure [kJ mol^{-1} K^{-1}]
c_s	surface concentration [kmol m^{-3}]
c_w	outer concentration: water vapour [kmol m^{-3}]
$c_{w,i}$	inner concentration: water vapour [kmol m^{-3}]
C	constant
C_1, C_2, C_3	constants to be determined
C_E	equilibrium constant
d	constant
	diameter [m]
	width [m]
d_H	hydraulic diameter [m]
D	diffusion coefficient [m^2 s^{-1}]
	thickness of metal (tube holder) [m]
D'	diffusivity: water vapour in porous product [m^2 s^{-1}]
D_{12}	mutual diffusion coefficient [m^2 s^{-1}]

D_{con} diffusion coefficient: concentrate [$m^2\ s^{-1}$]
$D_{g,w}$ diffusion coefficient: gas in water [$m^2\ s^{-1}$]
($D_{K,w}$) Knudsen diffusion coefficient: water vapour [$m^2\ s^{-1}$]
D'_K Knudsen diffusion coefficient (intercept value) [$m^2\ s^{-1}$]
D_{K_2} Knudsen diffusion coefficient: component 2 [$m^2\ s^{-1}$]
D_N diffusion coefficient: naphthalene vapour [$m^2\ s^{-1}$]
$D(p)$ diffusion coefficient (function of pressure) [$m^2\ s^{-1}$]
D_t diffusion coefficient (transition flow) [$m^2\ s^{-1}$]

e emissivity
exponential
E expansion: water vapour [$kJ\ m^{-3}\ s^{-1}$]
potential gradient [$V\ m^{-1}$]
E_A activation energy [J]
$E_i(q)$ exponential integral

f fraction: diffuse reflection of molecules at walls
frequency [s^{-1}]
f_1 net fraction: escaping molecules
f_2 net fraction: impinging molecules
f_c probability: water molecule reaches vapour trap or condenser
$f_s = d/x^n$, Stephenson's fractional drying factor
F conductivity [$m^3\ s^{-1}$]
F_1 geometrical viewing factor
F_2/A_2 external pressure: gas component 2 [$N\ m^{-2}$]
$F(r/\lambda)$ total fraction: molecules travelling distance r without collision
$1-F(r/\lambda)$ total fraction: colliding molecules
$F(\bar{t}) = (1/t^{1/2})\int_0^{\bar{t}} (\partial T/\partial \bar{y})\,d\bar{t}$, function normalized to unity

g gravitational acceleration [$9{\cdot}806\ m\ s^{-2}$]
G gas or water vapour flow rate [$kg\ m^{-2}\ s^{-1}$]
G' external gas flow rate (positive in negative x-direction) [$kg\ m^{-2}\ s^{-1}$]
G_1 potential energy: diffusing gas [$Pa\ m^3$]

h coefficient of heat transfer [$W\ m^{-2}\ s^{-1}$]
h' penetration depth [m]
h_1, h_2 surface heat transfer coefficients [$W\ m^{-2}\ s^{-1}$]
H heat gain [$kJ\ kg^{-1}$]
sensible heat [$kJ\ kg^{-1}$]
total heat [$kJ\ kg^{-1}$]

H_e effective heat [kJ kg^{-1}]
H_1, H_2 sensible heats before and during freezing [kJ kg^{-1}]
ΔH heat of crystallization [kJ kg^{-1}]
heat of evaporation [kJ kg^{-1}]
heat of fusion [kJ kg^{-1}]
heat of sorption [kJ kg^{-1}]
heat of sublimation [kJ kg^{-1}]
ΔH_0 molar heat of fusion at the freezing point of water [kJ kmol^{-1}]
ΔH_v latent heat of fusion per unit volume [kJ m^{-3}]

i vector sum
i_1 charging current vector [A]
I_0, I_1 modified Bessel functions

j grid line number

$k = 0.24n_1$
constant-ratio
freezing depression constant [1856·46 kg K kmol^{-1}]
k', k'' structural constants
k^* velocity constant
k_1, k_2 constants
k_1', k_2' shape factors
k_c external mass transfer coefficient [mol Pa^{-1} m^{-1} s^{-1}]
$k_c^* = (M/\rho_d)(\partial r'/\partial m)_T(p_w^\circ/RT)[k_c\beta'/(1+\beta')]$, effective mass transfer coefficient
k_j mass transfer coefficient: component j [mol Pa^{-1} m^{-1} s^{-1}]
k_w mass transfer coefficient: water [mol Pa^{-1} m^{-1} s^{-1}]
$K = \lambda_d/\lambda_f$
arbitrary constant
permeability [m^2 s^{-1} and kg m^{-1} Pa^{-1} s^{-1}]
resultant coefficient [m^2 s^{-1}]
$K' = DM/RT$
K_0, K_1 modified Bessel functions
K_1' permeability of water vapour [kg m^{-1} Pa^{-1} s^{-1}]
K_2, K_2' adjusted constants
K_∞ steady state permeability [kg m^{-1} Pa^{-1} s^{-1}]
K_j equilibrium constant: component j
$Kn = \lambda/d$ Knudsen number
K_v viscous permeability coefficient [m^2]

K_w	equilibrium constant: water
K_x	equilibrium constant
$\Delta K/\Delta P$	slope of asymptote
l	length [m]
	thickness [m]
L	annual charge to repay \$1 loan + interest
	length [m]
$(L_e/L)^2$	tortuosity factor
m	mass (water, vapour, ice and molecule) [kg]
	moisture content (dry basis) [kg kg^{-1}]
m_1	moisture content at some fixed point [kg kg^{-1}]
$m_{a,0}$	mass of aroma constituent a in feed [kg]
$m_{a,1}$	mass of aroma constituent a after evaporation [kg]
m_{cyc}	mass of ice sublimed per cycle [kg]
m_{tot}	total mass sublimed [kg]
$m_{w,0}$	mass of water in feed [kg]
$m_{w',1}$	mass of water after evaporation [kg]
m_u	mass of unfrozen water [kg]
M	molecular weight [kg $kmol^{-1}$]
M_2	molecular weight: gas component 2 [kg $kmol^{-1}$]
M_S	molecular weight: solute [kg $kmol^{-1}$]
n	number (molecules per unit volume, capillaries, cycles and years)
n_1	molecular density (gas component 1) [kg m^{-3}]
	number (accessible capillaries in first cycle)
n_2	molecular density (gas component 2) [kg m^{-3}]
n_{cyc}	effective number of accessible pores in *n*th cycle
n_D	number of dead organisms
n_g	molecular density: gas mixture [kg m^{-3}]
n_L	number of live organisms
nc_N	critical salt concentration [kmol m^{-3}]
N	molecular flow rate [kmol s^{-1}]
	mole fraction
	number of cycles
N_1	molecular flow rate (gas component 1) [kmol m^{-2} s^{-1}]
N_2	molecular flow rate (gas component 2) [kmol m^{-2} s^{-1}]
N_S	mole fraction: solute

$[O_2]$ oxygen concentration [kmol m^{-3}]

p Laplace transform variable
vapour pressure [Pa]
p' limiting pressure [Pa]
p_1 vapour pressure corresponding to surface concentration c_1 [Pa]
p_2 vapour pressure corresponding to surface concentration c_2 [Pa]
p_a vapour pressure at T_a (temperature of surrounding walls) [Pa]
p_a° saturated vapour pressure (constituent a) [Pa]
p_i' partial vapour pressure in boundary layer adjoining ice [Pa]
p_j° saturated vapour pressure (constituent j) [Pa]
p_{oil} oil manometer pressure [Pa]
p_s' partial water vapour pressure at the surface [Pa]
$p_{w,e}$ partial water vapour pressure in equilibrium [Pa]
$p_{w,s}$ partial water vapour pressure in equilibrium with T_s and m_s [Pa]
dp'/dx partial pressure gradient [Pa m^{-1}]
dp_2 partial pressure gradient: gas component 2 [Pa m^{-1}]
P fraction (nomogram Fig. 11.17)
pressure (absolute and operating) [Pa]
P^* residual gas pressure [Pa]
P_1 measured absolute pressure [Pa]
P_D probability that an organism dies
P_F flash pressure [Pa]
P_i initial pressure [Pa]
P/k_2 resistance due to diffusion [m^2 s kg^{-1}]

q number of increments to centre line
substitution term
Q fraction (nomogram Fig. 11.17)
heat flux [kJ m^{-2} s^{-1}]
total heat flux [kJ s^{-1}]
Q' external heat flux [kJ m^{-2} s^{-1}]
$Q_0 = (1/\tau)\oint \dot{m}\,\Delta H\, dt$
Q_1, Q_2 opposing heat fluxes [kJ m^{-2} s^{-1}]
Q_{max} maximum heat flux [kJ m^{-2} s^{-1}]
$Q(t)$ heat flux as a function of time (ice interface) [kJ m^{-2} s^{-1}]

r number of increments (porous region)
radius [m]
$= t/t'$, ratio (cyclic to normal process drying times)

$r' = p'_w/p^\circ_w$, vapour pressure ratio
r_0 interest rate [%]
r_{th} radius: tube holder [m]
R gas constant [8·314 m^3 Pa mol^{-1} K^{-1} and 8·314 J mol^{-1} K^{-1}]
length (spherical, cylindrical and slab geometries) [m]
Re Reynolds number

$s \equiv X$ (McCullough and Sunderland's notation)
thickness (slab) [m]
s_g width (gas space) [m]
s_t thickness (tray) [m]
s_p displacement of pump [m min^{-1}]
S mass fraction: solute
solubility [m^3 m^{-3}]
Sc Schmidt number

t time [s and h]
t' drying time (normal process) [h]
t_1 high-pressure period [min]
time for temperature to reach T'_s [h]
t_2 drying time[h]
low-pressure period [min]
$t_{\frac{1}{2}}$ half-equilibrium time [h]
t_c characteristic time [min]
time (surface tension forces overcoming gravity) [min]
t_{cyc} drying time (cyclic process) [h]
t_d drying time [h]
t_0 time: same weight loss at zero pressure [h]
t'_0 time intercept [h]
t_s time: surface to reach temperature T_s[h]
time: ice sublimation to thickness l [h]
T temperature (Celsius [°C] and absolute [K])
$\bar{T}$ Laplace transformation
T^* equilibrium temperature (solution containing large ice crystals) [°C]
T_0 freezing point of pure water [°C]
uniform surface temperature [°C]
T_1 temperature at some fixed point [°C]
temperature of vitrification [°C]
T_2 temperature of devitrification [°C]
T_{cs} maximum temperature: complete crystallization [°C]

T_F	temperature: freezing point [°C]
T_i'	temperature of vapour in boundary layer adjoining ice [°C]
T_{im}	minimum temperature: incipient melting [°C]
T_{im}'	maximum temperature: incipient melting [°C]
T_L	temperature: back face [°C]
$T(0)$	Laplace transform, $t=0$
T_s	surface temperature [°C]
T_s'	safe product temperature (temperature setting) [°C]
	temperature in boundary layer adjacent to surface [°C]
ΔT^*	lowering of equilibrium temperature: spherical crystals in solution [K]
u	new variable
	velocity [m s^{-1}]
u_0	uniform velocity: vapour space [m s^{-1}]
u_1	macroscopic velocity: gas component 1 [m s^{-1}]
u_2	macroscopic velocity: gas component 2 [m s^{-1}]
U_L	superficial liquid velocity [m s^{-1}]
U_s	velocity of diffusion: upward wash to downward liquid [m s^{-1}]
v	new variable
	velocity [m s^{-1}]
dv/v	fraction of molecules
V	specific volume [m^3 kg^{-1}]
	volume [m^3 or l]
V'	volume of vapour space [m^3]
V_1, V_2	volumes above mercury columns at pressures P_1 and P_2 respectively [m^3]
V_{ch}	volume: chamber [m^3]
V_i	viability
V_1/t	flow rate [m^3 h^{-1}]
V/V_{max}	variation: power supply intensity
w	mass fraction
	velocity [m s^{-1}]
w_f	mass fraction: frozen portion
w_{ins}	mass fraction: insolubles
w_{is}	mass fraction: ice slurry solids after separation
w_{ju}	mass fraction: juice
w_{mix}	mass fraction: frozen mixture before separation
$w_{m,l}$	mass fraction: mother liquor
w_S	mass fraction: solute

$w_{S,i}$ mass fraction: solute + ice
w_{unf} mass fraction: unfrozen portion
W moisture content (wet basis) [kg kg^{-1}]
$W' = G' c_g L/\lambda_d$, dimensionless constant
$W'' \simeq c(T_s - T_0)/y\,\Delta H$, dimensionless constant
$W_1 \simeq 1/(1+3l/8a)$, Clausing factor

x amount of water frozen out [kg]
depth (distance) [m]
substitution parameter
thickness (variable) [mm]
Δx spacing in porous region [mm]
Δx_i spacing in ice region [mm]
x_0 total thickness: slab [mm]
X depth of interface [mm]
X_c characteristic thickness [mm]
$[X]_0$ concentration: X at $t = 0$ [kmol m^{-3}]

$y = X/L$, constant
non-dimensional concentration
$y_1 = p_1/P$, ratio: partial to total pressures (component gas 1)
$y_2 = p_2/P$, ratio: partial to total pressures (component gas 2)

$z = x/L$
distance

Greek symbols

α angle [rad]
coefficient (evaporation, sublimation and condensation)
thermal diffusivity [m^2 s^{-1}]
α' constant defined by equation (4.II.7)
$\alpha_{a,w}$ relative volatility with respect to water: aroma constituent a
α_e effective relative volatility
$\alpha_{j,w}$ relative volatility: constituent j
$1/\alpha k_1$ kinetic resistance [m^2 s kg^{-1}]

$\beta = (M/2RT)^{1/2}$
fraction: absolute rate of sublimation at zero pressure
β' Bunsen coefficient [m^3 m^{-3}]
constant defined by equation (4.II.16)

$\beta_1 = \mu A'/mr$, time constant [s^{-1}]
thickness ratio (nomogram, Fig. 11.17)
β_2 thickness ratio (nomogram, Fig. 11.17)

γ activity coefficient
constant of proportionality
$\gamma_1 = 0{\cdot}5772$, Euler's constant
γ_a activity coefficient: component a
γ_j activity coefficient: component j
$\Gamma = \exp[-\beta^2 u_0^2 - \beta u_0 \pi \operatorname{erfc}(\beta u_0)]$, correction factor

δ increment
loss angle [rad]
overall fraction of water evaporated
thickness [m]
δ' reflection coefficient
δ_1 surface roughness factor, $=\pi/4$ (smooth pores), $=3\pi/16$ (rough pores)
$\tan\delta$ loss tangent
Δ difference

ε porosity
ε' new variable
phase angle [rad]

$\zeta = x/L$
new variable

η new variable
efficiency
η_{sep} efficiency: ice phase separation
θ angle of contact [rad]
polar angle [rad]
θ_1 angle of cone: escaping molecules [rad]
$\theta_d = (T_d - T_s)/(T_i - T_s)$, normalized temperature: dry layer
$\theta_f = (T_f - T_L)/(T_i - T_L)$, normalized temperature: frozen layer

κ variable factor
κ' loss factor
κ'' relative dielectric constant

λ thermal conductivity (W $m^{-1}K^{-1}$]
mean free path [mm]
λ_1 thermal conductivity (gas/solid in series) [W m^{-1} K^{-1}]
thermal conductivity (normal process) [W m^{-1} K^{-1}]
λ_2 isothermal thermal conductivity (gas/solid in series) [W m^{-1} K^{-1}]
$\lambda_{d,e}$ effective thermal conductivity: dry layer [W m^{-1} K^{-1}]
λ_e effective thermal conductivity [W m^{-1} K^{-1}]
$\lambda_{g,0}$ thermal conductivity: free gas at atmospheric pressure [W m^{-1} K^{-1}]
λ_L thermal conductivity: Lucite [W m^{-1} K^{-1}]
λ_0 thermal conductivity (low pressure) [W m^{-1} K^{-1}]
$\lambda(p)$ thermal conductivity as a function of pressure [W m^{-1} K^{-1}]
λ_{sp} wavelength in space [m]
λ_T thermal conductivity: turkey [W m^{-1} K^{-1}]
$\Lambda = x(\partial^2/\partial x\,\partial t) + v(\partial/\partial x) + \partial/\partial t$, operator [$s^{-1}$]

μ mass fraction (ice sublimed)
viscosity (cP)
$\mu^2 = 2h/\lambda_{me}D$, exponent (Bessel functions)
$\mu_1 = [Q(t)X(t)]/\alpha_f\rho_f\,\Delta H$, McCullough and Sunderland's notation

ν drying exponent
integer
$d\nu/\nu\, dA$ fraction of molecules reaching dA in half-cylindrical space

$\xi = u^2$, $u - u_0$ and x (new variables)

π partial air prssure [Pa]
pi (= 3.1416)

ρ density [kg m^{-3}]
$\rho' = mn$, vapour density [kg m^{-3}]
ρ_M density: mercury [kg m^{-3}]
$\rho_{s,1}$ density: solid–liquid phases [kg m^{-3}]
ρ_{oil} density: oil [kg m^{-3}]
$\rho/P = M/RT$, ratio: vapour density to vapour pressure

σ	diameter: water molecule [0·282 nm]
	Stefan's constant [0·0567 μW m^{-2} K^{-4}]
σ_{con}	surface tension: concentrate [mN m^{-1}]
σ_D	energy density [kJ m^{-3}]
σ_E	surface free energy [J m^{-2}]
σ_M	surface tension: mercury [mN m^{-1}]
Σ	summation
τ	new variable
	$= t_1 + t_2$, cycle period [min]
	time (actual, drying and initial) [h]
τ_a	actual drying time [h]
τ', $\bar{\tau}$	dimensionless drying time
ϕ	particle loss factor
	polar angle [rad]
ω	$= 2\pi/\tau$, angular frequency [s^{-1}]
$d\omega$, $d\omega'$	solid angles [sr]

Subscripts

a	air (external)
c	condenser (ice layer, cooling medium)
con	concentrate
cyc	cyclic process
d	dry layer
	dry product
do	downstream
e	effective
	equilibrium
f	final
	frozen (layer, ice at 0°C)
F	flash
	freezing point
g	gas
h	heater
i	ice interface
	ice (surface)
in	internal

j	constituent
K	Knudsen
m	molecular
	stagnant air
me	metal
n	normal process
0	initial
	reference
	zero
p	pump
s	sublimation
	surface
sol	solution
ss	steady state
up	upstream
v	viscous
w	water
	water vapour

Superscripts

$\cdot$	derivative: time
$^{\circ}$	saturated vapour
$'$	partial (vapour)
	saturated (lower temperature)
$-$	mean

PART 1
GENERAL PRINCIPLES

1 Introduction

1.1 Advent and aim of freeze-drying

The primary object of freeze-drying is to preserve biological materials without injury by freezing the contained water and then removing the ice by sublimation. Thus, one can expect to combine the advantages of both freezing and desiccation to obtain a more propitious state of preservation.

Preservation by freezing, desiccation and even "freeze-drying", occurs in nature because water so readily undergoes a change of state. Parkes (1962) noted several examples of such drying processes occurring in nature, and recalled that Arrhenius had thought living organisms in the dried state from outer space provided the genesis of life on earth. Becquerel (1951) found that some spores of bacteria and moulds survive just the sort of conditions which are imposed by freeze-drying, including even cooling, to around absolute zero. This observation has further given credence as to the origin of life on earth. Becquerel calculated from van t'Hoff's law that the reduction in metabolic rate for spores would give a survival of 70 billion years at −270°C if they survived one year at 10°C.

Systematic freeze-drying is a procedure which is applied to three broad categories of biological material:

1. Non-living matter such as blood plasma, serum, hormone solutions and foodstuffs.
2. Surgical transplants which are made non-viable so that the host cells can grow on them as the skeleton. Examples include artery, bone and skin.

3. Living cells destined to remain viable for long periods of time. Examples include bacteria, yeasts and viruses but not mammalian cells such as spermatozoa.

Freeze-drying is of paramount importance for blood plasma and for foodstuffs because it stops the growth of microorganisms, inhibits deleterious chemical reaction and facilitates distribution and storage. Whereas stored liquid plasma components deteriorate fairly rapidly with the loss of physiological and pharmacological properties, frozen and freeze-dried products deteriorate extremely slowly. Today, freeze-dried plasma is accepted not only as equivalent but even superior to stored liquid plasma. Freeze-drying of stored foodstuffs greatly reduces loss of quality due to deleterious chemical reactions caused by enzymic and non-enzymic browning and oxidative rancidity. Because foods are relatively cheap and of high water content it has often been remarked: "Freeze-drying for foods is no more than a vulgar application of a biological idea." With the exception of instant coffee and soup powders, many of the freeze-dried products produced commercially are used mainly for military and camper's rations.

1.2 Early history

The ancient Egyptians relied on the sun to dry food and on the dryness of the desert to assist mummification. Dry cold climates have been similarly utilized, e.g. the Vikings produced klip-fish in the Arctic and the Incas produced a potato flour, called *chuño*, in South America.

It is the capacity of cold dry air to evaporate moisture from biological materials that is so significant for the biologist and the food technologist. Such air produces desiccation by sublimation or "freeze-drying"—a change of state from ice to vapour without melting. Freeze-drying at great heights in mountains, instead of in air at sea level, has advantages as the Incas found in the cold rarefied air of the High Andes (Mellor, 1962a).

The role of desiccation in the process of inoculation has been known since Tissot (1792) described a method preparing and employing a thread impregnated with dried smallpox virus. However, a century had passed before Vansteenberghe (1903) succeeded in producing desiccation by freeze-drying rabies virus over sulphuric acid under vacuum. Bordas and d'Arsonval (1906) used a refrigerated condenser instead of sulphuric acid to remove the water vapour, as did Shackell (1909), who showed the advantages of pre-freezing as a first step to drying.

Altmann (1890) was the first to freeze-dry organs and tissues for histological examination under the microscope. His work established the practice of systematic freeze-drying of biological specimens.

Anticipating another use for freeze-drying, Carrel (1912) stated the need for preserving tissues for surgical transplants.

Thus the pattern of freeze-drying in medicine and biology was set for the next 50 years and there followed improvements in Altmann's method (Gersh, 1932); a theory for the design of the apparatus (Stephenson, 1953, 1954); centrifugal vacuum freezing for suspensions of bacteria inside the dryer (Greaves, 1944) and very low-temperature freeze-dryers for artery, bone and skin transplants intended for banking (Flosdorf and Hyatt, 1952). There were specific applications of the technology to electron microscopy (Wychoff, 1946).

Parallel with the early developments on the biological side there were investigations of the basic physical principles. Leslie (1811) was the first to publish an account of freezing by evaporation and subsequent sublimation of ice in an evacuated vessel containing sulphuric acid. Wollaston (1813) followed this with a demonstration at the Royal Society, London, using an evacuated cryophorus, or frost producer.

Rey (1960b, 1964b) describes other possibilities for lyophilization.

1.3 Developments

Freezing and freeze-drying would have been impossible without advances in vacuum pumping and in refrigeration machines. The first vacuum pump was constructed by Guericke (1672). Extremely low temperatures were attained later by liquefying gases such as air (Linde, 1895) and carbon dioxide (Thilorier, 1835). However, the "deep-freeze" for preservation purposes came about only after Linde (1873) had constructed a practical ammonia refrigeration compressor.

Early experimentalists were confronted by the problems of thermal and osmotic shock and cell disruption, all caused by freezing. There are two possible solutions: firstly, very rapid freezing, to obtain an amorphous state (Tamman, 1898; Luyet, 1939); and secondly, to capitalize on the protective properties afforded living cells by glycerol during fast freezing (Rostand, 1946), and during slow freezing (Polge and Rowson, 1952). The use of glycerol is today the standard method of preserving bull semen for artificial insemination of cattle.

To improve still further the viability of freeze-dried micro-organisms Fry and Greaves (1951) developed a suspending fluid, *Mist. Desiccans*, which protects organisms during freezing and drying. The composition of the fluid was adjusted so as to exclude carbonyl groups, which would have reacted with bacterial membranes (Scott, 1958), and the addition of a neutral polymer—for example, polyvinylpyrrolidone (PVP)—to aid reconstitution (Greaves, 1964). Dimethyl sulphoxide (DMSO) affords even better protection as, unlike other agents, it penetrates the cells so that the freezing rate is not so critical (Lovelock and Bishop, 1959).

Up to the time of World War II freeze-drying had been regarded as an occasional scientific tool and the opportunity for large-scale use had not arisen. Flosdorf (1949) in his book envisaged its use for foodstuffs and other bulky materials. When freeze-dried blood plasma was produced by Greaves (1942) for the armed services the potential of the process was fully realized and a number of patents followed, including an early British patent (Kidd, 1941) for freeze-drying foods using the heat-pump principle. Meanwhile, a large freeze-dryer with refrigerated and scraper condenser was built in Florida for drying fruit juices; it proved uneconomic because the formation of gummy substances on the surface of the product restricted further drying. One method investigated for overcoming this is surface-wetting the frozen juice prior to drying (Mellor, 1954). In addition, aromatic compounds essential for the flavour were lost during vacuum distillation and these had to be prepared separately and added to the freeze-dried product. Kallistratos and von Sengbusch (1964) found little retention of pure volatiles though Rey and Bastien (1962) showed retention in freeze-dried products to be better in the presence of glucose. These findings opened the way to a completely new field concerning quality considerations of dried products.

After World War II, work was resumed on dehydration of foodstuffs at the experimental factory of the British Ministry of Food Research Establishment at Aberdeen, Scotland. Vacuum contact plate-drying equipment was used and, in 1954, trials of initial freeze-drying followed by ordinary vacuum drying of fish fillets were carried out. In 1956 a novel idea for improving the freeze-drying rate for steaks, by impaling them on spiked heater plates, was put forward by Brynko and Smithies (1958) of Canada. The Ministry of Food (Gooding and Rolfe, 1957) adapted the idea to their existing plant (which had by that time been equipped with more efficient steam ejectors to complete the freeze-drying process) by interposing sheets of expanded metal, instead of spikes, between the drying surfaces of the frozen food and heating platens. Because of the substantially higher rates of freeze-drying thus achieved the process was named "accelerated freeze-drying" (AFD). In 1960 details of the new process were released to industry for commercial exploitation.

The measurement and control of product temperature in the course of freeze-drying are difficult to achieve because a constant sublimation temperature has to be maintained by means of a temperature gradient through the frozen layer. There is always the tendency for drying to take place (1) from the edges of the frozen layer and (2) at its surfaces in contact with the heater plate. The first difficulty can be overcome by controlling the sublimation pressure by periodically shutting off the condenser for short periods (Neumann, 1960) and the second by drying materials at lower temperatures in metal trays or by radiant heating.

An ingenious method of eliminating some or all of these difficulties and of reducing vapour flow resistance caused by the dry layer has been described by Greaves (1960b). It involves continuously scraping the dry layer during its formation.

Rieutord (1965) controlled the sublimation temperature in freeze-dryers by introducing short puffs of air into the vacuum system. An automatic control system, depending on the electrical conductivity of the frozen product during drying, has been devised by Rey (1959) and depends for its operation on the data obtained from a differential thermal analysis (DTA) technique for predetermining the temperature of incipient melting in a frozen solution (Rey, 1960a).

The possibility that freeze-drying could be accelerated by varying the vacuum pressure was foreshadowed when Ehlers *et al.* (1962) and Harper (1962) measured the heat-transfer coefficients in freeze-dried foods at different vacuum pressures and also when Harper and el Sahrigi (1964) made similar measurements with high-conductivity gases. They attributed reductions in drying time at higher pressures to increased heat transfer. Earlier, Görling (1956) had found that the diffusion of water vapour from products freezing under reduced pressure could be greater than the diffusion at normal pressures.

A cyclic-pressure method, combining both the heat and vapour transfer effects, was invented by Mellor (1967a), who obtained substantial reductions in drying time when the pressure cycle was optimized. Earlier, Mellor (1966b) had shown that the higher rates of drying depend on sudden reductions in pressure causing a flash of vapour in each cycle. Higher-pressure methods, controlled by small pressure variations, have been proposed (Hamilton, 1966), and two methods for improving the pressure in vacuum systems by raising the pressure several times to atmospheric have been patented (Takesaki, 1921; Kniepen, 1929).

1.4 Stages of the constant-pressure freeze-drying process

The aim of freeze-drying is to preserve moisture-laden materials, mostly of a labile biological nature, by freezing and sublimation and later to rehydrate them. Systematic freeze-drying requires several successive steps, as shown in Fig. 1.1.

1. *Prefreezing* The materials are prepared and frozen at a low temperature.
2. *Primary Drying* The ice crystals formed on freezing are sublimed by vigorous and then gentle heating, usually under vacuum.
3. *Secondary Drying* With the disappearnace of the ice the residual moisture is desorbed at around ambient temperature under high vacuum.
4. *Conditioning and Rehydration* At a sufficiently low moisture content the vacuum is broken with a dry inert gas, and the product is then packed and stored.

The dried product can be shelf-stored until it is required for rehydration. The operating conditions—temperature, pressure and time—for each step are critical and must be carefully controlled.

The final freeze-dried product is dry, light and porous with almost its original shape and texture and so disposed to have an evenly distributed residual moisture content. In the dry state it can be stored almost indefinitely provided there is no contact with moisture and oxygen from the air. Because of its "lyophile" nature—ability to

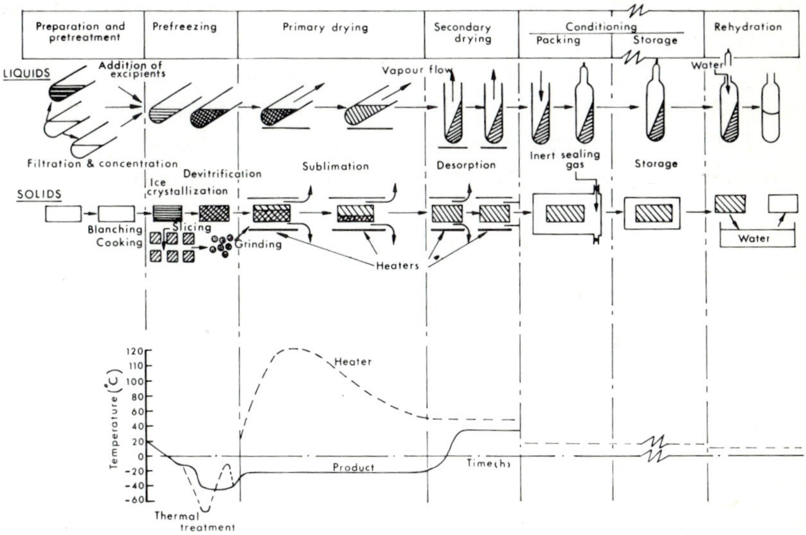

FIG. 1.1. Flow sheet for the successive steps in freeze-drying for liquid and solid materials and process temperatures (after Rey, 1964a).

absorb moisture—it quickly recovers most of its original physical, organoleptic and physiological properties.

1.4.1 PREPARATION

The materials should be of good quality and suitably prepared so as to suit the operational conditions of the drying apparatus or plant. To aid drying, pieces of material should have large surfaces and thin uniform cross-sections, or should have small spherical or thin wedge shapes. They should be distributed evenly throughout the drying chamber.

Very often materials to be dried are pretreated. Some foods are cut, ground, strained or concentrated whereas others are blanched or precooked. Others are pretreated with sulphur dioxide as an inhibiting agent to prevent the browning reaction. Vitamins and hormones are sterilized; serums and hormones are filtered; tissues are infused with ethylene dioxide and those employed as inert grafts are irradiated.

For microorganisms and tissue cells, excipients which can be added to the suspending fluid are important in pretreatment; they often serve a multiple purpose, sometimes to protect during freezing and drying, other times to buffer the eventual formation of carbonyl groups in storage, or to aid reconstitution. Excipients such as dextran, polyvinylpyrrolidone (PVP) and DMSO are used with bacteria and viruses for protection in freezing and sucrose and glucose for water fixation. Additionally, lactose, dextran and serum albumin are used to prevent suspensions of vitamins, hormones and bacteria from collapsing and to improve consistency during drying. Sodium glutamate is used for buffering.

1.4.2 PREFREEZING: ICE CRYSTALLIZATION

Prefreezing is an important step in the freeze-drying process only if the frozen material has not already been "denatured" during the formation of ice crystals.

All cellular materials contain large quantities of "free water", dissolved organic and mineral substances in the inter- and intra-cellular spaces; "remnant water"—described as "bound" to macromolecules either by electrostatic forces or by hydrogen bonds. On freezing, ice crystallizes out of the free water quite easily but generally not out of the bound water. The free water is therefore eliminated by sublimation in primary drying and, since hydrogen bonds are relatively weak, some of the bound water is also eliminated by evaporation at the beginning of secondary drying. By contrast, most of the water bound by electrostatic forces, being difficult to remove, remains in the dry product, e.g. probably about 1 per cent moisture in freeze-dried cells. During the process of crystallization the ice is surrounded by a continuous network of interstitial fluid containing all the dissolved and suspended compounds, which become more concentrated as freezing continues. In the end the whole mass should become rigid, forming a eutectic.

Unless a eutectic is formed there is a tendency towards denaturation, because the presence of concentrated electrolytes is damaging to cells and proteins. Ideally a rigid eutectic phase is necessary also to ensure that primary drying is by sublimation only and not combined with partial evaporation from a liquid state. However, this may mean going to very low temperatures, for example, if calcium chloride is present. Denaturation of protein may also be caused by concentrated solutions as temperatures are steadily lowered. Melting and inadequate freezing should therefore be avoided at all times; they manifest themselves by the formation of frothy and gummy substances in the final product when a vacuum is applied.

1.4.3 CRYSTALLIZATION IN COURSE OF SUBLIMATION

To control the course of sublimation Rey (1960a) determined the maximum temperature of complete crystallization, T_{cs}, in terms of the electrical conductivity of a frozen solution. In Rey's technique, changes in the resistance of the interstitial fluid to a low-frequency alternating current are measured. The resistance rises in proportion to the drop in freezing temperature until it levels off at about −45°C, when it is several million $\Omega\ cm^{-1}$ (Section 13.1).

The technique is the best devised so far for controlling freeze-drying of medical products and fruit juices; it is not often used for controlling the course of sublimation of materials in bulk. For these, other techniques are used in order to determine the sublimation temperature directly with thermocouples or the maximum safe product surface temperature by means of radiation pyrometers.

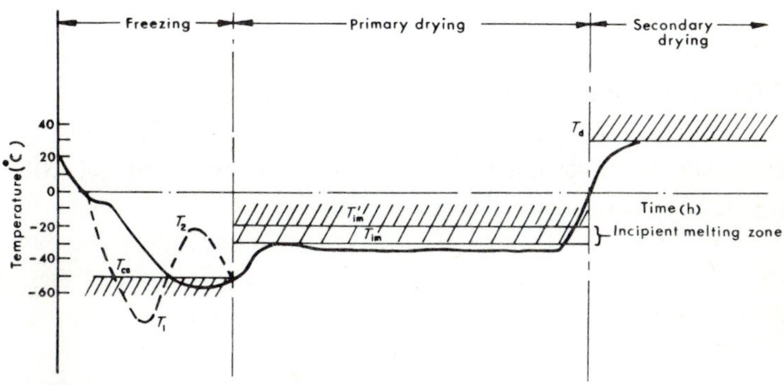

FIG. 1.2. Successive steps in freeze-drying. Maximum temperature of complete crystallization T_{cs}, incipient melting zone $T'_{im} - T_{im}$ and safe final product temperature T_d (after Rey, 1964a).

There are instances where freezing leads to a vitreous state instead of crystallization such as with fruit juices and aqueous solutions of B group vitamins. For these Rey (1964a) proposes a thermal treatment whereby the material is first cooled to a temperature T_1 (Fig. 1.2) to induce a vitreous state, then rewarmed slowly to a temperature T_2 where, because it is metastable, there is crystallization (devitrification); then the material is recooled to T_{cs}. Rey has determined precisely, by differential thermal analysis, the range of devitrified temperatures which he calls the minimum and maximum temperatures of incipient melting T_{im} and T'_{im} respectively. T_{im} is the prescribed sublimation temperature which is automatically controlled in terms of the resistance of the frozen product.

1.4.4 TECHNIQUE

The procedures for freezing vary widely. The choice between slow and rapid freezing will be determined according to the characteristics of the product. In general, the faster the freezing rate the longer it takes to freeze-dry and the better the quality of the final product; however, there are exceptions because circumstances vary enormously for freeze-drying processes and products. Usually freezing is carried out by contact with a cold air current or refrigerated plate, a relatively slow process. Freezing under vacuum (self-evaporation cooling) is quicker and here the size of a piece of material and the degree of vacuum are significant factors. This freezing technique is often damaging unless the material has excess surface moisture. For rapid freezing cryogenic fluids are employed, either by direct immersion or by sprays of liquid nitrogen. For the best results, small biological specimens require immersion in isopentane cooled by liquid nitrogen so there are no problems with the insulating effect of the spent nitrogen gas surrounding the specimens.

1.4.5 PRIMARY DRYING

The frozen material is placed in a closed chamber from which the air is evacuated by a vacuum pump. As a result, the ice sublimes slowly if the energy for latent heat is supplied to it. The vapour given off is prevented from returning to the product and from reaching the vacuum pump by imposing a cold trap or refrigerated condenser across its path.

The driving force for sublimation depends essentially on the difference in vapour pressure between the ice interface and the condenser. The vacuum ensures that the interface temperature is always very low. Energy for the requisite amount of latent heat can be supplied either by conduction through the ice layer and/or dry layer or by microwaves directly. The maintenance of a difference in temperature between the product and the condenser and the heat supply is therefore essential. However, the basic principle of primary drying is not as simple as this because other factors, such as resistances to heat and vapour flow in the dry product, are just as important. The problems concerning resistances are discussed in Sections 2.2 and 3.2, those concerning the risk of melting the subliming mass and overheating the surface of the product in Section 3.4, and those concerning the methods of heating in Section 3.7.

1.4.6 SECONDARY DRYING

This commences when the last crystal of ice stops subliming and moisture desorbs from the drying material which was once the site of the ice crystal. At the same time it is necessary to decrease the heating rate so as to retain the temperature below the limiting value T_d (Fig. 1.2) at which it is safe to hold the product. The temperature T_d is roughly 30°C for biological materials and 50°C for foods: and they can be automatically controlled. Meanwhile, the vacuum pressure in pascals[1] often falls from 13.3 Pa to about 1.3 Pa because of the inherently low vapour pressure of the water vapour in the chamber at this stage. For some biologicals secondary drying is often accomplished in a separate vacuum chamber with a phosphorus pentoxide trap.

Secondary drying sometimes takes up to a third of the total drying time to desorb the moisture from the large internal surface area within the porous dried product. Unless the end-point of drying can be measured during this stage (Section 13.3), over-drying to the detriment

[1] The SI unit "pascal" for pressure, symbol Pa, is used throughout the book. To convert to other pressure units still in use 1 Torr = 133.3 Pa and 1 mbar = 100 Pa.

of the product and to the cost of freeze-drying is inevitable. All that is required is an end-point in terms of the relative property of the measuring device. Absolute measurements of residual moisture level are best left until the drying chamber is opened and the product removed for packing; then the measurement should follow one or other of the standard methods (Section 13.2).

1.4.7 CONDITIONING AND REHYDRATION

A conditioning stage is sometimes necessary so that the residual moisture attains the same level throughout the entire mass of the product; such a stage is necessary in the operation of shelf dryers, where temperature gradients exist between top and bottom shelves.

It is obligatory to retain a certain quantity of residual moisture for storage purposes: for biologicals such as antibiotics and tissues 0.1–1 per cent, for microorganisms and viruses 1–3 per cent, and for foods 2–10 per cent have been recommended.

Breaking the vacuum with a dry inert gas such as nitrogen (for foods) or argon (for biological materials) is critical, otherwise the ingress of moisture-laden air and oxygen will damage the product in storage and render useless the work of desorption. Stopping, sealing and packing with inert gases are now done in the drying chamber (Rieutord, 1960; Hackenberg, 1965), but sealing under vacuum is done separately outside the chamber, mostly on small ampoules and vials.

Rehydration—the last stage in the whole freeze-drying process—is carried out only when the product is required in its original condition. Goldblith (1964) considers that the rate of rehydration is a good measure of quality for freeze-dried foods. The same cannot be said of freeze-dried microorganisms and viruses. Many techniques have been evolved for adding the water (Section 12.4); some, for instance, control the way in which the water penetrates the product at the same time expelling the air filling the pores.

1.5 Important analytical theories for the process

Theories which represent the heat and vapour transfer in the simplest of terms have been developed by Harper and Tappel (1957), and

Ginnette *et al.* (1958), using quasi-steady state models and these are the ones usually quoted in the literature. A more acceptable model—a moving boundary between a dry outer mantle and a semi-infinite frozen solid—has been developed by Mellor (1962b, 1966b) and Luikov *et al.* (1971), assuming a constant heating rate. In the latter, the theory applies to heat conduction through the ice layer or dry layer and a proportionality constant is implicit in terms of the surface temperature. A similar theory for the outer temperature of the frost layer on condensers is attributed to de Roissart and Laederich (1966).

The theory developed by McCulloch and Sunderland (1970) estimated the temperature within the mantle by a quadratic equation for two cases of radiant heating. Firstly, the surface temperature is considered to rise gradually to the heater temperature; secondly, it remains steady because the heater temperature is lowered. The proportionality constant appearing in McCulloch and Sunderland's analysis is then in terms of the depth of the ice interface below the surface of the product.

For prefreezing, prediction of the freezing rates for regularly shaped blocks of food began with the simple theory proposed by Plank (1913), and has since undergone change. Heiss (1933) appears to have been the first to determine the amount of water frozen out at any given temperature. Enthalpy–temperature relationships (Riedel, 1950) gave the amount of water that is bound to proteins.

2 Classical Theory of Sublimation

2.1 Introduction

In recent years the classical theory of sublimation has been transformed into an applied science of freeze-drying. The first step in this transformation was due to Carman (1948), who gave the basic theory in a paper to the Faraday Society. Carman's theory involved such parameters as the absolute rate of sublimation and the mean free path of the vapour molecules, but it is equally applicable to the more complicated parameters such as vapour-pressure driving force and actual rate of freeze-drying.

The next step was a series of experiments designed to verify Knudsen's equation for the absolute rate of sublimation of ice G_s:

$$G_s = \alpha p_s \left(\frac{M}{2\pi RT}\right)^{1/2}, \qquad 0 < \alpha < 1, \tag{2.1}$$

where α = coefficient described by Knudsen as the "coefficient of evaporation", p_s = saturated vapour pressure of ice, M = molecular weight of water vapour, R = gas constant, and T = absolute temperature.

In deriving this equation from kinetic theory two assumptions are made: (1) dynamic equilibrium between the surface of the ice and its vapour; (2) the number of vapour molecules escaping depends on the temperature of the ice, whereas the number returning to the ice surface and not escaping again depends on the pressure and temperature of the vapour.

The classical theory as far as water and ice are concerned had its origins in attempts to verify the equation for the absolute rate of evaporation, or for determining the exact value of α. It was thought that α should ideally have a value of unity.

Strasser *et al.* (1966) summarized the various investigations including the results for the sublimation of ice. Tschudin (1946) found $\alpha = 0{\cdot}96 \pm 0{\cdot}06$ at −60°C to −85°C; Kramers and Stemerding (1951) found $\alpha = 1{\cdot}0$ at −40°C to −60°C; and Strickland-Constable and Bruce (1954) found $\alpha = 0{\cdot}44$ to 0·63 at −40°C. Values less than unity were explained by these authors in terms of a higher working temperature and a small mean free path of 0·3 mm at −40°C compared with 3 mm at −60°C, at which temperature high α values have been obtained. A better explanation of their work is due to Burrows (1957).

A low value for α was obtained by Alty (1933). Later, Powell (1939–40) established a relationship between the sublimation temperature of ice spheres in a refrigerated atmosphere and the vapour pressure, but Tschudin (1946) found a value of 0·94 for the evaporation coefficient in the temperature range −60°C to −85°C *in vacuo* (explained, as already mentioned, in terms of a high working temperature and small mean free path). An alternative explanation, due to Burrows (1965), is concerned with the limitation imposed by Knusden's formula, namely that most experiments involve mass flow conditions rather than equilibria (the early work of Alty (1933) on ice at −8°C in this respect is of doubtful validity).

Schrage (1953) argued on similar lines and developed a correction factor for calculating the actual sublimation rate from Knusden's theory. Sublimation at the ice interface under a dry porous mantle, i.e. real freeze-drying, usually involves higher pressures with the resistance to diffusion of the escaping vapour becoming important. Consequently, sublimation rates are much less, by a factor of 1000 or so, than those predicted by Knusden's equation (Kramers, 1958). Cooke and Sherwood (1955) set out to measure and predict values of the evaporation coefficient under experimental conditions, but Carman (1948) and Ede (1947, 1949a) tackled the problem differently by considering the rate of diffusion in the vacuum space. Another approach has been by considering the rate of diffusion in the porous mantle.

Other investigators also have failed to obtain high values for α because of working at higher sublimation temperatures at which the mean free path of the vapour molecules was less than the dimensions of the vacuum space.

The theory for non-equilibrium conditions developed by Schrage (1953), on the basis of a Maxwellian velocity distribution function in the neighbourhood of the ice interface, has been discounted because it gives correction factors which differ only marginally from unity (Appendix 2.III).

Despite difficulties, attempts have been made to elucidate many of the problems of freeze-drying via the classical approach. Some success has been achieved but the results have been qualitative only—for example, a satisfactory explanation has been given of the behaviour of the vapour-pressure difference on which the driving force for freeze-drying depends (Section 2.2).

Kramers (1958) has pointed out that, to satisfy Knusden's equation, for ice sublimation at −30°C the temperature gradient below the ice surface would have to be about 50°C mm^{-1}, which is impossible. In spite of this, equation (2.1) has been modified to take account of the mean free path of the vapour molecules, the temperature and vapour pressure of ice at the condenser, the non-equilibrium sublimation conditions and several factors bearing on the evaporation coefficient. These factors will now be discussed.

The processes controlling the rate of sublimation were first studied in detail by Powell (1939–40), who sublimed 90 mm diameter ice spheres in stagnant air at −9°C. Later, Cooke and Sherwood (1955) sublimed naphthalene spheres 11·1 mm diameter in air streams *in vacuo* because they considered that if ice were chosen the sublimation mechanism would be affected by the transient temperature effects due to its large latent heat. They found that sublimation rates at low pressures could be predicted by adding the kinetic resistance due to the escaping vapour to the diffusive resistance of the vapour in the air stream towards the condenser, namely

$$\frac{1}{G}=\frac{1}{\alpha k_1}+\frac{P}{k_2}, \tag{2.2}$$

where $1/\alpha k_1$ is the first resistance, taken from equation (2.1) with $k_1 = p_s(M/2\pi RT)^{1/2}$ and P/k_2 is the second resistance with $P=$ absolute pressure, taken from the equation of Ranz and Marshall (1952) with $k_2 = 20\, D_N\, Mp_sP(1+0{\cdot}3\,\mathrm{Re}^{1/2}\,\mathrm{Sc}^{1/2})/\mathrm{RTd}$, where $D_N =$

diffusivity of naphthalene vapour, d = diameter of the sphere, Re = Reynolds number, Sc = Schmidt number.

Cooke and Sherwood found it convenient to correlate their data over the entire pressure range in terms of a coefficient β, defined by

$$G_s = \beta k_1, \qquad \alpha \neq \beta, \tag{2.3}$$

where β = fraction of the absolute sublimation rate G_s at zero pressure.

Substituting for βk_1, equation (2.2) yields

$$\frac{\alpha}{\beta} = 1 + \frac{\alpha k_1}{k_2} P. \tag{2.4}$$

Cooke and Sherwood found that the time-average value $\bar{\alpha} = 0{\cdot}1$ and is essentially constant and independent of pressure, and can be determined in a single test from the weight loss and time. Thus equation (2.4) may be written as

$$t = t_0\left(1 + \frac{\bar{\alpha} k_1}{k_2} P\right), \tag{2.5}$$

where t = time for a test at pressure P, and t_0 = time for the same weight loss at zero pressure.

Their results for the variation of β with pressure for stagnant and moving gas streams follow the curves in Fig. 2.1 and show that the value of α, measured in a stagnant gas, may be unreliable due to the accumulation of vapour from the subliming body.

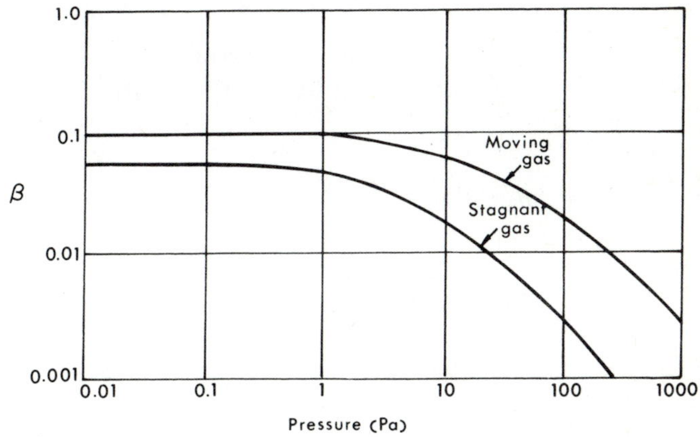

FIG. 2.1. Rate of sublimation expressed as a function of the absolute rate at zero pressure (after Cooke and Sherwood, 1955).

Carman (1948) considered the effect of higher pressures on sublimation in a different way, by invoking Stefan's diffusion equation. The subject has been dealt with in greater detail by Ede (1949a)—see Appendix 2.I.

The rate of diffusion G_m of the vapour through stagnant air to the condenser for plain ice sublimation is given by

$$G_m = \frac{MDP}{RTl} \ln \frac{P - p'_c}{P - p'}, \quad p' \neq p'_c \text{ when } P \neq p', \tag{2.6}$$

where D = diffusion coefficient of vapour in air, M = molecular weight of vapour, P = total pressure, p' = partial vapour pressure outside the interface, p'_c = partial vapour pressure outside the condenser, l = distance between the evaporator and condenser.

Taking average values of $\bar{D}$ and $\bar{T}$, since there is a temperature gradient and since DP is independent of P and proportional to $T^{3/2}$, equation (2.6) reduces to

$$G_m = \frac{k_3}{l} \ln \frac{P - p'_c}{P - p'}, \tag{2.7}$$

where $k_3 = M\bar{D}P/R\bar{T} \propto T^{1/2}$, but can be considered constant.

Many researchers (Carman, 1948; Kramers and Stemerding, 1951; de Boer, 1953) have considered the effect of a condenser in the vacuum system, maintaining a vapour pressure substantially below the sublimation pressure whatever the distance separating the interface and condenser $l < \lambda$, where λ = mean free path. Then vaporization may be considered to occur simultaneously from the interface and from the condenser, so that the rate of sublimation equals the absolute rate of evaporation less the rate of condensation. Therefore, it follows from equation (2.1) that

$$\begin{aligned} G_s &= (p_s - p'_c)\left(\frac{M}{2\pi RT}\right)^{1/2}, \quad l < \lambda, \ \alpha = 1 \\ &= \alpha(p_s - p'_c)\left(\frac{M}{2\pi RT}\right)^{1/2}, \quad l > \lambda. \end{aligned} \tag{2.8}$$

It also follows that, if the partial vapour pressure in the vapour adjacent to the ice interface is p', then

$$G_m = (p_s - p')\left(\frac{M}{2\pi RT}\right)^{1/2}, \quad l < \lambda. \tag{2.9}$$

Therefore, the rate of sublimation from equations (2.1) and (2.9) may be equated to the rate of diffusion, given by equation (2.7), to yield

$$G_m = G_s\left(1 - \frac{p'}{p_s}\right) = \frac{k_3}{l} \ln \frac{P - p'_c}{P - p'}. \tag{2.10}$$

Carman (1948) and Kramers and Stemerding (1951) considered two limiting conditions:

1. At a relatively low air pressure, $P \simeq p'_c$ and $p' \simeq P$, then $l \ll \lambda$, therefore

$$G_m = G_s\left(1 - \frac{p'_c}{p_s}\right). \tag{2.11}$$

2. If the heat supply is insufficient, $G_m \leqslant G_s$ and $p' \simeq p_s$; the temperatures of the interface and condenser become equal in the absence of air. In the presence of air equation (2.7) becomes

$$G_m = \frac{k_3}{l} \ln \frac{P - p'_c}{P - p_s}. \tag{2.12}$$

A more general result has been obtained by Luikov *et al.* (1971) by taking the kinetic resistance equal to the reciprocal of the evaporation rate in terms of the number of molecules instead of the mass rate of evaporation as taken by the previous authors, and the diffusive resistance as the reciprocal of the vapour transfer coefficient. This latter resistance is also inversely proportional to the product of the Nusselt and Knudsen numbers. Luikov *et al.* considered that the mean molecular velocity of the vapour is proportional to the ratio of the vapour diffusivity to the mean free path; this assumption leads similarly to equation (2.4).

2.2 Characteristics of sublimation

2.2.1 EVAPORATION COEFFICIENT

If the actual rate of freeze-drying is to be predicted from modifications to equation (2.1) it seems important to examine the evaporation coefficient α and its dependence on several factors, including the area for sublimation.

In determining the absolute rate of sublimation, the apparent surface of the ice interface is taken rather than the much larger microscopic surface resulting from the sublimation process. This course seems justified since the dimensions of the microscopic roughness can be considered to be substantially less than the mean free path of the escaping vapour molecules. The proof given by Strickland-Constable and Bruce (1954) assumes that the number of molecules crossing an imaginary plane (equal to the apparent area), without collisions in the saturated vapour at a distance less than a mean free pathway from a rough ice interface, will be given by $k_1 p$ (apparent area). On the other hand, if the contour of the ice interface had a larger size than the mean free path and projected beyond the imaginary plane, then it would be impossible to imagine an apparent area on which to calculate the number of molecules crossing it in the course of sublimation.

Another factor concerns the reflections of molecules striking the ice interface. In considering the absolute rate of sublimation there is no reflection of molecules at the surface but, as the actual rate of sublimation becomes smaller, it can therefore be assumed that part of the molecules striking the surface are reflected. Reflection is also a function of the size of the ice interface in a confined vapour space. With water molecules, which are polar, reflection is considered to play an important part at the surface in reducing the value of α. It is reported that the reflection coefficient for water near its freezing point is very high. On the other hand, Kramers and Stemerding (1951) have shown that there is practically no reflection between −40°C and −60°C.

The evaporation coefficient ranges from a maximum value of unity, where all the vapour molecules are fully reflected, to a minimum value of zero where the escape and return to the subliming surface are equal under equilibrium conditions. Between these limits it is necessary to take account of the probability of the distance x a molecule moves without collision and is given by $\exp(-x/\lambda_e)$, where λ_e is the mean free path in equilibrium vapour. If the molecules are not in the equilibrium vapour the effective mean free path may be expressed by $(b\lambda_e)$ where b is a suitable factor. The fraction of molecules colliding is, of course, given by $1-\exp(-x/\lambda_e)$. Burrows (1957) has corrected Strickland-Constable and Bruce's data by calculating α values on this basis for the type of sublimator that was used in their experiments (see Section 2.3).

It is noted (de Boer, 1953) that equation (2.1) for the absolute rate of sublimation might be used to determine the saturated vapour pressure of a substance at some particular ice temperature. This would depend on whether experimental conditions could be obtained without reflection or could be arranged so that the reflection coefficient could be accurately calculated. Ideal experimental surfaces for vapour pressure and sorption measurements have been obtained by an initial freeze-drying procedure with which many workers have obtained good results.

Katayama *et al.* (1969) have found that, in the pressure range 1·33 Pa to 113 Pa, the ice surface becomes very rough, and the maximum rate of sublimation is reached when the surface roughness increases to a maximum.

2.2.2 VAPOUR-PRESSURE DIFFERENCE

Vapour-pressure difference is an important concept in freeze-drying since the driving force for sublimation depends on it. Theoretically, the vapour-pressure difference can be expressed in terms of the sublimation rate and the various pressures included in the freeze-drying system (Mellor, 1966b).

Consider a simple type of freeze-dryer where p_s, p_c are the sublimation and condensation vapour pressures respectively, and p', p'_c the partial vapour pressures near the outer surfaces of the product and condenser respectively, as shown in Fig. 2.2.

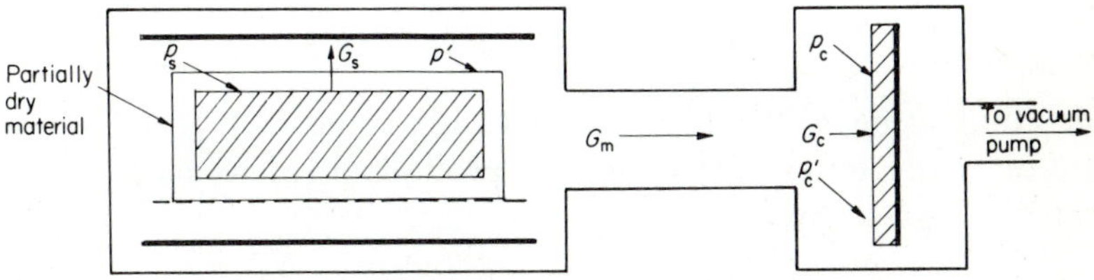

FIG. 2.2. Representation of a simple type freeze-dryer.

The rate of sublimation as given in Appendix 2.II, equation (2.II.10), shows that G_s is proportional to the difference between the vapour pressures at the ice interface and the surface of the product, that is

$$G_s = a(p_s - p'), \tag{2.13}$$

where $a = KM/XRT$ (constant).

Similarly, the rate of condensation is proportional to the differences between the vapour pressure in the vicinity of the condenser and its surface, that is,

$$G_c = a'(p'_c - p_c), \tag{2.14}$$

where a' = constant which is not necessarily equal to a.

The rate of migration of water vapour in the vacuum space between the product and condenser, in terms of Stefan's equation, is

$$G_m = b \ln \frac{P - p'_c}{P - p'}, \tag{2.15}$$

where $b = DMP/lRT$ (constant), D = diffusion coefficient, water vapour in air, M = molecular weight of water vapour, l = distance between product and condenser, R = gas constant, T = absolute temperature, P = total pressure.

In general, considering the freeze-dryer to be a closed system, the rates are equal, thus

$$G_s = G_c = G_m. \tag{2.16}$$

Substituting G_s for G_c in equation (2.14) gives

$$p'_c = p_c + G_s/a'. \tag{2.17}$$

Transposing equation (1.15) and substituting G_s for G_m gives

$$p' = P - (P - p'_c) \exp(-G_s/b). \tag{2.18}$$

After eliminating p'_c from equation (2.18) using (2.17), the vapour pressure at the outer surface of the product p' in terms of known quantities is given by

$$p' = P - (P - p_c - G_s/a') \exp(-G_s/b). \tag{2.19}$$

Now the vapour-pressure difference between the moving interface and the surface of the dry layer is given by $\Delta p = p_s - p'$; therefore, from equation (2.19), this becomes

$$\Delta p = p_s - P + (P - p_c - G_s/a') \exp(-G_s/b). \tag{2.20}$$

Equation (2.20) could have also been obtained by considering Knudsen's equation, namely

$$G_s = \alpha(p_s - p')\left(\frac{M}{2\pi R T_s}\right)^{1/2}, \tag{2.21}$$

where α = sublimation coefficient and T_s = absolute temperature of sublimation, because $\alpha(M/2\pi R T_s)^{1/2}$ is also constant. The absolute temperature of sublimation applies only to sublimation of ice under special conditions such that $\alpha = 1$. Knudsen's equation requires serious corrections in the case of sublimation of frozen materials since the process, even for pure ice, is regarded as one where non-equilibrium conditions prevail (Schrage, 1953).

In freeze-drying there is, first, a short constant-rate drying period during which the vapour flows hydrodynamically because the sublimation pressure is higher than the absolute vacuum pressure. This is followed by a long falling-rate period in which vapour transfer occurs by diffusion. The constant-rate period occurs principally when sublimation takes place from a free ice surface with a gradient in total pressure but, as the ice interface recedes below the surface of the material being dried, the resistance impeding the vapour flow becomes more and more effective, and so this period gives way to a falling-rate period with diffusive flow.

On the other hand, when the absolute pressure is large compared with the vapour pressure of ice and there is a negligible gradient in total pressure, diffusive flow predominates.

The vapour-pressure differences for each case follow from equation (2.20).

2.2.3 CONSTANT-RATE PERIOD

Beginning with freeze-drying and for a short period thereafter, a constant-rate period with hydrodynamic or bulk flow takes place. This manifests itself as a large vapour-pressure difference from the ice surface across a boundary layer to the immediate neighbourhood, extending beyond to the condenser where a small vapour-pressure difference exists. As the dry layer is not sufficiently developed and the absolute pressure high enough, resistance to bulk flow remains small and so the vapour flows unhindered.

The pressure curve (Fig. 2.3a) indicates that $p' = p'_c$ for the vacuum space, therefore, from equation (2.I.8) [Appendix 2.I], $\exp(-G_s/b) = 1$. Further, because $p'_c > p_c$, G_s/a' is not neglected—this follows from relationships (2.14) and (2.16). On substitution of these conditions in

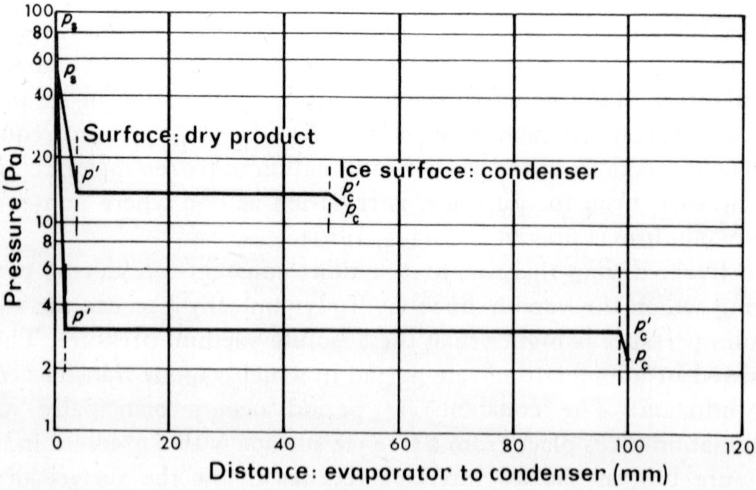

FIG. 2.3a. Variation of vapour pressure across the vacuum space during the constant-rate period.

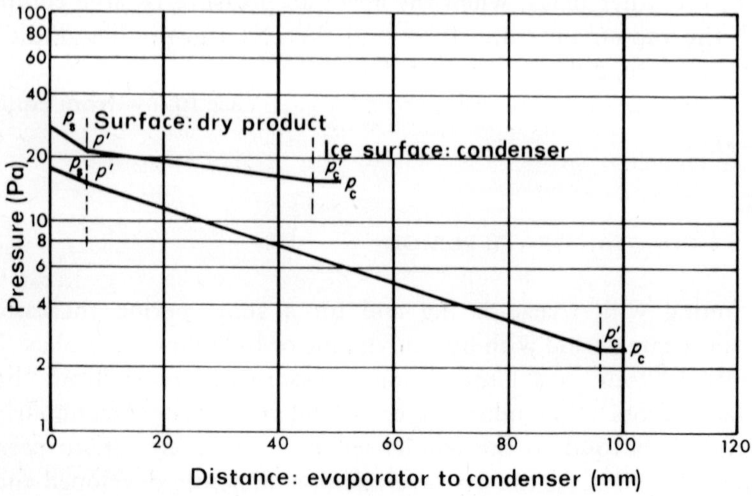

FIG. 2.3b. Variation of vapour pressure across the vacuum space during the falling-rate period.

equation (2.20) the vapour-pressure driving force for the constant-rate period is found to be given by

$$\Delta p = p_s - p_c - \frac{G_s}{a'}. \tag{2.22}$$

As there is no diffusive flow, it can be shown that the analysis given in Section 2.2.2 leads directly to the same result.

The term G_s/a' in equation (2.22) is evaluated from equation (2.14) by taking $G_s = G_c$, where $a' = \alpha(M/2\pi RT_c)^{1/2}$, and where $\alpha =$ condensation coefficient and $T_c =$ temperature of condensation. Thus, for water vapour at -40°C, and assuming $\alpha = 1$, it is found that $a' = 0.016$.

Table 2.1 indicates good agreement between calculated and measured values for the constant-rate period for orange juice; but for cooked meat the values agree only to within about 10 per cent.

This error may be reduced by applying the correction factor $1 - (p_i'/p_s)(T_s/T_i')^{1/2}\Gamma$ to obtain α (Appendix 2.III). Now, at -20°C the absolute sublimation rate G is $0.12\,\text{kg m}^{-2}\,\text{s}^{-1}$, so $G_s/b = 0.008$. Applying the temperature correction $(T_s/T_i')^{1/2} \simeq 0.9$, T_i' is taken equal to 273 K, and the correction factor $\Gamma = 0.98$ from Fig. 2.9 (Appendix 2.III), where it is found that $\alpha = 0.6$ and so $\Delta p = 101.7$. The example shows that changing the assumption from equilibrium conditions for freeze-drying to non-equilibrium conditions involves only a small correction. Actual rates of freeze-drying are such that a much bigger correction would be necessary in order to satisfy the classical theory of freeze-drying. If the correction ($\alpha = 0.1$) for a time-average value as is used to obtain equation (2.5) is applied, then the calculated value $\Delta p = 94{\cdot}2$, which is in good agreement with the measured value of $92{\cdot}1$.

2.2.4 FALLING-RATE PERIOD

Once the dry layer has become appreciably thick its thermal resistance increases and diffusive flow takes place during the remainder of the freeze-drying run. These are the circumstances corresponding to a falling-rate period where mass transfer is the controlling factor. As the flow rate falls off the term G_s/a' becomes small, and from relationship (2.14) the pressure difference $p_c' - p_c$ is negligibly small as seen in Fig.

TABLE 2.1

Calculated and measured values of vapour-pressure differences (after Mellor, 1966b)

Product	Time (h)	Sublimation Temperature (°C)	p_s (Pa)	Condensation Temperature (°C)	p_c (Pa)	Vacuum pressure P (Pa)	Measured sublimation rate G_s (kg m^{-2} s^{-1}) ×10^3	Vapour-pressure difference across boundary or dry layers: Measured Δp (Pa)	Calculated Δp (Pa)	Vapour-pressure difference across vacuum space: Ratio $\Delta p'/\bar{\pi}$	Ratio $\Delta p'/\bar{\pi}l$
Orange juice											
	0·5*	−26	57·3	−40	12·9	32·0	1·37	43·5	43·2		
	1·4*			(end of constant-rate period)							
(16-h drying period)	2	−32	30·9	∼−40	13·3	29·3	0·63			0·97	
	3	−33	28·0	−38·5	15·3	26·7	0·07	5·7	8·3	0·73	0·033
Cooked meat											
	0·25*	−19·8	105·4	−54	2·4	26·7	1·00	92·1	102·2		
(12-h drying period)	2·5	−37	18·1	−54	2·4	17·3	0·18	2·7	1·6	0·32	
(20-h drying period)	8	−26	57·3	−47	5·7	80·0	0·04			0·01	0·030

* Constant-rate period.

2.3b. Because diffusive flow takes place, equation (2.I.8) (Appendix 2.I) applies, and consequently the term G_s/b is retained in equation (2.20). The vapour pressure driving force for the falling-rate period is therefore given by:

$$\Delta p = p_s - P + (P - p_c) \exp(-G_s/b), \tag{2.23}$$

where $b = 1.013 \times 10^6 (DM/lRT)$, D = diffusion coefficient of water vapour in air, l = path length.

Comparison of the performance of two types of dryer shows that agreement between calculated and measured values for the vapour-pressure difference (Table 2.1) is at least qualitatively good. During the falling-rate period, a pressure drop occurs across the dry layer and another one across the vacuum space, as seen in Fig. 2.3a and b. By contrast, in the constant-rate period the main pressure-drop occurs across the boundary layer at the evaporator end. Values of G_s (Table 2.1) have been measured; they could also have been estimated from equation (2.II.11) (Appendix 2.II).

In conclusion, equation (2.20) is of more general application than equation (2.I.12) (Appendix 2.I) for diffusion, because it can be applied to both the contant-rate and the falling-rate periods to predict the vapour-pressure driving force. Thus, the whole sublimation process is covered by this theory and leads to equation (2.20). More research, including measurements of vapour-pressure gradients in the system, is needed as a basis of design for freeze-dryers.

2.2.5 VAPOUR FLOW FROM EVAPORATOR TO CONDENSER

Ede (1949a) observed the pressures in different parts of the drying chamber and the way in which ice builds up on the condenser and suggested that the vapour stream, passing from the evaporator to the condenser, occupies the whole cross-section of the drying chamber. To account for this, the theory proposed (Appendix 2.I) gives: $\Delta p'/\bar{\pi} = 0{\cdot}077$ (total drying rate) from equation (2.I.12), where the drying rate is expressed in terms of the whole cross-sectional area. This is in agreement with the experimental results at low drying rates, as shown in Fig. 2.4; at higher drying rates an increasing but unexplained discrepancy occurs.

The theory shows also that $\Delta p'/\bar{\pi} \propto l$, provided that the absolute pressure P and drying rate G_m remain constant. By applying a suitable correction, based on the areas of 900 cm^2 and 220 cm^2 for the drying chamber and evaporator respectively, Ede showed that for a drying

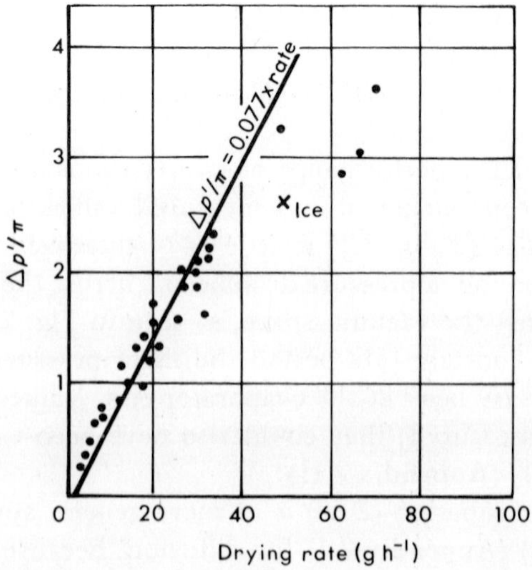

FIG. 2.4. Variation of the ratio vapour-pressure difference to the mean air pressure as a function of the drying rate (after Ede, 1949a).

rate of 8 g h^{-1}, where l ranges from 2.5 to 40 cm, the ratio $\Delta p'/\bar{\pi}l$ is a constant (=0.016), but for a higher rate of 50 g h^{-1} no such correction applies. Thus, there is a limit to which the condenser will work nearer to the evaporator.

Ede therefore concluded that the relations $\Delta p'/\bar{\pi} \propto G_m$ and $\Delta p'/\bar{\pi} \propto l$ might not hold at higher flow rates because of non-uniform gas flow.

Similar conclusions may be drawn from the data in Table 2.1 by Mellor (1966b) The basic parameters for Table 2.1 and Table 5.1 (Section 5.6) are the same. After 3 hours freeze-drying of orange juice in a similar dryer to Ede's, the rate was 11 g h^{-1} and the ratio $\Delta p'/\bar{\pi}l$ = 0·033, taking 860 cm^2 and 422 cm^2 for the central cross-sectional area of the chamber and the area of the evaporator (or condenser) respectively. Again, after 8 hours freeze-drying of cooked meat in a chamber having a cross-sectional area of 368 cm^2 and an external condenser with an area of 60 cm^2, the rate was 60 g h^{-1} and the ratio $\Delta p'/\bar{\pi}l = 0{\cdot}030$. Data for such systems are sparse; with the exception of Ede's results; these appear to be the only two meaningful values available for the ratio $\Delta p'/\bar{\pi}l$.

The design of freeze-dryers should be considered in relation to this aspect and equipment should be built and tested that conforms closely to the theoretical considerations outlined. This equipment should be used to obtain further data and confirmation of the theory.

2.3 Collision theory applied to sublimation problems

Limitations are imposed by the sublimator in so far as the classical theory of freeze-drying is concerned.

Mention has been made in Section 2.1 of the low values of the evaporation coefficient ($\alpha < 1$) obtained by Strickland-Constable and Bruce (1954) at a sublimation temperature of −40°C, and of the explanation for these results put forward by Burrows (1957). He points out that wall constrictions, such as when the level of the subliming surface falls below the top of the container, or the interposition of a condenser surface that interferes with the free motion of the molecules, alters their behaviour and thus modifies the value of α. This observation has been ignored by many researchers who have attempted to measure the evaporation coefficient. Another important point is that the effective mean free path, so far as collision theory is concerned, has to be expressed by ($b\lambda_e$) where b is a suitable factor and λ_e is the mean free path in equilibrium vapour. Though λ_e would be difficult to determine because of the concentration gradient in the vapour space and the "cross-fire" effect of re-evaporating molecules, in practice its value at such higher temperatures as −40°C is sufficiently small to render the fraction, $\exp(-a/b\lambda_e)$, of molecules moving without a collision negligible.

2.3.1 THEORY

The Clausing factor W_1 is the probability that a molecule will escape from a non-condensing tube and, for evaporation taking place in a tube, would be the ratio of the escape velocity at the outlet to the inlet velocity of the vapour at the evaporating surface. Clausing (1929) derived his factor on the assumption of free molecular flow, with molecules passing straight through the tube and those making collisions with the wall of the tube rebounding randomly according to the cosine law. Thus, assuming the total flow rate in a tube equals the sum of the free molecular and laminar viscous flow rate equations (Pollard and Present, 1948), the incorporation of factor W_1 into these equations yields

$$G_K + G_v = a^2 W_1 \,\Delta p\left(\frac{\pi RT}{2M}\right)^{1/2}\left(1+\frac{\pi a^2}{8l\lambda W_1}\right) \tag{2.24}$$

where W_1 can be determined by the approximate expression

$$W_1 \simeq \frac{1}{1+3l/8a}, \tag{2.25}$$

l being the length of the tube and a its radius. It will be noted that $W_1 = 1$ when $l = 0$ corresponds to an orifice, and that $W_1 \rightarrow 8a/3l$ for a very long tube; both limits were deduced by Clausing.

The term $\pi a^2/8l\lambda W_1$ represents the "reduced viscous flow"—for example, when an ice interface subliming at $-54{\cdot}8$°C reaches a level of 2 cm in a tube 1·5 cm in radius. It will be found from equation (2.24) that the flow will be 18 per cent viscous and 82 per cent molecular.

Imagine two groups of molecules resulting from the sublimation of a moving ice interface in the tube shown in Fig. 2.5(a). For a short

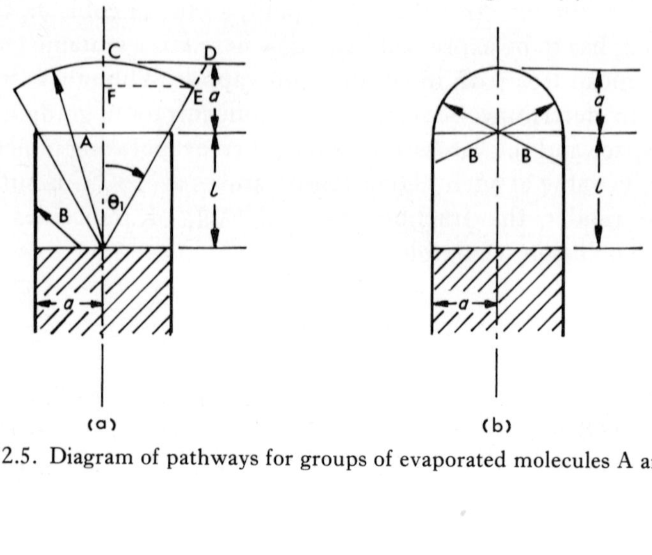

FIG. 2.5. Diagram of pathways for groups of evaporated molecules A and B.

tube, group A molecules would pass unhindered within a cone to a point outside the tube equal to its radius where they would then be in the equilibrium vapour. Another smaller group B would impinge on the wall of the tube.

It is shown in Appendix 4.I that the total fraction of molecules travelling a distance a without collision in a tube would be $F(a/\lambda)$ and the total fraction that collide with one another would therefore be $1-F(a/\lambda)$. The fractions in this case can be shown from equation (4.I.8) for group A molecules to be

$$F(a/b\lambda_e) = 2\exp[-(l+a)/b\lambda_e]\int_0^{\theta_1}\sin\theta\cos\theta\,d\theta$$
$$= \frac{a}{a^2+l^2}\exp[-(l+a)/b\lambda_e] \tag{2.26}$$

since, from Fig. 2.5(a), $\sin\theta_1 = a/(a^2+l^2)^{1/2}$ and

$$1-F(a/b\lambda_e) = [1-e^{-(l+a)/b\lambda_e}]/[1+(a^2+l^2)^{1/2}/a]. \tag{2.27}$$

where it is assumed that a few intermolecular collisions occur in the cone on the way to the tube exit, and that thereafter the motion is random. The fraction escaping after collision would be proportional to the ratio of the outlet area to the combined area of the outlet and conical surface. Thus the ratio equals

$$\frac{\pi a^2}{\pi a^2 + \pi a(a^2+l^2)^{1/2}}$$

since the conical surface can be shown to equal

$$\pi(\mathrm{CD})(\mathrm{FE}) = \frac{a(a+l)}{l}\,\frac{a(a+l)}{(a^2+l^2)^{1/2}}$$

and it can be assumed that for a short tube $l \simeq 2a$.

Combining equation (2.26) and (2.27) and the Clausing factor W_1, the net total fraction of molecules escaping is given by

$$f_1 = W_1 a^2/(a^2+l^2)\{e^{-(l+a)/b\lambda_e}$$
$$+[1-e^{-(l+a)/b\lambda_e}]/[1+(a^2+l^2)^{1/2}/a]\}. \tag{2.28}$$

Consider next group B molecules, some of which collide with the wall, rebound and thus contribute to total outward flow—a kind of slip flow until they escape eventually beyond the hemispherical region shown at the end of the tube in Fig. 2.5(b). By analogy with the analysis in Appendix 4.I the net total fraction of those escaping is given by

$$f_2 = [W_1 l^2/(a^2+l^2)][e^{-2a/\lambda_e} + (2/3)(1-e^{-2a/\lambda_e})] \tag{2.29}$$

since, of these, the total fraction escaping is

$$\left(1-\frac{a^2}{a^2+l^2}\right)\exp(-2a/\lambda_e)$$

and the total fraction colliding is

$$\frac{2}{3}\left(1-\frac{a^2}{a^2+l^2}\right)[1-\exp(-2a/\lambda_e)]$$

and of these latter molecules the fraction contains the ratio of the hemispherical surface to the outlet area, which has a value of 2/3. A mean free path equal to λ_e can be taken here, since the molecules are sufficiently random at the exit of the tube.

The evaporation coefficient α is then the total overall fraction of molecules escaping, that is, equations (2.28) and (2.29) combined lead to

$$\alpha = f_1 + f_2. \tag{2.30}$$

It is interesting to review the data obtained by Strickland-Constable and Bruce (1954), relating to the sublimation of ice, in the light of the present theory. In their apparatus a small condenser was located about 80 mm from the evaporator. Since

$$\lambda_e = \frac{0{\cdot}8\times10^{-17}\,T_s}{p_s\sigma^2} = 1{\cdot}43 \text{ mm}$$

where T_s is the sublimation temperature (−54·8°C), p_s the sublimation pressure and σ the diameter of a water vapour molecule (0·282 nm); the condenser has little effect on collisions between molecules in the neighbourhood of the evaporator. During sublimation the ice interface dropped a distance of 2 cm in the container.

Table 2.2 gives the theoretical evaporation coefficient calculated from equations (2.28) to (2.30). The average value is 0·43, which is in good agreement with the value of 0·44 obtained by Strickland-Constable and Bruce at −54·8°C. Their results did not, however, support the prediction (in Table 2.2) of a lowering of evaporation rate in the course of sublimation.

Burrows considers that molecules rebounding from the top of the wall of the evaporator container (see Fig. 2.5(b)) may affect the

TABLE 2.2.
Values of the theoretical evaporation coefficient for sublimation of ice

l (mm)	l/a	Sublimation temperature (°C)	λ_e (mm)	W_1	f_1	f_2	α
0	0	−54·8	1·43	1·00	0·500	0·000	0·500
5	0·33	−54·8	1·43	0·89	0·391	0·059	0·450
10	0·67	−54·8	1·43	0·80	0·251	0·164	0·415
15	1·0	−54·8	1·43	0·73	0·146	0·243	0·389
20	1·33	−54·8	1·43	0·67	0·090	0·285	0·375

evaporation rate. This suggests that it would be desirable to use a thin-wall container to give the molecules a better chance to escape.

2.4 Analysis of the freezable water in freezing and drying

Sometimes in the freezing and drying of substances of a medical and chemical nature for which physical data are scanty, there arises a need for a theoretical prediction of the freezing temperature and water activity in terms of the "freezable" water in the substance. Freeze-drying will primarily depend on these factors in relation to: (1) initial freezing where the approach to the eutectic point governs the amount of water frozen out; (2) sublimation where optimal temperatures consistent with the reduced amount of water determine the risk of melting. These factors can either be calculated by thermodynamic analysis or determined experimentally (but with some difficulty).

The analysis requires knowing the exact water content and the freezing point, or, alternatively, the water content and the amount of solute from which a theoretical freezing point can be found. There are also the insoluble solids and a small non-freezable portion of the water content of biological materials which is a more important

TABLE 2.3
The influence of temperature on the amount of water frozen out for various foodstuffs (after Kuprianoff, 1964)

Foodstuff	Total water content (%)	Percentage of water frozen out in relation to the freezable water at different temperatures (°C)					Non-freezable water (kg kg^{-1} solids)
		−5	−10	−15	−20	−30	
Lean beef	74	83	93	97	99	100	0·35
Haddock	83·5	87	94	97	98	100	0·39
Cod	80·5	85	94	97	98	100	0·39
Perch	79	84	94	97	98	100	0·39
Whole egg	74	90	95	98	99	100	0·20
Egg yolk	50	89	94	97	98	100	0·40
Egg white	86·5	93	96	98	99	100	0·40
Fruit juice	88	75	87	93	96	100	0·20
Whitebread	46	50	87	97	99	100	0·30
Peas	78	68	86	92	96	100	0·2–0·3
Beans	89	84	92	96	98	100	0·2–0·3
Spinach	93	95	97	98	99	100	0·20

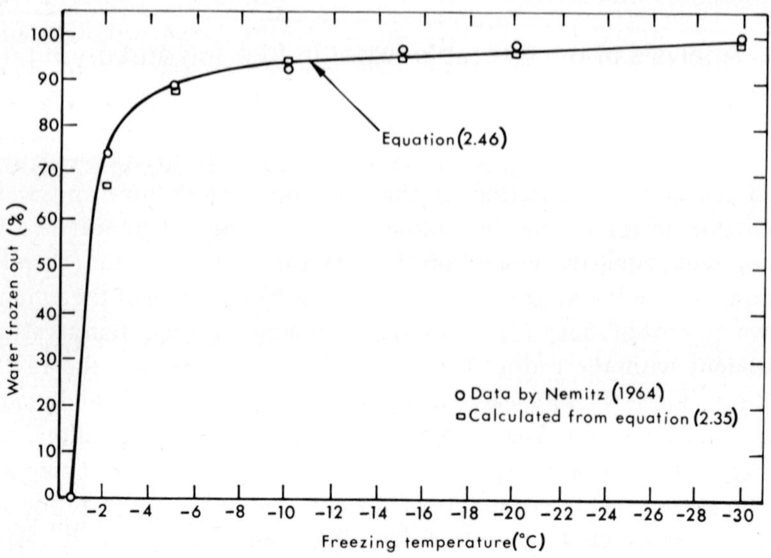

FIG. 2.6. Percentage of water frozen out in relation to the "freezable" water at different temperatures for egg albumen.

consideration at temperatures of −25°C to −30°C, and can only be determined experimentally. The present author has used a theoretical approach to predict sublimation temperatures for the satisfactory production of freeze-dried fruit purées (Anet and Reynolds, 1957), ox spleen (Cobbin and Thorp, 1957) or other miscellaneous preparations.

Hirsch (1946) has stated that freezing and drying are intimately related processes involving changes in phase equilibria of water–ice mixtures. Graphical illustration has been employed by him to coordinate existing data on various foodstuffs and solutions isopiestic with them. The freezing of food has been treated thermodynamically by Bartlett (1944), and data obtained by Short (1944) for raw beef and orange juice tend to confirm Bartlett's theory. There are other data due to Heiss (1933) and to Kuprianoff (1964), the latter having tabulated his results according to the measurements of Riedel (Table 2.3). Data published by Nemitz (1964) for egg albumen (Fig. 2.6) seem important in the present context.

2.4.1 FREEZING AND RELATIVE LOWERING OF VAPOUR PRESSURE

Biological materials in a partially frozen state may be regarded as heterogeneous mixtures of insoluble solids and two phases involving water: an ice phase consisting of water frozen out of solution and a solution phase of unfrozen water and solutes. In thermodynamic equilibrium at a constant temperature above the eutectic point and at a constant absolute pressure the two phases can be assumed to have equal vapour pressures, so that

$$p_i = p_{sol} \qquad (2.31)$$

where p_i = vapour pressure of ice and p_{sol} = vapour pressure of the solution.

Equal vapour pressures at a constant temperature below the eutectic point can also be assumed for a metastable state present in some substances. For example, the process of formation of sucrose hydrate crystals is extremely slow, and consequently the metastable state will exist at all temperatures below the eutectic point of −13·0°C (Young and Jones, 1949), as shown in Fig. 2.7. Therefore, if temperature equilibrium is reached, whether in freezing or in the course of freeze-drying, the vapour pressure of the residual phase, comprising the remaining water and solutes, equals the vapour pressure of ice.

Applying a modified Raoult's law: "The ratio of the vapour pressure of the solution p_{sol} to that of the pure solvent (water) p_w at the same temperature and pressure equals the mole fraction of the solvent", then

$$N_w = \frac{p_{sol}}{p_w} = \frac{W/M_w}{W/M_w + S/M_S}, \tag{2.32}$$

where N_w = mole fraction of water, W = amount of water, S = amount of solute, M_w = molecular weight of water, M_S = molecular weight of solute. In general, the water activity a_w is given by

$$a_w = \frac{p_i}{p_w} = \frac{W}{W + S(M_w/M_S)}. \tag{2.33}$$

This follows from equations (2.31) and (2.32).

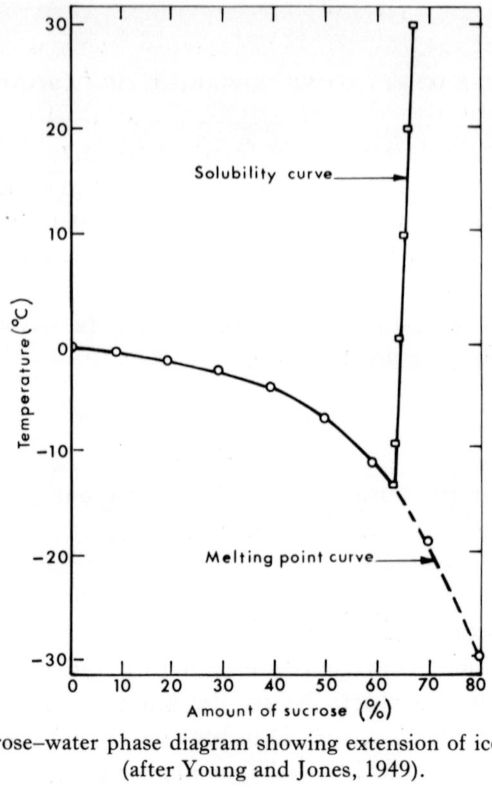

FIG. 2.7. Sucrose–water phase diagram showing extension of ice equilibrium curve (after Young and Jones, 1949).

Consequently, at the freezing point the water activity is dependent on temperature and may be determined by taking the ratio of the vapour pressure over ice to that over water at the same temperature; for example, for an aqueous sucrose solution with a freezing point of −0·9°C the water activity $a'_w = 0{\cdot}991$. It is also dependent on solute concentration (equation 2.33).

If W is the "freezable" water content and x the amount of ice (water frozen out), then $W - x$ is the amount of water unfrozen, and so the activity at the freezing point a'_w can be assumed to be given by equation (2.33); likewise, the activity at a lower temperature T is given by

$$a_w = \left(\frac{p_i}{p_w}\right)_T = \frac{W - x}{W - x + S(M_w/M_S)} \qquad (2.34)$$

Below the freezing point the concentration increases as water is frozen out and, providing the percentage frozen out in relation to the total freezable water is known, the water activity also can be estimated in this way. However, since most biological substances contain always a small percentage of non-freezable water bound to macromolecules this water may be regarded as part of the solute concentration for purposes of estimation. For example, for fruit juice (Table 2.3) the non-freezable water amounts to 2·4 g per 100 g solution; thus this amount when added to the amount of solids gives an adjusted value for S of 14·4. Reference to the data in Table 2.3 and equation (2.34) then gives the water activity $a_w = 0{\cdot}818$ at a freezing temperature of −20°C, or $(p_i/p_w)_{-20°C} = 0{\cdot}822$.

Eliminating the term $S(M_w/M_S)$ from equations (2.33) and (2.34) yields

$$\frac{x}{W} = \frac{1 - (a_w/a'_w)}{1 - a_w} \qquad (2.35)$$

Alternatively, equation (2.35) written in terms of vapour pressure becomes

$$\frac{W - x}{W} = \frac{(p_w - p_i)/p_i}{(p'_w - p'_i)/p'_i}, \qquad (2.36)$$

where the superscript prime denotes the ice and water vapour pressures at the lower temperature. Thus, the amount of water frozen out can be expressed in terms of the relative lowering of vapour pressure at the freezing point to that at any other temperature within the freezing range.

2.4.2 FREEZING POINT ACTIVITY

The accuracy of equation (2.35) is only fair in the temperature range down to $-10°C$ (Fig. 2.6). The freezing point is very often unknown and, in order to improve the accuracy, the activities a_w, a'_w and the freezing point may be determined thermodynamically in the following way.

Consider the variation of the activity a_w, or the mole fraction of water N_w, with temperature T within the freezing range (freezing point to eutectic point) at constant pressure P, which is given by van t'Hoff's law of freezing point lowering (Glasstone, 1947) as

$$\left(\frac{\partial \ln a_w}{\partial T}\right)_P = \left(\frac{\partial \ln N_w}{\partial T}\right)_P = \frac{\Delta H_m}{RT^2}, \qquad (2.37)$$

where R = gas constant, ΔH_m = molar latent heat of fusion of ice, and $a_w = N_w$ (by definition).

Since the solution is becoming more concentrated as the temperature is lowered below the freezing point, ΔH_m must vary with temperature, otherwise, ΔH_m could only be regarded as constant for dilute solutions (high freezing points). The variation of ΔH_m with absolute temperature T may be expressed by Kirchhoff's equation (Glasstone, 1947) as

$$\left(\frac{\partial(\Delta H_m)}{\partial T}\right)_P = (c_P)_w - (c_P)_i = \Delta c_P, \qquad (2.38)$$

where $(c_P)_w$ = molar specific heat of water at constant pressure, $(c_P)_i$ = molar specific heat of ice at constant pressure. The difference in specific heats Δc_P can be considered constant over a short temperature range, so that integration of equation (2.38) gives

$$\Delta H_m = \Delta H_0 - \Delta c_P (T_0 - T), \qquad (2.39)$$

where ΔH_0 = molar latent heat of fusion at T_0; and T_0 = freezing point of water.

Equation (2.37) combined with (2.39) becomes

$$\mathrm{d} \ln a_{\mathrm{w}} = \frac{\Delta H_0 - \Delta c_p (T_0 - T)}{RT^2} \mathrm{d}T, \tag{2.40}$$

which, after integration and with an integration constant evaluated at $a_{\mathrm{w}} = 1$ for the freezing point of water ($T = T_0$), yields

$$a_{\mathrm{w}} = \exp(-A + B/T_0 + C \log T_0). \tag{2.41}$$

And also at the freezing point T_{F}

$$a'_{\mathrm{w}} = \exp(-A + B/T_{\mathrm{F}} + C \log T_{\mathrm{F}}), \tag{2.42}$$

where the derived constants are given by

$$A = \frac{\Delta c_P T_0 - \Delta H_0}{RT_0} + 2{\cdot}303 \frac{\Delta c_P}{R} \log T_0,$$

$$B = \frac{\Delta c_P T_0 - \Delta H_0}{R},$$

$$C = 2{\cdot}303 \frac{\Delta c_P}{R}.$$

The values of these constants, using the following data for ice and water, are: $\Delta H_0 = 601{\cdot}08\ \mathrm{J\ mol^{-1}}$, $(c_P)_{\mathrm{w}} = 75{\cdot}24$ and $(c_P)_{\mathrm{i}} = 37{\cdot}62\ \mathrm{J\ mol^{-1}\ K^{-1}}$, $T_0 = 273\ \mathrm{K}$ and $R = 8{\cdot}31\ \mathrm{J\ mol^{-1}\ K^{-1}}$, become $A = 27{\cdot}306$, $B = 513{\cdot}848$, $C = 10{\cdot}435$.

2.4.3 FREEZING POINT DEPRESSION

The freezing point depression for biological materials is comparatively small (between 0·5°C and 2°C), which means it can be obtained by assuming a freezing point T_F for a dilute solution, and using equation (2.40) with the mole fraction of water N_{w} expressed simply by

$$\mathrm{d} \ln N_{\mathrm{w}} = \frac{\Delta H_0}{RT^2} \mathrm{d}T. \tag{2.43}$$

Integration with the same conditions ($a_w = 1$, $T = T_0$) as were used to derive equation (2.41) gives

$$\ln N_w = \frac{\Delta H_0}{RT_0} - \frac{\Delta H_0}{RT_F} = \ln(1 - N_S) \tag{2.44}$$

where N_S is the mole fraction of the solute.

Now, since N_S is small for dilute solutions, $\ln(1 - N_S) \simeq N_S$ and also

$$\frac{\Delta H_0}{RT_0} - \frac{\Delta H_0}{RT_F} = \left(\frac{\Delta H_0}{RT_0^2}\right)(T_0 - T_F)$$

since the freezing point depression $\Delta T = (T_0 - T_F)$ is small; whence

$$\Delta T = N_S \frac{RT_0^2}{\Delta H_0} = k \frac{m}{M_S}, \tag{2.45}$$

where

$$k = M_w \frac{RT_0^2}{\Delta H_0} = 1856{\cdot}46 \text{ kg K mol}^{-1},$$

$$m = \frac{S}{W} \quad \text{(fractional amount of solute in solution)},$$

$$N_S \simeq \frac{S/M_S}{W/M_w} \quad (S/M_S \ll W/M_w \text{ for dilute solutions}).$$

2.4.4 AMOUNT OF WATER FROZEN OUT

The amount of water frozen out x in relation to "freezable" water content W at any temperature T for a biological material with a freezing point T_F then follows, by combining equations (2.35) and (2.42) to give

$$\frac{x}{W} = \frac{1 - a_w \exp(A - B/T_F - C \log T_F)}{1 - a_w}, \tag{2.46}$$

where the activity $a_w = \exp(-A + B/T + C \log T)$ for freezing temperatures ($T < T_F$). And the value of T_F is given by equation (2.45).

The adequacy of the above theory is clearly demonstrated by the curve of percentage water frozen out for egg albumen with the lowering of freezing temperature (Fig. 2.6).

2.4.5 NON-FREEZABLE WATER

It has already been stated (in Section 2.4) that the non-freezable water can only be determined experimentally. However, one possibility of calculating it for biological materials would be to apply a correction to the present theory for the insoluble substances, and to include among these the non-freezable part of the total water content. This water might be considered as bound and remaining solid, like the insolubles at very low temperatures, whether or not the eutectic point is well defined.

Rearranging the freezing equation of Kramers (1958) for the mole fraction of the water frozen out in the presence of insolubles gives

$$w_{\text{ins}} = \frac{w_{\text{S,i}}(1 - w_{\text{ins}}) - w_{\text{S}}}{N_{\text{w}}} - w_{\text{S,i}}(1 - w_{\text{ins}}) \tag{2.47}$$

where w_{S} = mass fraction of solute in the material, w_{ins} = mass fraction of insolubles in the material, $w_{\text{S,i}}$ = mass fraction of solids (solute + ice) in the pure solution at the eutectic mixture.

For unit mass of eutectic mixture, writing w_{f} = mass fraction of the frozen portion and w_{unf} = mass fraction of the unfrozen (including insolubles) portion, equation (2.47) becomes

$$w_{\text{unf}} = \frac{1 - N_{\text{w}}}{N_{\text{w}}} w_{\text{f}} - w_{\text{S}}, \tag{2.48}$$

where $w_{\text{S,i}} = (w_{\text{S}} + w_{\text{f}})/(1 - w_{\text{unf}})$, $w_{\text{ins}} = w_{\text{unf}}$ and $1 - w_{\text{unf}}$ = mass fraction for the pure solution.

For example, $w_{\text{unf}} = 0{\cdot}065$ for fruit and vegetable juices with 12 per cent solids content in the temperature range −25°C to −30°C, using the data from the nomogram of Riedel (1950), and his enthalpy relationship gives the fraction of insolubles $w_{\text{ins}} = 0{\cdot}036$. Thus, the fraction of non-freezable water is $0{\cdot}065 - 0{\cdot}036 = 0{\cdot}029$, or 2·9 per cent, which agrees reasonably well with the experimental determination of 2·4 per cent (Table 2.3). Maltini (1975) gives values of 2·0 and 2·8 per cent for natural orange and lemon juices respectively.

As the total solids content is 12 per cent, the amount of dissolved solute is 8·2 per cent, which is in agreement with juices having an approximate freezing point of 1°C from equation (2.45) and also from Fig. 2.7 and Riedel's nomogram.

Appendix 2.I Vapour space diffusion

Mean air pressure

Ede (1947) assumed that the partial air pressure π increased exponentially from the surface of the drying product to the condenser according to the equation $\pi = a\,e^{bx}$, where a, b are constants and x is the distance from the surface to the condenser.

The mean air pressure is given by

$$\bar{\pi} = \frac{\int_0^l \pi\, dx}{\int_0^l dx}, \tag{2.I.1}$$

where l = distance between the drying surface and condenser.

Substitution for $dx = (1/b)(d\pi/\pi)$ and insertion of the boundary conditions: $\pi = P - p'$ at $x = 0$ and $\pi = P - p'_c$ at $x = l$ in equation (2.I.1), and integration give

$$\bar{\pi} = \frac{p' - p'_c}{\ln (P - p')/(P - p'_c)} \qquad 2.I.2)$$

Rate of diffusion

The rate of diffusion in the vapour space (Ede, 1949a) is found by considering the reciprocal penetration of air and vapour at a distance x from the ice interface during freeze-drying, which is linearly and uniformly distributed according to the Stefan–Nusselt equations:

$$-\frac{DM}{RT}\frac{dp}{dx} + \frac{v_g M}{RT} p = G_m, \tag{2.I.3}$$

$$-\frac{DM_a}{RT}\frac{d(P-p)}{dx}+\frac{v_g M_a}{RT}(P-p)=0, \tag{2.I.4}$$

where D = diffusivity of either gas, M = molecular weight of water vapour, M_a = molecular weight of air, v_g = velocity of the gas mixture towards the condenser.

Assuming that the absolute pressure P is constant everywhere and that temperature variations are ignored, elimination of v_g from equations (2.I.3) and (2.I.4) gives

$$\frac{v_g}{D}=\frac{1}{p}\frac{dp}{dx}+\frac{RTG_m}{DMp}=\frac{1}{P-p}\frac{d(P-p)}{dx}. \tag{2.I.5}$$

Substitution of $\pi = P - p$ for the partial air pressure in equation (2.I.5) yields

$$-\frac{1}{P-\pi}\frac{d\pi}{dx}+\frac{RTG_m}{DM(P-\pi)}=\frac{1}{\pi}\frac{d\pi}{dx}, \tag{2.I.6}$$

with $dP/dx = 0$, since P is constant.

Integration of equation (2.I.6) gives

$$\ln\frac{\pi}{c}=\frac{RTG_m x}{DMP}, \tag{2.I.7}$$

where c = constant.

Substitution from the boundary conditions $\pi = P - p'_c$ at $x = l$ and $\pi = P - p'$ at $x = 0$ in equation (2.I.7) gives the vapour flow rate due to diffusion as

$$G_m=\frac{DMP}{RTl}\ln\frac{P-p'_c}{P-p'}. \tag{2.I.8}$$

Vapour-pressure difference

Ede (1949a) extended the theory to include the vapour-pressure difference and distribution of air pressure in the vacuum space. In general, since $p \propto x$, equation (2.I.8) becomes

$$\frac{P-p}{P-p'}=\exp\left(\frac{G_m RTx}{DMP}\right). \tag{2.I.9}$$

By dedendum, with $p = p'_c$ at $x = l$ (condenser face), this leads to

$$\frac{p' - p'_c}{P - p'} = \exp\left(\frac{G_m RTl}{DMP}\right) - 1 \tag{2.I.10}$$

Consider next the logarithmic expansion of the original expression (2.I.8), by applying dedendum to $(P - p'_c)/(P - p')$ giving $(p' - p'_c)/(P - p')$ and addendum giving $(2P - p' - p'_c)/(P - p')$. Since the ratio of the two, viz. $(p' - p_c)/(2P - p' - p'_c)$, is a comparatively small value, the first term of the expansion yields

$$\ln\frac{P - p'_c}{P - p'} \simeq \frac{2(p' - p'_c)}{2P - p' - p'_c} \tag{2.I.11}$$

or

$$\frac{\Delta p'}{\pi} = \frac{G_m RTl}{DMP}, \tag{2.I.12}$$

where the mean air pressure across the vacuum space is given by $\bar{\pi} = (2P - p' - p'_c)/2$ and the vapour-pressure difference across the vacuum space is given by $\Delta p' = p' - p'_c$.

Ede showed also, by combining equations (2.I.9), (2.I.10) and (2.I.12), that the distribution of air pressure is given by

$$\pi = \bar{\pi}\frac{G_m RTl}{DMP}\frac{e^{-G_m RTx/DMP}}{e^{-G_m RTl/DMP} - 1} \tag{2.I.13}$$

which is in quantitative agreement with the original assumption that $\pi = a\,e^{bx}$.

Appendix 2.II Vapour-pressure equation

To obtain the basic equation for vapour transfer, Mellor (1966b) considered the flow of water vapour from a moving ice interface at a depth $x = X(t)$ below the surface of a semi-infinite solid (Fig. 2.8) with a saturated vapour pressure p_s of ice which remains constant; the vapour pressure at the surface of the product being zero.

The vapour pressure p at a depth x and time t in the dry layer satisfies the fundamental equation

$$\frac{\partial^2 p}{\partial x^2} = \frac{1}{K}\frac{\partial p}{\partial t}, \qquad 0 < x < X, \tag{2.II.1}$$

where K = permeability for water vapour in the dry product. The vapour pressure also satisfies two boundary conditions:

Surface: $p = 0, \quad x = 0, \quad t > 0$ (2.II.2)

Interface: $p = p_s, \quad x = X.$ (2.II.3)

Equations (2.II.1) and (2.II.2) have the solution

$p = A \operatorname{erf}\{x/[2(Kt)^{1/2}]\}$ where A is a constant to be determined.

From equation (2.II.3), $p_s = A \operatorname{erf}\{X/[2(Kt)^{1/2}]\}$.

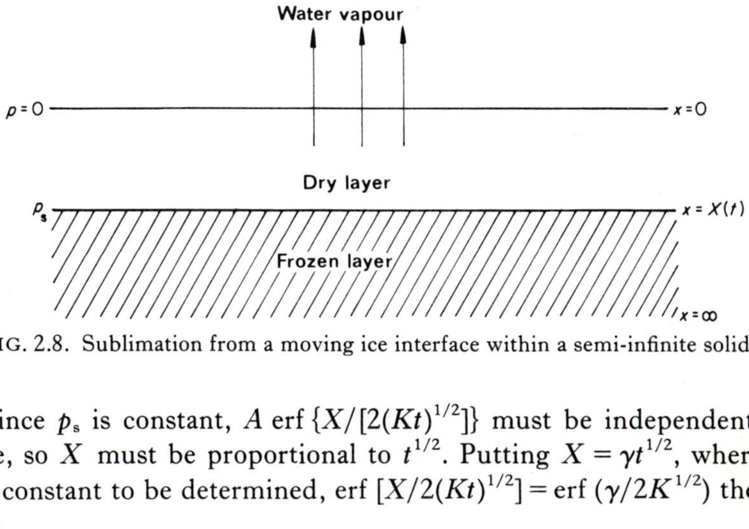

FIG. 2.8. Sublimation from a moving ice interface within a semi-infinite solid.

Since p_s is constant, $A \operatorname{erf}\{X/[2(Kt)^{1/2}]\}$ must be independent of time, so X must be proportional to $t^{1/2}$. Putting $X = \gamma t^{1/2}$, where γ is a constant to be determined, $\operatorname{erf}[X/2(Kt)^{1/2}] = \operatorname{erf}(\gamma/2K^{1/2})$ therefore

$$\frac{p}{p_s} = \frac{\operatorname{erf}\{x/[2(Kt)^{1/2}]\}}{\operatorname{erf}(\gamma/2K^{1/2})} \tag{2.II.4}$$

For thin sections the value of the exponent $\gamma/2K^{1/2}$ has been found to be small. Expanding it as a power series,

$$\operatorname{erf}\frac{\gamma}{2K^{1/2}} = \frac{2}{\pi^{1/2}}\sum_{\nu=0}^{\infty}\frac{(-1)^{\nu}(\gamma/2K^{1/2})^{2\nu+1}}{(2\nu+1)\nu!} \simeq \frac{\gamma}{(\pi K)^{1/2}}. \tag{2.II.5}$$

Substituting $\gamma/(\pi K)^{1/2}$ for erf $[X/2(Kt)^{1/2}]$ in equation (2.II.4) and changing the zero of the vapour-pressure scale to p' (for the surface of the product), that is, p_s is equivalent to $p_s - p' = \Delta p$, gives

$$p = \frac{(\pi K)^{1/2}\Delta p}{\gamma} \operatorname{erf} \frac{x}{2(Kt)^{1/2}}. \tag{2.II.6}$$

As the interface moves, the rate G_s at which the vapour escapes is given by

$$G_s = \rho_i \left(\frac{dX}{dt}\right), \tag{2.II.7}$$

where ρ_i = density of ice. Further, from the definition of K,

$$G_s = \frac{KM}{RT}\left(\frac{\partial p}{\partial x}\right)_{x=X} \tag{2.II.8}$$

Differentiation of equation (2.II.6) with respect to x and of equation $X = \gamma t^{1/2}$ with respect to t, and substitution in equations (2.II.7) and (2.II.8) yields

$$G_s = \frac{KM\,\Delta p}{RT\gamma t^{1/2}} \exp(-x^2/4Kt)|_{x=X} = \rho_i \frac{\gamma}{2t^{1/2}}. \tag{2.II.9}$$

As t becomes larger, equation (2.II.9) leads to

$$\gamma^2 = \frac{2KM\,\Delta p}{\rho_i RT}. \tag{2.II.10}$$

In equation (2.II.8), substitution of $\Delta p/x$ for $(\partial p/\partial x)_{x=X}$ and $X = \gamma t^{1/2}$ gives

$$G_s = \frac{KM\,\Delta p}{RT\gamma t^{1/2}} = \frac{KM\,\Delta p}{XRT}. \tag{2.II.11}$$

Appendix 2.III Kinetic theory of sublimation and condensation

In the theory for a non-equilibrium condition developed by Schrage (1953), starting with Maxwell's law of the distribution of velocities, the number of molecules lying in the range du dv dw with velocities u, v, w is

$$dn = n(\beta^3/\pi^{3/2}) \exp[-\beta^2(u^2+v^2+w^2)] \, du \, dv \, dw, \tag{2.III.1}$$

where n = total number of molecules per unit volume, and $\beta = (M/2RT)^{1/2}$.

Consider net sublimation as the sum of the rates of sublimation away from the ice–vapour interface ($u > 0$) and of recondensation towards the interface ($u < 0$); then Maxwell's law can be applied to a vapour space having mean velocities $\bar{u}$, $\bar{v}$ and $\bar{w}$ as the frame of reference.

The absolute rate of sublimation is given by

$$\begin{aligned} G_s &= \text{mu } dn_s \\ &= \rho_s \frac{\beta_s}{\pi^{3/2}} \int_0^\infty \int_{-\infty}^\infty \int_{-\infty}^\infty u \exp[-\beta_s^2(u^2+v^2+w^2)] \, du \, dv \, dw \end{aligned} \tag{2.III.2}$$

where m = mass of a molecule, $\rho_s = mn_s$ is the vapour density, and subscript s denotes sublimation (saturated conditions). Therefore

$$G_s = \frac{\rho_s \beta_s^3}{\pi^{3/2}} \frac{\pi}{\beta_s^2} \int_0^\infty u \exp(-\beta_s^2 u^2) \, du. \tag{2.III.3}$$

Putting $u^2 = \xi$, $2u \, du = d\xi$, equation (2.III.2) yields

$$G_s = \frac{\rho_s \beta_s}{2\pi^{1/2}} \int_0^\infty \exp(-\beta_s^2 \xi) \, d\xi = \frac{\rho_s \beta_s}{2\pi^{1/2}} \left[-\frac{\exp(-\beta_s^2 \xi)}{\beta_s^2} \right]_0^\infty = \frac{\rho_s}{2\beta_s \pi^{1/2}}. \tag{2.III.4}$$

The absolute rate of recondensation is given by

$$\begin{aligned} G_c &= mu \, du \\ &= mun \frac{\beta^3}{\pi^{3/2}} \int_{-\infty}^\infty \int_{-\infty}^\infty \int_{-\infty}^\infty \exp\{-\beta^2[(u-\bar{u})^2 + (v-\bar{v})^2 \\ &\quad + (w-\bar{w})^2]\} \, du \, dv \, dw. \end{aligned} \tag{2.III.5}$$

If there is a uniform velocity u_0 in the vapour space and there are no transverse velocities, $\bar{u} = u_0$, $\bar{v} = \bar{w} = 0$; thus equation (2.III.5) becomes

$$G_c = \frac{\rho'\beta^3}{\pi^{3/2}} \int_{-\infty}^{\infty} u \exp[-\beta^2(u-u_0)^2]\,du$$

$$\cdot \int_{-\infty}^{\infty} \exp(-\beta^2 v^2)\,dv \int_{-\infty}^{\infty} \exp(-\beta^2 w^2)\,dw$$

$$= \frac{\rho'\beta}{\pi^{1/2}} \int_{-\infty}^{0} u \exp[-\beta^2(u-u_0)^2]\,du, \qquad (2.III.6)$$

where $\rho' = mn$ is the vapour density.

Putting $u - u_0 = \xi$, $du = d(\xi + u_o) = d\xi$, the integral in equation (2.III.6) yields

$$\int_{-\infty}^{0} u \exp[-\beta^2(u-u_0)^2]\,du = \int_{-\infty}^{-u_0} (\xi + u_0)\exp(-\beta^2\xi^2)\,d\xi$$

$$= \int_{-\infty}^{-u_0} \xi \exp(-\beta^2\xi^2)\,d\xi$$

$$+ u_0 \int_{-\infty}^{-u_0} \exp(-\beta^2\xi^2)\,d\xi. \qquad (2.III.7)$$

Putting $\xi^2 = \eta$, $2\xi d\xi = d\eta$ in the first integral, and $\beta\xi = \zeta$, $\beta\, d\xi = d\zeta$ in the second, the above integrals become

$$\frac{1}{2}\int_{-\infty}^{u_0^2} \exp(-\beta^2\eta)\,d\eta + \frac{u_0}{\beta}\int_{-\infty}^{-\beta u_0} \exp(-\zeta^2)\,d\zeta$$

$$= \left[-\frac{\exp(-\beta^2\eta)}{2\beta^2}\right]_{\infty}^{u_0^2} + \frac{u_0}{\beta}\int_{\beta u_0}^{\infty} \exp(-\zeta^2)\,d\zeta$$

$$= -\frac{\exp(-\beta^2 u_0^2)}{2\beta^2} + \frac{\pi^{1/2} u_0}{2\beta}\,\mathrm{erfc}\,(\beta u_0). \qquad (2.III.8)$$

Since by definition

$$\frac{\pi^{1/2}}{2}\,\mathrm{erfc}\,(\beta u_0) = \int_{\beta u_0}^{\infty} \exp(-\zeta^2)\,d\zeta,$$

therefore

$$\frac{1}{2}\int_{\infty}^{u_0^2} \exp(-\beta^2\eta)\,d\eta + \frac{u_0}{\beta}\int_{-\infty}^{-\beta u_0} \exp(-\zeta^2)\,d\zeta$$

$$= -\frac{1}{2\beta^2}[\exp(-\beta^2 u_0^2) - \beta u_0 \pi^{1/2}\,\mathrm{erfc}\,(\beta u_0)].$$

Whence equation (2.III.6) yields

$$G_c = -\frac{\rho'}{2\beta\pi^{1/2}}[\exp(-\beta^2 u_0^2) - \beta u_0 \pi^{1/2} \operatorname{erfc}(\beta u_0)] = -\frac{\rho'\Gamma}{2\beta\pi^{1/2}} \tag{2.III.9}$$

where $\Gamma = \exp(-\beta^2 u_0^2) - \beta u_0 \pi^{1/2} \operatorname{erfc}(\beta u_0)$ is a correction factor.

Taking the ratio of the equations (2.III.9) to (2.III.4) gives

$$\frac{G_c}{G_s} = -\frac{\rho'}{\rho_s}\frac{\beta_s}{\beta}\Gamma = \frac{p_i'}{p_s}\left(\frac{T_s}{T_i'}\right)^{1/2}\Gamma. \tag{2.III.10}$$

And since $p_i' \propto \rho'$ from the gas law $p_i'V = RT'$, equation (2.III.10) becomes

$$\frac{G_c}{G_s} = -\frac{p_i'}{p_s}\left(\frac{T_s}{T_i'}\right)^{1/2}\Gamma. \tag{2.III.11}$$

But the actual rate of sublimation is given by $G = \alpha(G_s + G_c)$, where α = sublimation–condensation coefficient; therefore, finally, equation (2.III.11) becomes

$$G = \alpha G_s\left[1 - \frac{p_i'}{p_s}\left(\frac{T_s}{T_i'}\right)^{1/2}\Gamma\right],$$

where p_s = saturated vapour pressure of ice, p_i' = vapour pressure in the boundary layer adjoining the ice; T_s = sublimation temperature, T_i' = temperature of the vapour in the boundary layer adjoining the ice. The variation of Γ with G_s/G is plotted in Fig. 2.9.

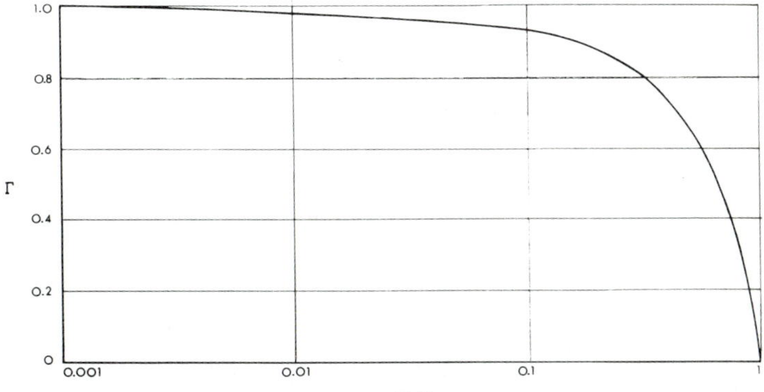

FIG. 2.9. Variation of the non-equilibrium correction factor Γ with the reduced sublimation rate G_s/G, assuming both α and T_s/T_i' equal unity.

3 Heat transfer

3.1 Internal and surface heat transfer in sublimation

Heat transfer is generally accepted as the rate-controlling step in freeze-drying when it takes place through an outer layer of dry product to the ice crystals. Drying occurs outside the product mainly by radiation from a heat source to the product surface and inside the product mainly by conduction. However, it may occur by convection with some of the heating methods which have been examined (Oetjen *et al.*, 1963; Meffert, 1965; and Neumann, 1963). In a typical case of heating in freeze-drying, Fig. 3.1 shows the course of sublimation, product surface and heating platen temperatures. Steady radiant heating causes the ice to sublime completely at almost a constant temperature whilst the product surface temperature rises; then, as the heating is reduced during desorption of residual moisture, the product approaches everywhere a safe temperature nearly equal to the final temperature of the heating platen.

The factors determining whether freeze-drying is either heat or mass transfer controlling are discussed with reference to drying rate in Chapter 5 (Sections 5.2, 5.3). Strasser *et al.* (1966) have examined the course of sublimation temperature for both alternatives, as shown in Fig. 3.1 (the lower curve is for mass transfer controlling). Usually, mass transfer is the controlling factor when heat is conducted directly through the ice layer instead of through the dry layer.

3.2 Steady-state model of heat transfer

3.2.1 RADIANT HEATING

Explanations of the two basic types of heat transfer have been given by simple models for the steady state proposed by Mink and Sachsel

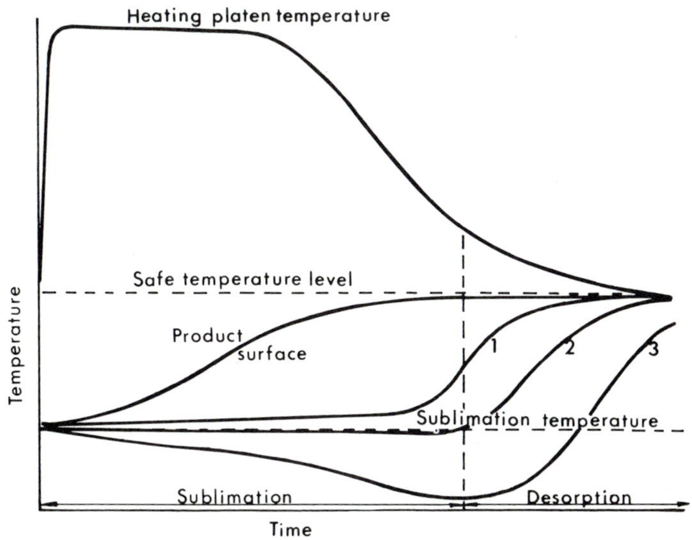

FIG. 3.1. Course of sublimation, product surface and heating platen temperatures: (1) heat transfer controlling; (2) neither controlling; (3) mass transfer controlling.

(1961), Kan and de Winter (1968) and King (1970). Figure 3.2 shows a typical model for the progress of freeze-drying with a radiant heat-source temperature T_h, and ice interface temperature T_i. If h_s and h_{in} are, respectively, the surface and internal heat-transfer coefficients of the material, then the heat flux per unit area Q is given by

$$Q = \frac{T_h - T_i}{1/h_s + 1/h_{in}} \tag{3.1}$$

The heat flux may also be related to the surface temperature T_s, through the equation

$$Q = h_{in}(T_s - T_i). \tag{3.2}$$

Eliminating Q from both equations yields

$$\frac{T_h - T_s}{T_s - T_i} = \frac{h_{in}}{h_s}. \tag{3.3}$$

Therefore, if the heating-platen and sublimation temperatures remain constant during sublimation (Fig. 3.2) the ratio of the internal to the

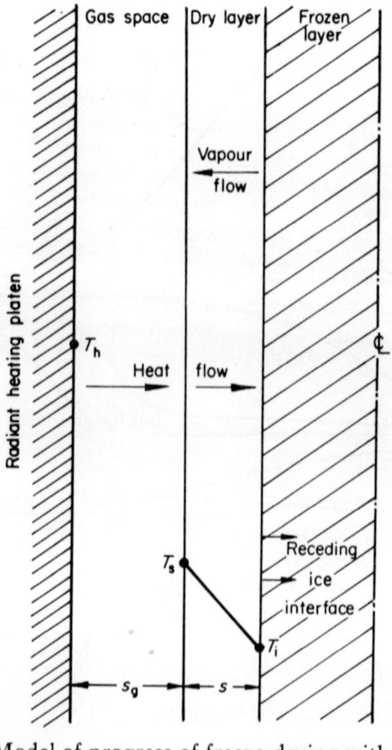

FIG. 3.2. Model of progress of freeze-drying with radiant heating.

surface heat-transfer coefficients determines the product surface temperature.

The internal heat-transfer coefficient h_{in} is obtained by assuming a slab of material, thickness s, freeze-drying with a uniform retreating ice interface; thus $h_{in} = \lambda_d/s$ where the thermal conductivity of the dry material λ_d is constant. On the other hand, the surface heat-transfer coefficient h_s is more difficult to determine but can be deduced if the surface is assumed to be heated by radiation and, in addition, by conduction in the gas space in contact with it. Assuming (1) a constant heat-source temperature T_h, (2) that the thermal conductivity of the gas space λ_g is independent of pressure, and (3) a linear temperature profile exists in the dry material, Kan and de Winter (1968) have given the following analysis. The total flux is given by

$$Q=\frac{\lambda_g}{s_g}(T_h-T_s)+\frac{\sigma F_1(T_h^4-T_s^4)}{1/e_h+1/e_s-1}=(\lambda_g/s_g+h_s)(T_h-T_s), \quad (3.4)$$

where $h_s=\sigma F_1(T_h+T_s)(T_h^2+T_s^2)/(1/e_h+1/e_s-1)$, s_g = width of the gas space, σ = Stefan's constant, F_1 = geometrical viewing factor ($F_1 \leqslant 1$), e_h, e_s = emissivities of the heat source and product surfaces.

In equation (3.4) the first term on the right is due to gas space conduction and the second is due to radiation, which has also been given by Rowe (1964a). For freeze-drying, the difference in absolute temperatures between the heat source and product surface is fairly small, therefore h_s may be approximated by

$$h_s \simeq \frac{4\sigma F_1 \bar{T}^3}{1/e_h+1/e_s-1}, \quad (3.5)$$

where $\bar{T}=(T_h+T_s)/2$.

Combining equations (3.2), (3.4) and $h_{in}=\lambda_d/s$ yields

$$Q=\frac{T_h-T_i}{s/\lambda_d+1/(\lambda_g/s_g+h_s)}. \quad (3.6)$$

In equation (3.6) the denominator represents the heat-transfer resistance between the heat source and the ice interface, from which h_s can be obtained.

3.2.2 CONDUCTION HEATING

Where conditions are such that a liquid or semi-rigid material can be frozen in contact with the heat source, heat may be conducted directly from one side of the frozen layer and the water vapour removed from the other (Harper and Tappel, 1957; and Lambert and Marshall, 1962). The surface temperature T_s refers then to the frozen layer in contact with the heat source which is adjusted to avoid melting, and h

becomes the overall heat-transfer coefficient through the frozen layer. Thus equation (3.2) simplifies to

$$Q = h(T_s - T_i). \tag{3.7}$$

Otherwise, for a slab of frozen material making good contact with a metal tray resting on a heating platen (Fig. 3.3), the equation for heat conduction through the different layers is given by

$$Q = \frac{T_s - T_i}{1/h_h + s_{me}/\lambda_{me} + 1/h + s/\lambda}, \tag{3.8}$$

where h_h, h_i = heat-transfer coefficients for the platen–tray and tray–frozen layer interfaces, s_{me}, s = thickness of the tray and frozen layer,

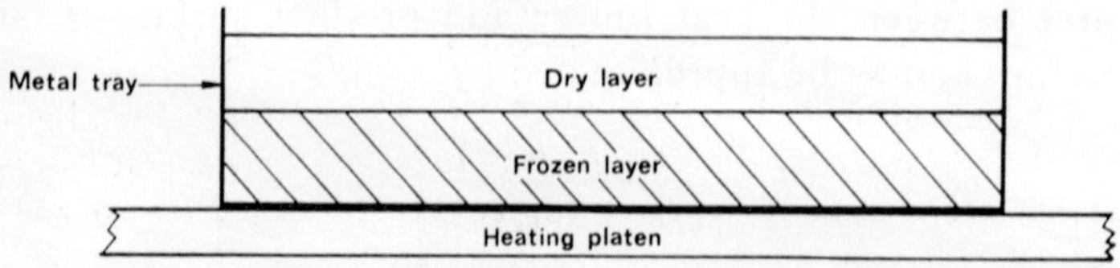

FIG. 3.3. Conduction heating of a frozen slab in contact with a metal tray.

λ_{me}, λ = thermal conductivities of the tray metal and the frozen material.

Often with this type of heating the frozen layer detaches itself from the tray, at first around the edges and then towards the centre where underside drying occurs. Consequently, a thermal and vapour barrier builds up, limiting further drying of isolated portions of the frozen layer. This melts when the vacuum is broken.

It is seen from equation (3.8) that interfacial contact of the intervening surfaces is very important. Freezing the material in contact with the tray presents no difficulty, but maintaining flat metal-to-metal surfaces to obtain good thermal contact is always difficult because the large temperature gradients which occur in freezing cause the tray to buckle. As the thermal conductivity of the frozen material is about 45 times that of the dry material there is a great incentive to improve thermal contact and increase further the thermal conductivity of the frozen material. Such improvements assist conductivity as well as dielectric heating.

3.3 Transient model of heat transfer

3.3.1 THEORY OF CONDUCTION WITH A MOVING INTERFACE

The theory of the sublimation process may be considered as a heat conduction problem involving (1) the frozen material as the outer layer of a dry semi-infinite solid and (2) a moving interface between the frozen and dry parts of the material (Mellor, 1962b). Classical theory dictates that the semi-infinite solid is the only one for which there are exact solvable analytical solutions, using the Stefan or Neumann approach (Carslaw and Jaeger, 1959). A number of limiting boundary conditions can be imposed, and it is proposed to show that, assuming as a boundary condition a constant surface heat flux, the problem can be solved along classical lines without Stefan's assumption that the interface depth varies as the square root of time.

The thermophysical properties (thermal conductivity and specific heat) of the ice phase and the dry porous phase are different, but, by taking the case of back-face heating instead of heating through the dry layer and the accompanying small temperature gradients in the ice phase, mean values for these quantities can be used.

The interface is not necessarily continuous in most materials that have been freeze-dried but may be considered as such in layers of reasonable thickness and homogeneity (uniform retreating ice front). When heat is conducted through the frozen layer to the interface a constant temperature gradient is usually maintained across it, even though the thickness of the layer is diminishing. The problem of heat being conducted through the dry layer instead of the frozen layer might perhaps be treated in a similar way.

From the theoretical standpoint two modes of heat transfer may be distinguished, providing different boundary conditions: (i) constant surface temperature and (ii) constant heat flux. A third (variable surface flux) will be examined in the light of McCulloch and Sunderland's (1970) analysis for a variable heat flux.

i. *Constant surface temperature*

Let us assume that the back face of the frozen outer layer, initially at the sublimation temperature T_i, is raised to a constant temperature

T'_s. The temperature profile at some particular point and time in the frozen layer is derived in Appendix 3.I, equation (3.I.10), namely:

$$\frac{T - T_i}{T'_s - T_i} = 1 - \frac{\operatorname{erf}\,[x/2(\alpha_f t)^{1/2}]}{\operatorname{erf}\,\gamma} \tag{3.9}$$

During sublimation, T'_s is usually regarded as the maximum safe temperature of the frozen material, T_i as a constant, X as the depth of the interface at any time t, α_f the thermal diffusivity of the frozen material and γ as a proportionality constant from Stefan's expression: $X = 2\gamma(\alpha_f t)^{1/2}$.

Differentiation of equation (3.9) with respect to x and substitution of $t^{1/2}$ from $X = 2\gamma(\alpha_f t)^{1/2}$, the heat flux $\lambda_f(\partial T/\partial x)_{x=0}$ into the frozen layer becomes

$$\lambda_f\left(\frac{\partial T}{\partial x}\right)_{x=0} = \frac{2\gamma\lambda_f(T'_s - T_i)}{\pi^{1/2}X\operatorname{erf}\,\gamma} \simeq \frac{\lambda_f(T'_s - T_i)}{X}, \qquad \gamma \to 0. \tag{3.10}$$

Usually, with freeze-drying, the approximation (equation 3.10) may be used, since $\operatorname{erf}\,\gamma \simeq 2\gamma/\pi^{1/2}$ where γ is small.

Consider, as the interface boundary condition, that heat quantity Q per unit area per unit time is removed by sublimation. Then, in terms of the latent heat of sublimation ΔH and the rate the interface moves, $-\mathrm{d}X/\mathrm{d}t$, Q is given by

$$Q = \lambda_f\left(\frac{\partial T}{\partial x}\right)_{x=X} = -\rho_f \Delta H \frac{\mathrm{d}X}{\mathrm{d}t}, \tag{3.11}$$

where λ_f is the thermal conductivity of the frozen material and ρ_f is its density.

Differentiation of equation (3.9) and substitution from equation (3.11) with $\gamma = X/2(\alpha_f t)^{1/2}$ yields

$$Q = \frac{2\gamma\lambda_f(T'_s - T_i)\exp(-\gamma^2)}{\pi^{1/2}X\operatorname{erf}\,\gamma} \simeq \frac{2\rho_f \Delta H \gamma^2 \alpha_f}{X}. \tag{3.12}$$

Again, where γ is small, a power series expansion for the exponential and error functions in equation (3.12) yields a second approximation for γ, given by

$$\gamma^2 = \frac{\lambda_f(T'_s - T_i)}{\lambda_f(T'_s - T_i) + 2\alpha_f\rho_f\,\Delta H} \tag{3.13}$$

Or a first approximation: $\gamma^2 = \lambda_f(T'_s - T_i)/2\alpha_f\rho_f\Delta H$ obtained from equations (3.10) and (3.12) if $\lambda_f(\partial T/\partial x)_{x=0} \equiv -Q$ is assumed.

Therefore, if the back face of a semi-infinite frozen solid is maintained at a maximum safe temperature, the heat flux into the frozen layer is given by equation (3.10) and the sublimation heating rate in terms of γ is given by equation (3.12). This shows that the flux into the frozen layer and the sublimation heating rate differ only by a second-order term in γ, i.e. $\lambda_f(\partial T/\partial x)_{x=0} \equiv -Q$ is a reasonable assumption; consequently, it may be assumed that virtually all the heat input is lost to sublimation and none by heat transfer to the surroundings.

A later theory, developed by Luikov *et al.* (1971), leads to similar conclusions after taking into account their specific heat flux equation in terms of the present terminology, viz. equation (3.12). Moreover, the conclusions drawn from their general gas dynamics problem for sublimation supports the moving-interface theory that the interface velocity is directly proportional to the specific heat flux, i.e. equation (3.11).

As the heat-transfer rate varies with distance between the surface and the interface, Mellor (1962b) deduced from equation (3.12) that the average sublimation heating rate $\bar{Q}$ for a slab may be obtained as an approximation by integrating $2\rho_f\,\Delta H\gamma^2\alpha_f/X$ over the total thickness of the slab x_0. Thus

$$\bar{Q} = \frac{2\rho_f\Delta H\gamma^2\alpha_f}{x_0}\ln x_0. \tag{3.14}$$

ii. Constant heat flux

In general, for constant heat flux Q, the interface depth $X(t)$ varies with time t (Appendix 3.I) according to

$$X(t) = \frac{Q}{\rho_f\Delta H}t - \frac{3}{8\alpha_f}\left(\frac{Q}{\rho_f\Delta H}\right)^3 t^2 + \cdots \tag{3.15}$$

When the ice phase is gently heated at a constant rate the interface depth may be expected to vary linearly with time according to the relation

$$X(t) = \frac{Q}{\rho_f\Delta H}t, \tag{3.16}$$

which is a first approximation to equation (3.15). The term $Q/(\rho_f \Delta H)$ may be regarded as a small quantity in the case of gentle heating since $\rho_f \Delta H$ is large; this is also obvious from equation (3.12), since γ^2 is normally very small.

Consequently, a certain triviality is manifest in the solutions of the heat conduction equations for freeze-drying unless, of course, contrary to normal practice, there are strong heat loads.

Given safeguards, stronger-than-normal heat loads seem necessary for economical freeze-drying of foodstuffs. It is hoped that the theory presented will throw some light on this important aspect.

iii. *Variable surface flux*

During radiant heating the surface temperature of the product always rises and, in reality, ice sublimes whilst the surface heat flux decreases with time. Two cases of variable heat flux at the product surface are discernible; either (1) the surface temperature rises towards the heating platen temperature or (2) it remains steady as the heating platen temperature falls. At no point is the surface temperature of the product allowed to rise beyond a safe limit.

McCulloch and Sunderland (1970) have attempted an analysis for both cases, using Goodman's integral energy balance and a quadratic equation for the temperature profile. However, their analytical predictions would appear to be of little value for most of the current freeze-drying problems involving heat transfer by radiation to the surface of the product and hence heat conduction in the dry layer.

They graphed the interface position and the course of surface temperature for some hypothetical freeze-drying runs, taking the heating platen temperatures at 67°C, 80°C, 93°C and 121°C and the usual properties of freeze-dried beef, except for the porosity, which was assumed in turn to take values of 0.6, 0.7 and 0.8. Assuming that the temperature of the dried beef never exceeds 38°C (safe temperature) and the heating platen temperature equals 93°C, their calculations show that a piece of beef 16 mm thick could be safely freeze-dried in 22 hours. The interface position (Fig. 3.4a) and the surface temperatures (Fig. 3.4b) have been obtained from equation (3.18), taking μ as constant. As their calculated drying times extend to

32 h for primary drying of products only 16 mm thick, it would appear that their curves are of little practical value.

For the heat flux $Q(t)$, McCulloch and Sunderland assumed a radiation boundary condition expressed by $Q(t) = \sigma\varepsilon[T_h^4 - T_s'^4(t)]$. And for the temperature profile they obtained

$$T - T_i = \frac{A\alpha_f}{2s}[1-(1+4\mu_1)^{1/2}][x-s] + \frac{A\alpha_f}{4s^2}[1-(1+4\mu_1)^{1/2}+2\mu_1][x-s]^2, \quad (3.17)$$

which upon reduction gives

$$T - T_i = \frac{A\alpha_f}{s}[-\mu_1(x-s)+\mu_1^2(x-s)+\frac{\mu_1^2}{2s}(x^2-2xs+s^2)]. \quad (3.18)$$

And

$$\frac{T-T_i}{T_s'-T_i} = 1 - \frac{x}{X} - \frac{\mu_1}{2X^2}(X^2-x^2),$$

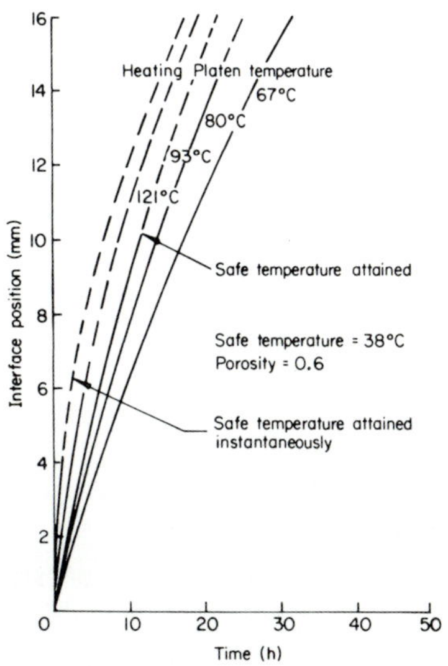

FIG. 3.4a. Variation of interface position with time for different heating platen temperatures.

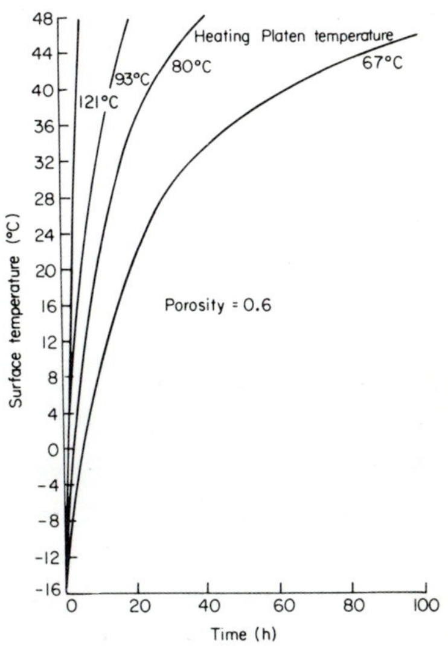

FIG. 3.4b. Variation of surface temperature with time for different heating platen temperatures.

where $A\alpha_f/s \simeq (T'_s - T_i)\mu_1 s$ and $s \equiv X$, $A \equiv \rho_f \Delta H/\lambda_f$ and $\mu_1 = Q(t)X(t)/\alpha_f\rho_f\Delta H$ using the present notation.

Mellor (1974) attempted to correlate the dimensionless variable μ_1 with the proportionality constant γ of equation (3.9), but found that this involved incorporating a variable factor κ which was unjustified.

On the other hand, if $\mu_1(=Q(t)X(t)/\alpha_f\rho_f\Delta H)$ could be made constant, $X(t)$ and $Q(t)$ would have to be related, which would mean a special and possibly impracticable means of heating. Thus, it seems that McCulloch and Sunderland's analysis is only useful with iterative procedures for obtaining the temperature profile and interface depth. It is therefore not an analytical solution, as they point out; although, in case (2) (constant surface temperature) of their analysis where they make μ_1 constant, the solution could be in line with the present analysis (Appendix 3.I).

3.4 Control of the heat supply

In drying labile materials, the quality of the finished product and the time to completely dry depend on the choice of a method for controlling the heat flux to the material. Usually, the surface is monitored to a fixed temperature with thermocouples at a number of locations, but this imposes difficulties because of piece-to-piece variations and the manual operations involved. Recognizing the advantages an automatic control system might have for sublimation drying, Krylov *et al.* (1967) obtained an analytical function that provides effective control for high-temperature radiating platens.

To obtain this function they divided the entire drying period into three stages. The first (heat flux off, vacuum being drawn) is unnecessary as far as the analysis is concerned, and the following simpler analysis involving two stages will therefore be discussed:

(1) Maximum heating to raise the surface to the temperature setting, $0 \leq t \leq t_1$, $Q = Q_{max}$ (constant).
(2) Decreased heating to control the surface at the temperature setting, $t_1 \leq t \leq t_2$, $Q = Q(t)$.

Figure 3.5 shows the two stages and the relationship the sublimation and surface temperatures bear to the intensity of the heat flux.

Other control systems are discussed in Section 10.4.

In their derivation of equations for these two heating stages Krylov *et al.* assumed all the radiant heat to be expended on sublimation whilst the surface temperature rises from a minimum value T_i to the temperature setting T'_s, beyond which the quality of the product would be lost. The following simpler approach begins with the first

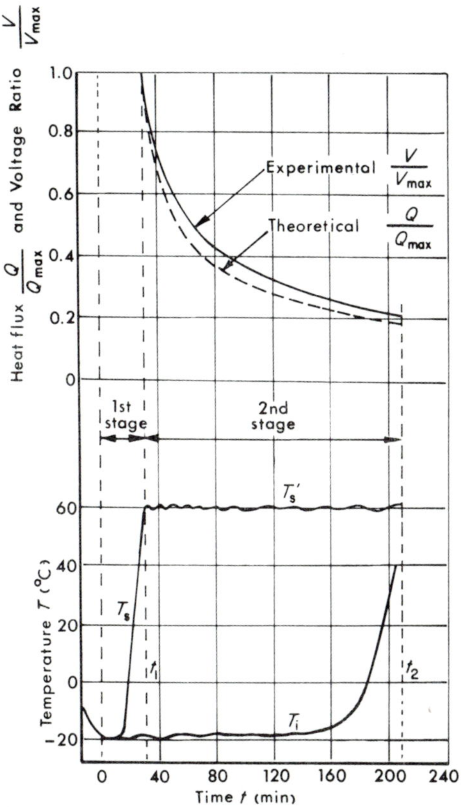

FIG. 3.5. Relationship of the sublimation and surface temperatures to variations in heat flux and power supply.

transcendental term of equation (3.I.7) (Appendix 3.I). In the first stage a negligible thickness for the drying surface layer is assumed, so that $x=0$; and, with $Q=Q_{max}$, the first term of equation (3.I.7) may be written as

$$T-T_i=\frac{Q_{max}}{\lambda_d}\left(\frac{\alpha_d}{\pi}\right)^{1/2}\int_0^t \frac{d\tau}{(t-\tau)^{1/2}}. \tag{3.19}$$

Integration gives

$$T - T_i = \frac{2Q_{max}}{\lambda_d}\left(\frac{\alpha_d t}{\pi}\right)^{1/2}. \tag{3.20}$$

Thus the time t_1 required for the surface to reach the temperature setting T'_s is given by

$$t_1 = \frac{\lambda_d^2 \pi (T'_s - T_i)^2}{4\alpha_d Q_{max}^2}. \tag{3.21}$$

In the second stage $t_1 \leqslant t \leqslant t_2$, $Q = Q(t)$, and therefore the first term of equation (3.17) becomes

$$T'_s - T_i = \frac{1}{\lambda_d}\left(\frac{\alpha_d}{\pi}\right)^{1/2} \int_{t_1}^{t} \frac{Q(\tau)}{(t-\tau)^{1/2}}\,d\tau. \tag{3.22}$$

Integration ought to provide the theoretical basis on which a control system depends. Applying Abel's integral equation to the heat flux $Q(\tau)$ yields

$$\begin{aligned} Q(\tau) &= \frac{\sin(\pi/2)}{\pi}\frac{d}{d\tau}\int_0^t \frac{Q(t)}{(\tau - t)^{1/2}}\,dt \\ &= \frac{Q_{max}}{\pi}\left[\frac{1}{(\tau - t_1)^{1/2}} - \frac{1}{\tau^{1/2}}\right], \end{aligned} \tag{3.23}$$

where $Q(t) = Q_{max}$, $0 \leqslant t \leqslant t_1$.
Thus the integral in equation (3.22) becomes

$$Q(t) = \frac{Q_{max}}{\pi}\int_{t_1}^{t} \frac{d\tau}{(\tau - t_1)^{1/2}(t-\tau)^{1/2}} - \frac{Q_{max}}{\pi}\int_{t_1}^{t}\frac{d\tau}{\tau^{1/2}(t-\tau)^{1/2}}. \tag{3.24}$$

Introducing new variables u and v, such that $\tau = t\sin^2 u + t_1 \cos^2 u$ and $\tau = t \sin^2 v$ in the first and second integrals respectively, yields

$$\begin{aligned} Q(t) &= \frac{2Q_{max}}{\pi}\int_0^{\pi/2} du - \frac{2Q_{max}}{\pi}\int_{\arcsin (t_1/t)^{1/2}}^{\pi/2} dv \\ &= \frac{2Q_{max}}{\pi}\arcsin (t_1/t)^{1/2}. \end{aligned} \tag{3.25}$$

The equation shows the variation of the heat flux required for sublimation during the second stage in order to maintain the surface at the temperature setting. The time t_1 for this equation can be determined from equation (3.21) and Q_{max} found experimentally. Disregarding the moisture loss whilst the vacuum is being drawn, Krylov *et al.* estimated Q_{max} from the moisture loss from a surface layer 1 mm thick during the first stage, which was the rate at which the dry layer advanced. They used an equation which is similar to equation (3.11).

Figure 3.5 (upper curves) shows their experimental curve for the variation of the intensity of the power supply V/V_{max} and the theoretical variation of heat flux $Q(t)/Q_{max}$ expended on sublimation. There is a similar trend in the shapes of the curves; consequently, to stabilize the surface temperature during the second stage, it is necessary to vary the power supply according to equation (3.25).

Equation (3.I.7) might appear doubtful since it represents plain (inward) heat conduction in the drying layer and not the resultant of inward as well as outward conduction (liberated sublimation heat or plane continuous heat source at the interface). The latter is observed as a cooling effect in the dry porous layer as the vapours escape from it. Understandably, this cooling effect would be less marked in either thin slabs or granulated products under intense heating conditions (which this theory sets out to cover). Moreover, the theory assumes that the product surface temperature T_s reaches the temperature setting in the shortest possible time, as shown in Fig. 3.5, and the presence of this high temperature would therefore weaken the cooling effect.

The theory of Krylov *et al.* then throws some light on a simple theory that ought to follow for the freeze-drying of small particles.

Designing a controller to follow a smooth curve exactly (Fig. 3.5, upper curve), for which equation (3.25) involves a transcendental function, might not be the only way in practice, since programmed control in a number of steps could be made to follow up and down the smooth curve and yet give greater flexibility. Such a controller is described in Section 8.1.

3.5 Heat conduction in condensation

The heat-flow resistance due to the ice layer on the condenser is proportional to its thickness and inversely proportional to its thermal conductivity. Limitations are imposed by making the area for condensation larger than for sublimation, so that the ice layer never exceeds the thickness of the frozen material at the beginning of freeze-drying. The layer will develop as snow instead of clear solid ice having a thermal conductivity of about 2·39 $\mathrm{Wm^{-1}K^{-1}}$, and its density and thermal conductivity will therefore be lower unless there is a good vacuum.

Condensers are designed largely on the basis of the area allowed for condensation, without considering the heat conducted through the ice layer. It would be true to say that the theory of heat flow in condensers for freeze-dryers has been largely neglected or overlooked. This is in direct contrast to the many theoretical and experimental studies of heat flow in different heating systems that are used in freeze-drying. First, Shumsky (1958) attempted a theory for "sublimation" condensers by presenting formulae for calculating the required surface area, ice distribution and overall heat transfer where there is appreciably no content of permanent gases, as in a good vacuum. He also extended the theory by approximating to actual conditions involving the partial pressure effect of the permanent gases. Later, Ciborowski and Surgiewiez (1961) also examined the problem, to determine the amount of ice deposited, especially for the case of freeze-drying in fluidized beds.

During freeze-drying, the partial air pressure near the condenser is always relatively high and so a large space, screened by it to accommodate a substantial proportion of the air liberated during sublimation and by leakage into the chamber, should be allowed for in design. The partial air pressure in the rest of the chamber will then be reduced. This aspect, which influences the efficiency of drying, has been examined by several workers and is dealt with in Chapter 2.

Finally, de Roissart and Laederich (1966) attacked the problem, treating it as a moving boundary between the ice front and the air–vapour mixture, to determine the thickness, capacity and temperature drop for ice condensers cooled by mechanical refrigeration, by dry ice, or by liquid nitrogen. Their theory will now be examined.

Based on equation (3.I.10) (see Appendix 3.I) but, because the temperature gradient in the ice layer is positive, and not negative as before, the equation may be written

$$T - T_s = (T_i - T_s)\frac{\operatorname{erf}\,[x/2(\alpha_i t)^{1/2}]}{\operatorname{erf}\gamma}, \qquad (3.26)$$

where T_s = surface temperature of the condenser, T_i = ice surface temperature, $\gamma = X/2(\alpha_i t)^{1/2}$. For a thin ice layer where the values of the exponents (equation 3.26) are small, the thickness, on substituting in equations (3.10) and (3.11), see Section 3.3, becomes

$$X \simeq \frac{1}{\alpha_i^{1/2}\gamma}\frac{\lambda_i t^{1/2}}{\rho_i \Delta H}(T_i - T_s) = \frac{2t_c^{1/2}}{X_c}\frac{\lambda_i t^{1/2}}{\rho_i \Delta H}(T_i - T_s) \qquad (3.27)$$

where X_c is the characteristic thickness and t_c the characteristic time required to reach that thickness. In de Roissart and Laederich's derivation, equation (3.27) contains also an auxiliary function:

$$F(\bar{t}) = \frac{1}{t^{1/2}}\int_0^{\bar{t}} \frac{\partial T}{\partial \bar{y}}\, \mathrm{d}\bar{t},$$

which they normalize to unity.

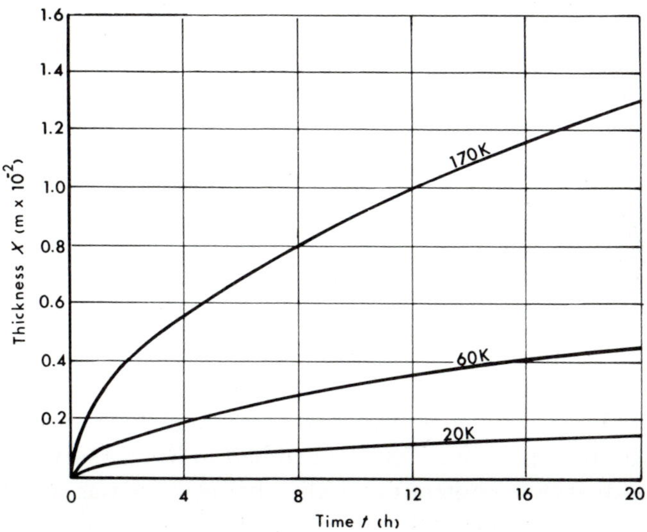

FIG. 3.6. Estimated ice layer thickness for a condenser operating at various temperature deficits.

Plots of the theoretical ice thickness X as a function of $T_i - T_s$ for different times are shown in Fig. 3.6. The authors consider that the temperature deficit $T_i - T_s$ should take the values: 20, 60 and 170 K for Freon, dry ice and liquid nitrogen refrigerants respectively.

3.6 Convection heating—atmospheric freeze-drying

The tendency for unwrapped frozen foods to dry out is due to atmospheric freeze-drying, whereby cold dry air from the refrigerated coil causes ice to sublime from the food by convection. Other examples of this occur especially in nature. Atmospheric freeze-drying is not so attractive, however, as some researchers have supposed. Experiment has shown it to involve a drying time greater than 60 hours with products reduced in size to 1 cm cube, higher sublimation temperatures far in excess of eutectic temperatures and the need for a high circulation of cold dry air. Nevertheless, consideration of some of the activity in this area of research indicated the potential of atmospheric freeze-drying.

Treffenberg (1953), Jensen (1954) and Jensen and Kavaljian (1957) reported experiments on the use of air for freeze-drying plant tissue. Later, Meryman (1959, 1962) suggested that the method could be improved by circulating the air stream through a molecular sieve or over a refrigerated coil, to remove the water vapour. Lewin and Mateles (1962) found that carrot and chicken pieces 8 mm thick could be dried in 30–40 hours in recirculated cold air regenerated by silica gel. Woodward (1961, 1963) revealed that diced roast beef dried more rapidly at temperatures above −10°C than below that temperature and he produced an economic analysis for the process. They all found that atmospheric freeze-drying is independent of air velocity above 0.5 m s^{-1}. To increase the drying rate Malecki *et al.* (1969) attempted atmospheric fluidized-bed freeze-drying of apple juice and egg white. Their work supports the conclusion of Dunoyer and Larousse (1961) that the drying rates in atmosphere can almost reach those in vacuum if the particle size is sufficiently small. However, Malecki *et al.* found, for particles of apple juice, that the bed had to be at −34°C to prevent sticking and, at this temperature, only 1 per cent ice was sublimed per hour, which was too low.

Atmospheric freeze-drying is a case of mass-transfer control involving diffusion of water vapour in free air as well as in air within the porous solid, and success depends on the partial pressure of water vapour of the convective gas being kept low enough to provide the driving force.

Heldman and Hohner (1972) analysed the process completely with a mathematical model and verified their results experimentally. A conclusion from their analysis is that the rate of drying can be increased most effectively by reducing particle size and increasing the surface mass transfer coefficient, the latter being best expressed by a dimensionless number as the ratio of the external to internal mass transfer. Using a one-dimensional model, as shown in Fig. 3.7, the ice

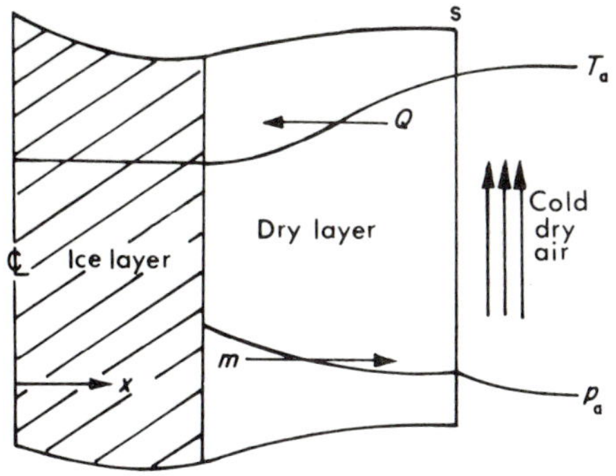

FIG. 3.7. Schematic diagram of atmospheric freeze-drying.

interface recedes to the centre line as heat of sublimation flows from the surface s to the interface in response to a temperature gradient represented by the upper curve. Simultaneously, water vapour flows through the dry layer in response to the water vapour pressure driving force represented by the lower curve.

Heldman and Hohner derived the heat-transfer equation from the conservation of energy in volume dV of the dry layer:

$$\begin{matrix}\text{Internal} \\ \text{energy}\end{matrix} = \begin{matrix}\text{Heat} \\ \text{conduction}\end{matrix} + \begin{matrix}\text{Energy input} \\ \text{(vapour)}\end{matrix} + \begin{matrix}\text{Sublimation} \\ \text{heat}\end{matrix}$$

$$\bar{\rho}_d(c_d + c_w)\frac{\partial T}{\partial t} = \lambda_d \frac{\partial^2 T}{\partial x^2} + K_e \frac{\partial p}{\partial x}\left(c_w \frac{\partial T}{\partial x}\right) + p_i \,\Delta H \frac{\partial m}{\partial t}, \qquad (3.28)$$

where c_d = specific heat of dry product, c_w = specific heat of water, K_e = effective water-vapour permeability, similar to equation (4.3), Section 4.2, m = moisture content, dry basis, $\bar{\rho}_d$ = mean density of dry product, p = vapour pressure, λ_d = thermal conductivity of dry product, ρ_i = density of ice, ΔH = latent heat of sublimation of ice, T = temperature in the dry layer, x = distance, t = time.

With the boundary conditions:

$$\begin{aligned} &\text{Surface:} \quad -\lambda_d\left(\frac{\partial T}{\partial x}\right)_{x=s} = h(T - T_a) \\ &\text{Interface:} \quad \rho_i \Delta H(m_0 - m_f)\frac{dX}{dt} = -\lambda_d\left(\frac{\partial T}{\partial x}\right)_{x=X} + \lambda_i\left(\frac{\partial T_i}{\partial x}\right)_{x=X}, \end{aligned} \tag{3.29}$$

where m_0 = initial moisture content, dry basis, m_f = final moisture content, dry basis, λ_i = thermal conductivity of ice, T_i = ice temperature, T_a = external temperature, h = surface heat-transfer coefficient, subscript X = interface, and subscript s = surface.

In order to evaluate the boundary condition at the interface, since $T_i = T_i(x, t)$, it can be written

$$\lambda_i \frac{\partial T_i}{\partial x} = \lambda_i \frac{\partial T_i}{\partial t}\left(\frac{\partial t}{\partial x}\right)_{x=X} + \lambda_i \frac{\partial T_i}{\partial x}\left(\frac{\partial x}{\partial x}\right)_{x=X} \tag{3.30}$$

As the temperature gradient $\partial T_i/\partial x$ in the ice is small the term containing this can be neglected and by definition $X(\partial x/\partial t)_{x=X} = \lambda_i/c_i\rho_i$. This leaves

$$\lambda_i \frac{\partial T_i}{\partial x} = \rho_i X(1 + m_0)c_i \frac{\partial T_i}{\partial t}. \tag{3.31}$$

Substitution therefore from equations (3.29) and (3.31) gives

$$\rho_i \Delta H(m_0 - m_f)\frac{dX}{dt} = -\lambda_d\left(\frac{\partial T}{\partial x}\right)_{x=X} + \rho_i X c_i(1 + m_0)\frac{\partial T_i}{\partial t}, \tag{3.32}$$

where c_i = specific heat of ice and $(1 + m_0)\rho_i$ accounts for the ice density.

Equation (3.32) expresses the drying rate in terms of heat conduction in the dry layer together with a small correction term for the change in heat capacity of the ice; it transcends equation (3.28).

The surface boundary condition may be used instead of the interface condition (equation 3.29) leading to Charm's (1963) equation for the drying rate based on normal air drying. Estimated drying times for these equations are about four times as long as vacuum freeze-drying times or about 2 days for freeze-drying slabs of beef 12·5 mm thick with an air-stream temperature of −20°C.

3.7 Heating methods

The many ways of applying heat in freeze-drying may be classified under three headings:

1. Thermal contact heating of the frozen layer.
2. Radiation or plate heating of the dry layer.
3. Internal heating by microwave energy.

A fourth method concerns mainly atmospheric freeze-drying, namely:

4. Convection heating.

In practice, some of the above methods occur in combination with one another and the fourth lends itself to fluidization of granulated as well as atomized products.

Essentially, a large quantity of heat, amounting to 2.84 MJ kg^{-1}, needs to be supplied as latent heat for sublimation to the ice interface between the frozen and dried layers, and it must be applied effectively by one or other of the four methods of heating.

3.7.1 THERMAL CONTACT HEATING

In the classical method of freeze-drying, heat is conducted to the ice interface by thermal contact with the underside of the frozen layer and maintained at a constant rate for the sublimation phase. However, near the end of sublimation, a gradual reduction in heating rate is necessary to avoid melting unless thermal contact heating is limited either to thin layers or to slow drying. There is also a tendency for drying to take place from the edges of the frozen layer and at surfaces in contact with the heating platen. The first difficulty may be solved by controlling the temperatures of the heater and drying boundary automatically, and the second can generally be overcome by drying at lower temperatures in metal trays or by superimposing thin plastic sheets between the bottom of the tray and the underside of the frozen layer, as Rowe *et al.* (1963) has done.

Heat is introduced as shown in Fig. 3.8, but the boundary and surface temperatures are limited by the maximum safe temperature for the frozen material in question. In some cases this will be its melting point; in others it will be necessary to maintain lower temperatures near the eutectic point, to prevent denaturation.

The theory behind the method follows in Sections 3.2 and 3.3.

In order to approach the ideal, as when drying a plug or shell of liquid in a vial, Mason (1963) devised a method of thermal contact

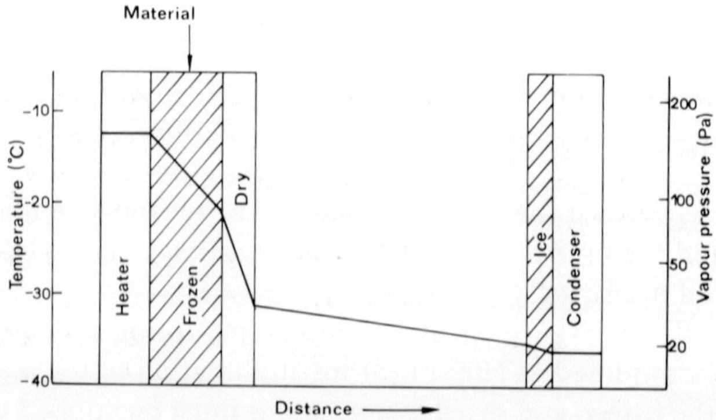

FIG. 3.8. Temperature and vapour-pressure gradients in freeze-drying with heating through the frozen layer.

which he called "inner" heating. The liquid material is frozen in the form of a shell on the outside of an array of vertical tubes through which the refrigerant circulates. The reverse arrangement was adopted by Seffinga Engineering Co. (1964). The frozen product is then subjected to vacuum in a drying chamber, and the tubes gently warmed by circulating a warm fluid through them at a temperature a little above the eutectic point of the liquid. When the product has been freeze dried it is scraped into a hopper feeding a vacuum lock, before being removed.

The same effect is possible with tray freeze-drying if the product remains intact throughout the drying process. To ensure that it does, della Porta (1967) demonstrated that a liquid could be frozen on to the evaporator plate of a domestic refrigerator and subsequently freeze dried from it by circulating a warm fluid through the plate. Good contact was maintained during drying since the frozen liquid was bonded to the perforations in the plate between the fluid channels.

It is possible that an impermeable surface layer forms during freezing in the "inner" heating method, as has been observed by Quast and Karel (1968), which hinders drying. However, the formation of radial vapour paths against the freezing tubes left by the sublimation of needle-shaped ice crystals would tend to alleviate this problem.

All these methods lend themselves to heat supply by thermal recycling from the refrigerator or from waste heat.

3.7.2 RADIATION AND PLATE HEATING

The second approach to the heat input problem is the application of radiant heat to the drying surface, or by the accelerated freeze-drying (AFD) technique (Hanson, 1961). One important advantage of these methods is that drying can be induced from both sides instead of from a single face. Both methods of heating, however, are far from ideal since it is necessary for the heat to traverse a large thermal gradient because of the thermal insulating effect of the dry layer in a vacuum—for example, freeze-dried beef has a thermal conductivity of $0{\cdot}045\ \mathrm{W\ m^{-1}\ K^{-1}}$ (13·3 Pa to 133·3 Pa). The gradient must be quite steep in order to deliver the heat through an appreciable thickness of dry material, but its steepness will be limited by the maximum permissible temperature at the surface, that is, the safe temperature below which denaturation does not occur.

In the AFD process developed by the British Ministry of Food Research Establishment, heat is conducted to the upper and lower surfaces of pieces of food and not to the frozen portion only, as in the classical method of heating. The problem of heating and at the same time allowing water vapour to escape is overcome by placing expanded metal sheets in the space between the surfaces of the food and the heating platens, and pressing the whole assembly together hydraulically (Forrest, 1959, 1961).

The sheets serve to conduct heat through the metal mesh and through the escaping vapour. The faster drying obtained outweighed the disadvantages of (1) high product surface temperatures (up to 60°C), (2) the need for high-capacity heating and pumping units for a peak (3 h) sublimation rate in an 8-h drying cycle, and (3) for a heating system under hydraulic pressure. Originally, steam ejectors, instead of a refrigeration plant, were used for direct pumping of the large volume of water vapour evolved, and for producing the vacuum. No doubt the location in Aberdeen (Scotland) with an abundance of water for steam production and condensing purposes greatly influenced this choice.

Temperature gradients for heating by radiant heat and accelerated freeze-drying are shown in Fig. 3.9. It should be noted that radiant heat can only be applied to surfaces of the drying material in a direct line of sight from the heating source.

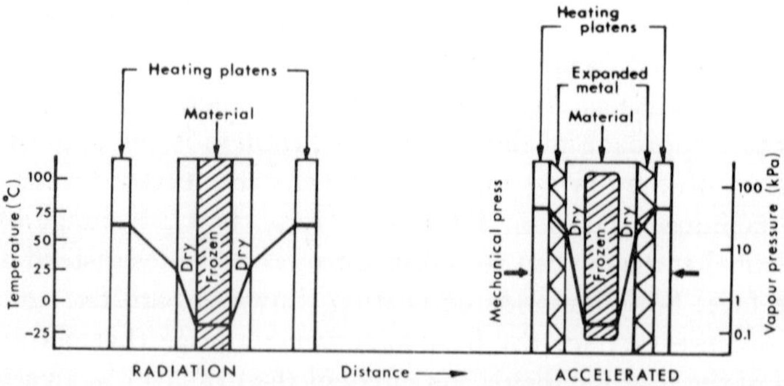

FIG. 3.9. Temperature and vapour-pressure gradients in freeze-drying with heating through the dry layers.

The heating method normally used in freeze-drying is by radiation, for many of the conduction methods by heating plate depend largely on heat being applied by radiation, the theory of which is given in Section 3.2. When the drying surface is heated directly by radiation a distinction is often made between black-body and infrared radiant heating. In the former, there is no significant heat penetration by radiation into the dry layer itself but, theoretically, blackened heating platens could be expected to give a higher heat transfer rate for a given heater temperature. Temperatures up to 100°C are used. With platens 1 m^2 and a small gap of 2 cm between them and the product surfaces to minimize edge effects, the heating rate is independent of gap size (Smithies and Blakley, 1959). On the other hand, Ginsburg *et al.* (1969) found that the temperature and wavelength range of infrared radiators may be chosen to provide the maximum ray penetration through the dry layer. However, for a complete freeze-drying operation the heating conditions need to be changed three times; initially the radiation is chosen in accordance with the optical characteristics of the frozen surface, then it is chosen so that the dry layer will not be overheated and the heat rays will reach the receding ice front and, finally, it is changed again for the evaporation of residual moisture in accordance with the optical properties of the dry layer. They recommended infrared radiators giving wavelengths of 1·1–1·2 (2527°C filament temperature) and 3·5–6·0 (227–527°C) for the first and second stages of drying respectively. Generally, heater temperatures do not exceed 100°C, but may reach 240°C for a short time at the beginning of drying. In another application, Zamzow and Marshall (1952) have tested and analysed infrared radiation applied directly to the frozen layer in freeze-drying of liquids.

If the dry powder layer as it forms is abraded away then high-temperature radiations can be used continuously. This was the method adopted by Greaves (1960b), who froze a liquid into a cylindrical block and mounted it on a rotating turntable fitted with a scraper that removed the dry powder. In another version of this method the liquid was spun frozen in a tube fitted with an axially mounted heater and scraper. The practical drawback was entrainment of fines by the water vapour which finished up on the condenser. Removal of the fines by means of an electrostatic precipitator has been suggested by Hackenberg (1966), and Dalgleish (1963) has proposed intermittent operation of Greaves' second method so that the powder can be scraped away when the condenser has been isolated from the drying chamber.

Radiant heating permits fairly rapid drying of irregularly shaped pieces of product; it overcomes the difficulty of the poor contact associated with conduction heating without involving sophisticated equipment such as is needed in accelerated freeze-drying (AFD).

The loading density for the chamber in the accelerated freeze-drying method is low because the hydraulic press, metal plates and heavy heating platens occupy much space; nevertheless, the throughput is higher than with most other heating methods (Rowe *et al.*, 1963). Another disadvantage is the need for precooling the shelves before loading, to prevent melting. Mellor (1962b) suggested two improvements: (1) integrated heating platens that would supply heat differentially to the slabs of material and to the escaping vapour; (2) placing the whole system of plates and slabs under pneumatic pressure.

Rowe *et al.* (1963) has proposed a modified AFD assembly (Fig. 3.10) using a pneumatic bag inflated with air from outside the chamber to obtain the pressure, and Oldencamp and Small (1965) and Mehrlich and Haugh (1965) have proposed means for improving the heating system. But, perhaps, the most novel idea for heating differentially is by impaling slabs of the frozen product material on spiked heating platens (Brynko and Smithies, 1958). Jeppson (1962) has suggested a variant in which warm dry gas is permitted to escape

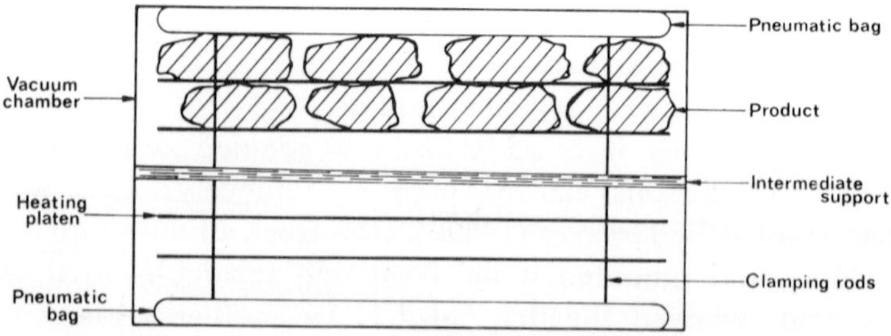

FIG. 3.10. A modified AFD assembly.

from the tips of the spikes, which only partly penetrate the product. The theory of heat transfer in these processes has been given by Forrest *et al.* (1962).

Accelerated drying rates can be explained in terms of a superheating effect in the vapour (Section 5.7). Certainly, superheating is likely with these methods, as the vapour may travel more than 0·5 m in a path close to the heater at 60–70°C.

3.7.3 INTERNAL HEATING BY MICROWAVES

A third method of heating, namely microwave heating (Fig. 3.11), has been frequently proposed and occasionally tried (Harper *et al.*, 1962; Copson, 1962; Decareau, 1962; Meryman, 1964). In this method, short-wave high-frequency alternating currents, 100 cm (300 MHz) to 3 cm (10 000 MHz), are employed and the heating effect is due to the "loss current" produced in the dielectric when the molecular dipoles, polarized by the field current, relax after reversal of the electric charge. Some heating is also due to frictional losses during polarization.

In the dielectric the loss current i_1 lags the charging current i by 90°, and so the ratio of the magnitudes of the two currents is expressed by the loss tangent ($\tan \delta$), namely

$$\tan \delta = \frac{i_1}{i}, \tag{3.33}$$

where δ is the loss angle between the charging current vector and the vector sum of i_1 and i. The loss tangent is equal to the ratio of the loss factor κ'' to the relative dielectric constant κ':

$$\tan \delta = \frac{\kappa''}{\kappa'}. \tag{3.34}$$

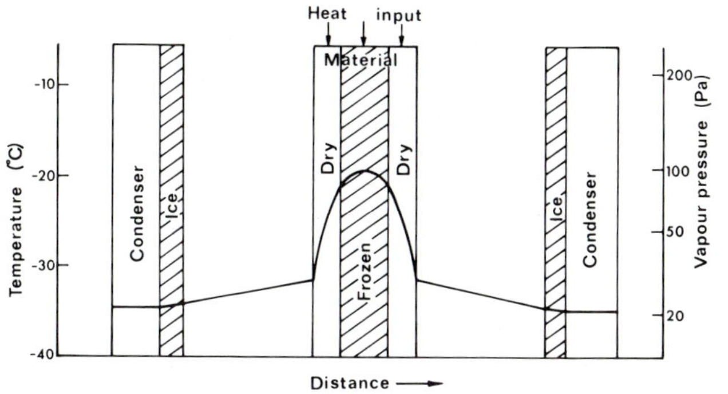

FIG. 3.11. Temperature and vapour-pressure gradients with microwave heating.

Since it is related to the mobility of the dipoles it tends to increase with increasing temperature. A material undergoing freeze-drying is itself a dielectric, with the dry layer having a low loss factor as compared with the frozen or moist layer. The rationale of the method is that the dry layer is not heated, only the frozen layer.

For the energy dissipation in the frozen layer

$$\sigma_D = 0{\cdot}556 \times 10^{-12} f E^2 \kappa'', \tag{3.35}$$

where σ_D = energy density, f = frequency, E = potential gradient.

Loss factors for frozen materials are listed in Table 3.1.

There are several disadvantages with this method of heating. In the first place, feedback of energy occurs, causing melting, because $\kappa' = 80$ for water compared with $\kappa' = 4$ for ice in the region below 100 Mc. Consequently, if any portion of a frozen specimen should melt there will be a sudden increase in energy dissipation, and since water absorbs 4000 times more energy than ice, pockets of steam will appear with explosive effects. It would appear therefore that microwave heating is more effective than elsewhere in areas of the product where there is bound water left after sublimation of the remaining ice. It is this factor which makes microwave heating attractive for removing residual moisture during the terminal stage of drying.

TABLE 3.1
Dielectric properties of frozen materials

Material	Temperature (°C)	Frequency f (MHz)	Dielectric constant κ'	Loss tangent $\tan\delta$	Loss factor κ''
Ice	−12	3000	3·2	0·0009	0·0029
Raw beef	−15	1000	5·0	0·15	0·75
Raw pork	−15	1000	6·8	0·12	0·82
Fat (average values)	all	all	2·4	0·1	0·24
Boiled peas	−15	1000	2·5	0·2	0·50
Boiled spinach	−15	1000	4·5	0·2	0·90
Baked squash	−15	1000	5·0	0·3	1·5
Peach	−15	1000	1·77	0·105	0·186
Pear	−15	2000	4·17	0·27	1·12
Boiled potato	−15	1000	4·5	0·2	6·5
Porridge	−15	1000	5·0	0·74	3·7
Suet	25	3000	2·5	0·07	0·175
Steak	25	3000	40	0·3	12·0

A second disadvantage is that at low frequencies there is limited penetration of the applied microwave field into raw food samples, and thus heating occurs only at a relatively small depth below the surface, whereas for pure ice, penetration is very much greater. The amount of penetration depends on κ''. Using microwave energy, the theoretical depth of penetration for ice at 3000 MHz is more than 10 m and for frozen beef at 1000 MHz it is nearly 30 cm. The depth of penetration is expressed by the relation

$$h' = \frac{\lambda_{sp}}{2\pi}\left[\frac{2}{\kappa'(1+\tan^2\delta)-1}\right]^{1/2}, \tag{3.36}$$

where h' = depth at which the field strength is $1/e$ of its original value, and λ_{sp} = wavelength in space.

At operating pressures of several hundred pascals and less there is maximum gas ionization. Experience with microwave heating at a frequency of 27 MHz shows that ionization occurs, resulting in discoloration of the product. Calculations based on equation (3.35) for a low field strength, thus preventing ionization, would indicate that a frequency of 1000 MHz should give better results. If the loss factor or breakdown voltage decreases with increasing frequency, going to a higher frequency will not be the solution. Data on the loss factor show that it is almost independent of frequency in the working range. On breakdown voltages the evidence is sparse. However, since the breakdown voltage is higher for water vapour than for air it might be an advantage to use a steam ejector pump instead of a refrigerating condenser to pump water vapour, thereby replacing the residual air in the chamber by water vapour. Better sealing against air leakage into the chamber and de-aeration of the product would, however, be necessary.

During freeze-drying the chamber impedance changes continuously, making matching of it with the output of the microwave generator necessary.

3.8 Pressure dependence of gas thermal conductivity

The possibility that freeze-drying could be accelerated by raising the vacuum pressure was anticipated when Ehlers *et al.* (1962) and Harper (1962) measured higher thermal conductivities in freeze-dried foods with increasing pressure and, moreover, when Harper and el Sahrigi (1964) made similar measurements with high-conductivity gases. Improvements in heat transfer, causing reductions in freeze-drying time which had been observed at higher pressures, may be deduced from Fig. 3.12 showing the variation of thermal conductivity with air and helium pressure for freeze-dried beef and apple. The curves show that the thermal conductivity in air-filled products increases with pressure by more than 50 per cent in the range 26·7 Pa to 2·7 kPa, and nearly 4000 per cent in the range 26·7 Pa to 13·3 kPa in helium-filled products. Without the gas, each product would have the lowest thermal conductivity equal to that of the material, comprising the structure alone of the product, in the range 0·026–0·040 W m^{-1} K^{-1}. This might be the magnitude of the thermal conductivity of those products which are normally freeze dried at a pressure of 13·3 Pa to 133·3 Pa. Also, it is seen that the variation of thermal conductivity with log pressure is almost linear in a part of their conductivity range. Various theories which have been proposed to fit the experimental results have not been entirely satisfactory; the one due to Harper (1962) is the best known.

3.8.1 EMPIRICAL THEORY

According to Harper, the effective thermal conductivity of any gas-filled porous material has two components, one associated with a solid

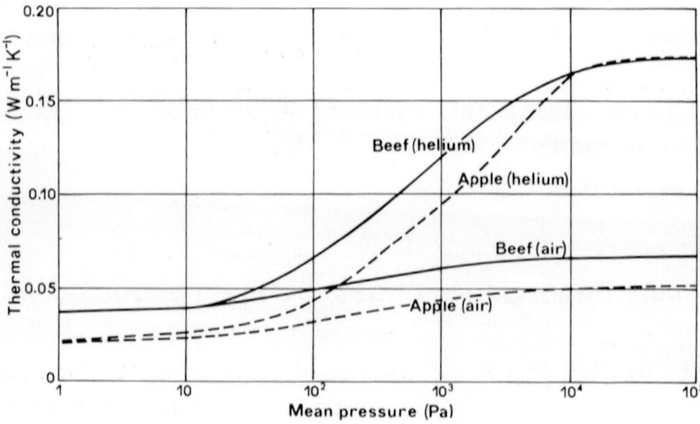

FIG. 3.12. Variation of thermal conductivity with air and helium pressure in freeze-dried beef and apple.

network and the other with the gas that fills the pores; he concluded that the experimental data can be adequately correlated by the empirical equation

$$\lambda_g = \frac{\lambda_{g,0}}{1 + C/P}, \tag{3.37}$$

where λ_g = thermal conductivity of the confined gas in the porous solid, $\lambda_{g,0}$ = thermal conductivity of the free gas at atmospheric pressure, C = constant, P = pressure.

For the confined gas, from equation (3.37) $\lambda_{g,0}(1 + C/P)^{-1}|_{\text{high}} \to \lambda_{g,0}$ at the high-pressure limit and $\lambda_{g,0}/(1 + C/P)^{-1}|_{\text{low}} \to (P/C)\lambda_{g,0} \simeq 0$ at the low-pressure limit. If λ_d = thermal conductivity of the porous solid (which is invariant), and λ_e = effective thermal conductivity of the material between the high- and low-pressure limits, then consider that

λ_e has plateaux given by

$$\begin{aligned} \lambda_e &= (1-\varepsilon)\lambda_d \quad \text{(low pressure)}, \\ \lambda_e &= (1-\varepsilon)\lambda_d + \lambda_{g,0} \quad \text{(high pressure)}, \end{aligned} \tag{3.38}$$

where ε = porosity, and $\lambda_{g,0}$ is taken without the volume fraction ε since λ_e would be independent of pressure in the pores at the high-pressure plateau.

Taking a weighted mean thermal conductivity $\bar{\lambda}_e$ with respect to the low- and high-pressure expressions for the confined gas yields:

$$\begin{aligned} \bar{\lambda}_e &= \frac{(1-\varepsilon)\lambda_d \left.\frac{\lambda_{g,0}}{1+C/P}\right|_{\text{low}} + [(1-\varepsilon)\lambda_d + \lambda_{g,0}] \left.\frac{\lambda_{g,0}}{1+C/P}\right|_{\text{high}}}{\left.\frac{\lambda_{g,0}}{1+C/P}\right|_{\text{low}} + \left.\frac{\lambda_{g,0}}{1+C/P}\right|_{\text{high}}} \\ &= (1-\varepsilon)\lambda_d + \frac{\lambda_{g,0} \left.\frac{\lambda_{g,0}}{1+C/P}\right|_{\text{high}}}{\left.\frac{\lambda_{g,0}}{1+C/P}\right|_{\text{low}} + \left.\frac{\lambda_{g,0}}{1+C/P}\right|_{\text{high}}} \end{aligned} \tag{3.39}$$

And, for the limit $\lambda_{g,0}$ in the denominator, equation (3.39) becomes

$$\bar{\lambda}_e = (1-\varepsilon)\lambda_d + \frac{\lambda_{g,0}}{1+C/P}. \tag{3.40}$$

Equation (3.40) gives the effective thermal conductivity for the region between low and high pressures and is the basis on which Harper summarized his data as shown in Table 3.2. It can also be written, in consideration of equation (3.38), as

$$\frac{\lambda_e - \lambda_{\text{low}}}{\lambda_{\text{high}} - \lambda_{\text{low}}} = \frac{1}{1+C/P}. \tag{3.41}$$

TABLE 3.2
Thermal conductivities of freeze-dried foods (W m^{-1} K^{-1}) (after Harper, 1962)

Pressure	Beef	Apple	Peach
Maximum	0·066	0·043	0·044
Minimum	0·038	0·015	0·016
Difference	0·028	0·028	0·028
Constant C	200·0	213·3	239·9

The thermal conductivity of the freeze-dried meat measured in the direction of the grain would explain a higher zero-pressure conductivity than for fruit. The intrinsic values for the solid $\lambda_d = (\lambda_e)_{P=0}/(1-\varepsilon)$ are 0·105, 0·123 and 0·175 W m^{-1} K^{-1} for beef, apple and peach respectively. Harper explained that these values are about the same as for similar porous materials. On this simple theory the value of C is between 200 and 240 for air but is 10 times greater for helium. Equation (3.41) is seen to be roughly linear between the low- and high-pressure plateaux when the effective conductivity λ_e is plotted against log mean pressure $\bar{P}$.

To take advantage of the higher conductivities so as to improve the heating rate for freeze-drying would mean operating at pressures which are above the range 13·3 Pa to 133·3 Pa. At pressures above 133·3 Pa this would be detrimental to the product since it would raise sublimation temperatures and produce incipient melting of the frozen layer. Higher-than-normal pressure operation has been advocated by Neumann (1963) using careful pressure control and by de Buhr (1967) with gas recirculation within the vacuum chamber. However, better results are obtained with a cyclic-pressure method due to Mellor (1967b), which combines the improved heat-transfer effect and a "flash" vapour-transfer effect with the lowering of pressure. The principle and operation of this method are discussed in Chapters 6 and 7.

3.8.2 THEORIES BASED ON ELECTRICAL ANALOGUE

In a fundamental approach, Kessler (1962) showed that a series-parallel model of the gaseous–porous solid of width d, as shown in Fig. 3.13, predicts the pressure dependence of the thermal conductivity of certain systems. He derived the following formula by analogy with electrical conduction for a simple series-parallel network.

$$\lambda_e = \frac{1}{(1-d)/[(1-\varepsilon)\lambda_d + \varepsilon\lambda_g] + d(1-\varepsilon)/\lambda_d + d\varepsilon/\lambda_g} \qquad (3.42)$$

Triebes and King (1966) simplified Harper's and el Sahrigi's model of a unit cell for heterogeneous media, which has been used previously by Wyllie and Southwick in connection with studies on electrical conduction. As shown in Fig. 3.14, they considered the heat flow along a conduction path which is both parallel and series for the gas and solid phases.

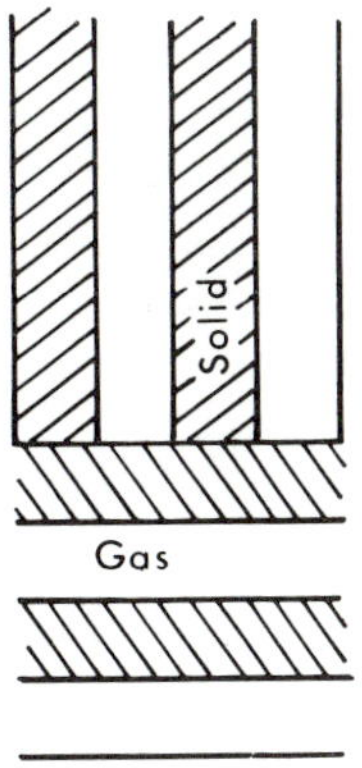

FIG. 3.13. Heat transfer model for gas filled porous material as used by Kessler (1962).

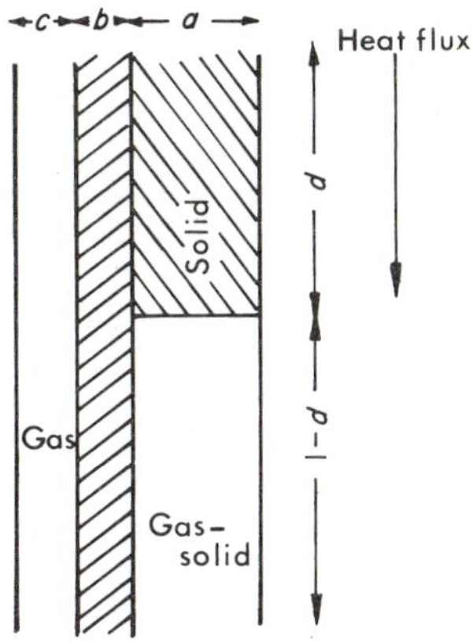

FIG. 3.14. Heat transfer model for gas filled porous material as used by Triebes and King (1966).

In addition to assuming no lateral conductivity, as had been done previously, i.e. vertical lines of heat flux only, Triebes and King assumed that the thermal conductivity λ_1 for the solid and gas in series with width d is given by

$$\frac{1}{\lambda_1}=\frac{d}{\lambda_d}+\frac{1-d}{\lambda_g}$$

or

$$\lambda_1=\frac{\lambda_d\lambda_g}{d\lambda_g+(1-d)\lambda_d}.$$

Hence the effective conductivity of the sections with dimensions a, b and c in parallel becomes

$$\lambda_e=\frac{a\lambda_g\lambda_d}{d\lambda_g+(1-d)\lambda_d}+b\lambda_d+c\lambda_g. \tag{3.43}$$

Triebes and King considered that for infinite lateral conductivity the plane separating the solid from the gas at a distance $1-d$ must be isothermal; and, therefore, it must be above the isothermal plane

$\lambda_1=(a+b)\lambda_d+c\lambda_g$ for distance d, whereas below it the thermal conductivity for the solid and gas in series is $\lambda_2=(a+c)\lambda_g+b\lambda_d$, for distance $1-d$. The effective conductivity is then

$$\frac{1}{\lambda_e}=\frac{d}{\lambda_1}+\frac{1-d}{\lambda_2}$$

$$=\frac{d}{(a+b)\lambda_d+c\lambda_g}+\frac{1-d}{(a+c)\lambda_g+b\lambda_d}. \quad (3.44)$$

From Fig. 3.14 it is seen that the conditions for the four constants are

$$a+b+c=1,$$
$$b+ad=1-\varepsilon. \quad (3.45)$$

At the low-pressure limit, $\lambda_g\to 0$ and, therefore, from equation (3.43) $\lambda_e\to b\lambda_d$ and, at the high-pressure limit, $\lambda_g=\lambda_{g,0}$. By plotting λ_e against λ_g as in Fig. 3.15, λ_s may be found for the solid material where it equals the apparent conductivity at the point $\lambda_e=\lambda_g$.

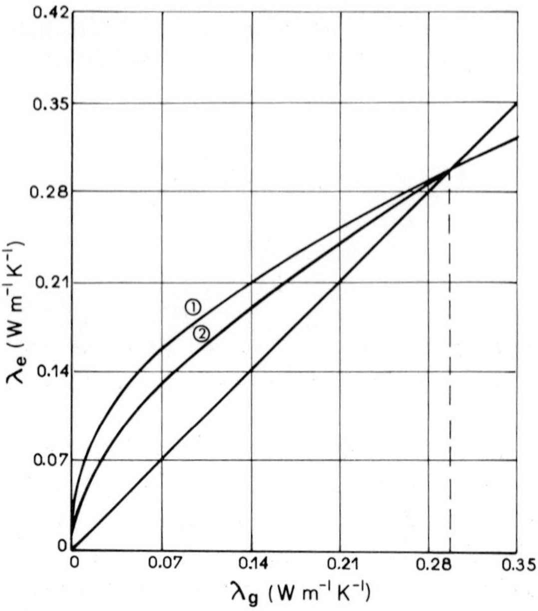

FIG. 3.15. Variation of effective thermal conductivity with the thermal conductivity of the confined gas for the two samples of freeze-dried turkey.

The four constants—a, b, c and d—were determined by Triebes and King (1966) from a solution of the above equations for a sufficient set of conditions pertaining to freeze-dried turkey meat, but for obvious reasons the solution was tedious. However, Magnussen (1969) preferred to assume zero lateral thermal conductivity, and therefore used equation (3.43) with four constants to be determined. Two constants may be found by combining the set of equations (3.45) and two other constants by taking the limiting values of λ_e at low pressure and atmospheric pressure. C in equation (3.41) may be determined by using the value of λ_e at the inflection point of the experimental curve.

3.9. Measurement of thermal conductivity

For steady-state measurement of thermal conductivity, Triebes and King (1966) employed a guarded hot-plate method and reference material consisting of "Lucite" in series with a sample of freeze-dried cooked turkey. They used two independent heat-conduction equations to derive the conductivity of the test sample: (1) from a derived heat flux to the Lucite in terms of its temperature gradient and conductivity and, (2) with a known heat flux Q. With subscripts T and L for turkey and Lucite respectively, the equations are

$$\lambda_T = \lambda_L \frac{\Delta T}{\Delta x}\bigg|_L \frac{\Delta x}{\Delta T}\bigg|_T,$$
$$\lambda_T = Q \frac{\Delta x}{\Delta T}\bigg|_T. \qquad (3.46)$$

Apparently Lucite and freeze-dried turkey gave comparable results when the pieces of Lucite were carefully aligned in the direction of heat flow.

Saravacos and Pilsworth (1965) also employed a guarded hot plate and reference material for measurements on freeze-dried model food gels. In use, these methods come to equilibrium slowly.

On the other hand, to obtain quicker measurements, Harper (1962) and Harper and el Sahrigi (1964) employed a straightforward conduction method with a single test sample and insulated holder between two heat-flow transducers, one on either side, for checking the two heat fluxes, Q_1 and Q_2 as in the equation

$$\lambda = Q_1 \frac{\Delta x}{\Delta T} \equiv Q_2 \frac{\Delta x}{\Delta T}. \tag{3.47}$$

The following description of a similar method to the above (Greenfield, 1972) uses two test samples instead of one, and is fairly rapid in use since there is no reference material. The apparatus (Fig. 3.16), based on a typical one for measuring thermal conductivities of

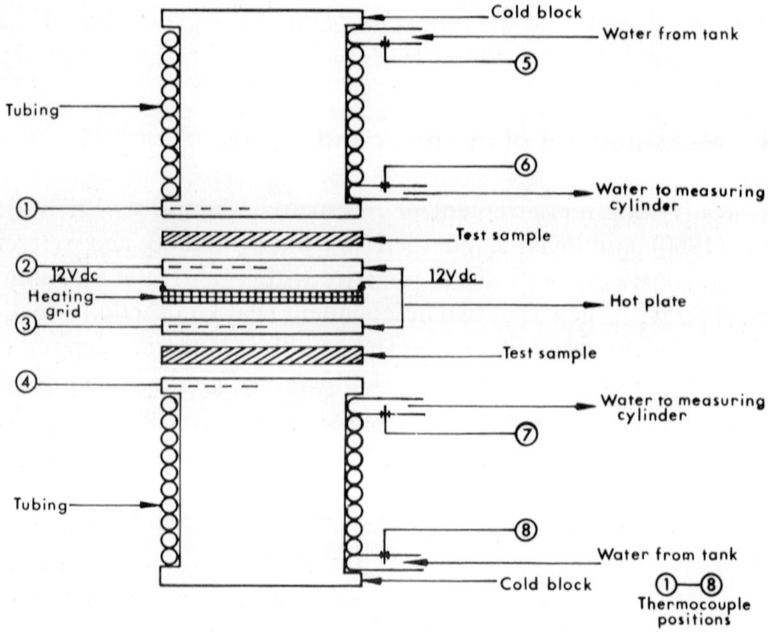

FIG. 3.16. Expanded view of thermal conductivity apparatus.

insulating materials, permits measurements on freeze-dried samples under vacuum. It consists of a hot-plate and two parallel cold blocks with samples of material 3·5 mm thick in the spaces between the hot and cold surfaces. The hot-plate is made of two copper disks, each 44·5 mm diameter by 3·5 mm thick, and a heating grid mounted between them made of nichrome ribbon wound on a thin mica former insulated from the plates by slightly larger mica disks. The grid was heated by a 12-volt d.c. battery and glycerine was used to improve the thermal contact between the grid and the copper disks. The two cold blocks (44·5 mm diameter by 44·5 mm) were solid copper cylinders with a spiral of 3·5 mm O.D. copper tubing as an integral part of their round surfaces. All the hot and cold metal surfaces in contact with the samples were machined flat to a tolerance of 0·0025 mm and blackened using zinc sulphide and sulphuric acid, and small holes were drilled to the central axis to within 2 mm from these flat surfaces. Fine No. 24 S.W.G. calibrated thermocouples were inserted in the holes to measure the temperature and others, consisting of five multi-junction thermocouples, were used to measure the water temperatures. Water from a constant-head tank circulated through the copper tubing. The entire assembly was mounted vertically inside a cylindrical chamber in which the pressure could be controlled at different levels between 0·133 Pa and 101·3 kPa.

The electric current to the heater and voltage were determined to check the heat input as measured from the water temperature and flow rate. Temperature recordings of the hot-plate and of the water were taken until they became steady (after 1–2 hours), and water flow rates were obtained by collecting 200 ml of water in measuring cylinders.

The adequacy of the apparatus was tested using cellular polystyrene and cork samples, the conductivities of which have been published by the National Physical Laboratory, England. The experimental value for the cork agreed to within 3 per cent, but a large error in the value obtained for the polystyrene sample suggests that the test material was of a different kind. In addition, the test showed that it was necessary to insulate the inlet water tubes from the tank to the cold blocks with cotton wool so as to reduce temperature fluctuations in the water and, by means of a temperature recorder, to determine the steady temperature when these fluctuations were minimal. Also, insulation around the entire assembly was necessary.

The freeze-dried samples were conditioned before the test in a desiccator containing phosphorus pentoxide. There were no problems in cutting the beef under atmospheric conditions, but the fruits were very hygroscopic and it was difficult to cut flat parallel surfaces without crumbling and without absorbing moisture. Dry specimens could support the weight of the copper block without damage to the surface pores of the sample. Granular products, e.g. freeze-dried rice, were supported using a thin cardboard ring. The thickness of the sample was measured with a vernier caliper and its area by means of a planimeter.

An improved method measures the heat flux directly by using a heat-flow transducer consisting of a multi-junction thermopile in a thin layer of protective material. This arrangement gave sufficiently accurate results in a shorter time than with the measurement of output flux from the water temperature and flow rate, when they were separately interposed between a test sample and a cold block. Philips' heat-flow transducers, which give an output potential difference of approximately 1 mV for a heat flux of 14·7 W m^{-2}, could be used without further calibrations. The thermal conductivity of the test sample was determined, as would be the case using equation (3.47).

As shown in Table 3.3, the expected behaviour of the thermal conductivity varies from a constant value at high pressure with an almost linear transition to a constant value at low pressure for freeze-dried sucrose and gelatin solutions. Similar results have been obtained for freeze-dried pear and rice samples.

TABLE 3.3
Variation of thermal conductivity with air pressure for freeze-dried sucrose and gelatin solutions (after Greenfield, 1972).

10% Sucrose		20% Sucrose		10% Gelatin	
Pressure (kPa)	Thermal conductivity (W m^{-1} K^{-1}) $\times 10^3$	Pressure (kPa)	Thermal conductivity (W m^{-1} K^{-1}) $\times 10^3$	Pressure (kPa)	Thermal conductivity (W m^{-1} K^{-1}) $\times 10^3$
0·0036	9·2	0·0033	15·5	0·0027	28·2
0·0039	9·2	0·0035	16·0	0·0031	26·0
0·0044	9·7	0·0080	17·6	0·0053	27·7
0·0120	10·9	0·0105	16·8	0·0059	28·2
0·0160	12·2	0·0207	20·2	0·0267	29·8
0·0173	13·0	0·0240	19·3	0·0293	32·3
0·0240	14·7	0·0700	25·2	0·0667	31·5
0·0533	18·5	0·0706	24·4	0·0733	34·4
0·0573	18·9	0·1040	27·7	0·1000	35·7
0·1320	24·8	0·1106	28·1	0·1226	35·7
0·1373	23·1	0·2533	34·4	0·2667	38·6
0·9998	34·9	0·7332	36·1	0·3999	39·9
1·200	35·7	1·333	39·9	1·333	46·2
13·33	37·0	4·532	40·7	6·665	52·5
26·67	37·0	12·53	41·2	13·33	54·2
47·99	37·0	26·66	41·6	39·99	53·8
80·65	36·5	53·99	42·4	93·31	54·6
101·31	37·0	101·31	42·8	101·31	57·6

Greenfield (1972) has observed that the results for 10 per cent gelatin pre-frozen in air at −20°C are higher than those obtained by Saravacos and Pilsworth (1965) who used dry ice to freeze the samples. This probably reflects the different porous structures resulting from the different freezing conditions.

Appendix 3.I Moving-interface boundary theory of heat conduction

Constant surface temperature

As already mentioned in Chapter 3, Section 3.3, Mellor (1962b) derived equations for the temperature profile and the heat flux for a moving interface boundary problem, for sublimation. These are now rigorously derived for the back-face heating of a semi-infinite frozen solid where the ice sublimation temperature at the surface is raised from zero to a constant value T_s; and where the sublimation temperature T_i is assumed to be temporarily at zero temperature scale.

Heat conduction in the ice phase causes the interface between the frozen and dry parts of the material to move with velocity $\dot{X}$ as shown in Fig. 3.17. In the first instance the heat conduction equation is

$$\frac{\partial T}{\partial t} = \alpha_f \frac{\partial^2 T}{\partial x^2}, \quad T(x, t), \tag{3.I.1}$$

where α_f = thermal diffusivity of the ice phase, and $T(x, t)$ = temperature in the ice phase.

The initial and boundary conditions are:

$$\begin{aligned} &\text{i.} \quad T = 0, \quad \text{all } x, \quad t = 0, \text{ initially,} \\ &\text{ii.} \quad T = T_s, \quad x = 0, \quad \text{all } t, \\ &\text{iii.} \quad \lambda_f(\partial T/\partial x)_{x=X} = -\rho_f\, \Delta H\, (\mathrm{d}X/\mathrm{d}t), \end{aligned} \tag{3.I.2}$$

where λ_f = thermal conductivity of the ice phase, ρ_f = density of ice phase, ΔH = latent heat of sublimation of ice, $X(t)$ = interface depth.

Applying the Laplace transformation to equation (3.I.1) yields

$$\frac{\partial^2 \bar{T}}{\partial x^2} - \frac{p}{\alpha_f}\bar{T} = -\frac{c_f \rho_f T(0)}{\lambda_f}. \tag{3.I.3}$$

Now the quantity $-c_f\rho_f T(0)$ is the heat gained per unit volume; it could equally be the latent heat $\rho_f \Delta H$ liberated by sublimation per unit volume, so that it seems appropriate to write

$$\frac{\partial^2 \bar{T}}{\partial x^2} - \frac{p}{\alpha_f}\bar{T} = \frac{\rho_f \Delta H}{\lambda_f}. \tag{3.I.4}$$

Note that Evans *et al.* (1950) obtained this equation without proof as the starting point for a recrystallization problem; however, they used

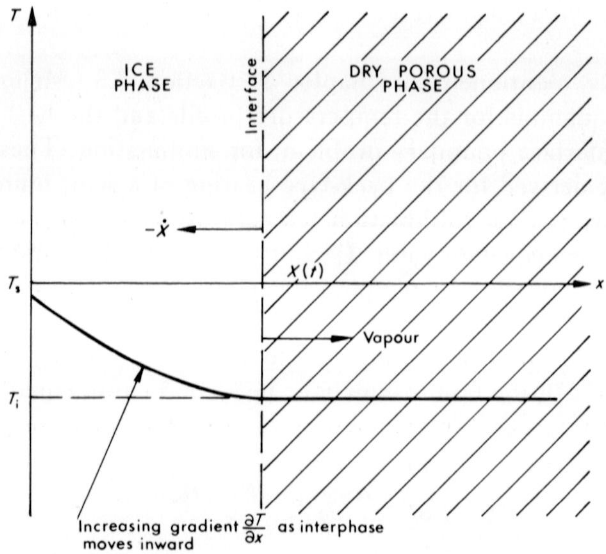

FIG. 3.17. Moving interface model for back-face heating.

the right-hand side of the equation with a decay function, viz. e^{-pt}. Another type of decay function is discussed by Carslaw and Jaeger (1959).

Also, it is worth noting that interface condition (iii) is utilized only after finding a solution by the classical proof for evaluating Stefan's proportionality constant γ. In equation (3.I.4), however, it is incorporated in the conduction equation.

The solution of equation (3.I.4) is given by

$$\bar{T} = \frac{\rho_f \Delta H}{2\lambda_f}\left(\frac{\alpha_f}{p}\right)^{1/2} e^{(p/\alpha_f)^{1/2}x} \int e^{-(p/\alpha_f)^{1/2}x}\, dx - \frac{\rho_f\, \Delta H}{2\lambda_f}\left(\frac{\alpha_f}{p}\right)^{1/2} \cdot e^{-(p/\alpha_f)^{1/2}x} \int e^{(p/\alpha_f)^{1/2}x}\, dx + C\, e^{-(p/\alpha_f)^{1/2}x}, \tag{3.I.5}$$

where $C = \alpha_f \rho_f\, \Delta H / p\lambda_f$ is a constant obtained from condition (i).

Changing x to ξ in equation (3.I.5) under the integration gives

$$\bar{T} = \frac{\rho_f\, \Delta H}{2\lambda_f}\left(\frac{\alpha_f}{p}\right)^{1/2} \int e^{-(p/\alpha_f)^{1/2}(\xi - x)}\, d\xi - \frac{\rho_f\, \Delta H}{2\lambda_f}\left(\frac{\alpha_f}{p}\right)^{1/2} \int e^{-(p/\alpha_f)^{1/2}(x-\xi)}\, d\xi + \frac{\alpha_f \rho_f\, \Delta H}{p\lambda_f} e^{-(p/\alpha_f)^{1/2}x}$$

$$= -\frac{\rho_f\, \Delta H}{\lambda_f}\left(\frac{\alpha_f}{p}\right)^{1/2} \int e^{-(p/\alpha_f)^{1/2}(x-\xi)}\, d\xi + \frac{\alpha_f \rho_f\, \Delta H}{p\lambda_f} e^{-(p/\alpha_f)^{1/2}x}, \tag{3.I.6}$$

since $\int e^{-(p/\alpha_f)^{1/2}(\xi - x)}\, d\xi = -\int e^{-(p/\alpha_f)^{1/2}(x-\xi)}\, d\xi$ after integrating and reversing the role of x and ξ.

The inverse transform of equation (3.I.6) is then

$$T = -\frac{\rho_f\, \Delta H}{\lambda_f}\left(\frac{\alpha_f}{\pi}\right)^{1/2} \int \frac{e^{-(x-\xi)^2/4\alpha_f t}}{t^{1/2}}\, d\xi + \frac{\alpha_f \rho_f\, \Delta H}{\lambda_f} \operatorname{erfc} \frac{x}{2(\alpha_f t)^{1/2}}$$

$$= -\frac{2\, \rho_f\, \Delta H}{\lambda_f}\left(\frac{\alpha_f t}{\pi}\right)^{1/2} \dot{X} + \frac{\alpha_f \rho_f\, \Delta H}{\lambda_f} \operatorname{erfc} \frac{x}{2(\alpha_f t)^{1/2}}. \tag{3.I.7}$$

where $\dot{X} = d\xi/dt$, $x - \xi = 0$ and so $e^{-(x-\xi)^2/4\alpha_f t} = 1$.

From the surface boundary condition (ii), and equation (3.I.7)

$$T_s = -\frac{2\rho_f\, \Delta H}{\lambda_f}\left(\frac{\alpha_f t}{\pi}\right)^{1/2} \dot{X} + \frac{\alpha_f \rho_f\, \Delta H}{\lambda_f}, \tag{3.I.8}$$

$$T_s - T = \frac{\alpha_f \rho_f\, \Delta H}{\lambda_f}\left[1 - \operatorname{erfc} \frac{x}{2(\alpha_f t)^{1/2}}\right]. \tag{3.I.9}$$

Now Stefan's criterion that the interface depth varies as the square root of time applies to this case—one of the few possible cases. That is: $T = T_i$ and $\gamma = X/2(\alpha_f t)^{1/2}$. Thus, finally, the solution becomes

$$\frac{T - T_i}{T_s - T_i} = 1 - \frac{\operatorname{erf}\, [x/2(\alpha_f t)^{1/2}]}{\operatorname{erf}\, \gamma}. \tag{3.I.10}$$

Constant heat flux

Consider a similar problem to the previous one but with a constant heat flux applied to the ice surface instead of a constant surface temperature. The starting point is again equations (3.I.2 and 3.I.3), with condition (ii) replaced by

$$-\lambda_f\left(\frac{\partial T}{\partial x}\right)_{x=0} = Q, \tag{3.I.11}$$

where Q = applied heat flux. Substitution from equation (3.I.11) in equation (3.I.5) yields constant $C = (\bar{Q}/\lambda_f)(\alpha_f/p)^{1/2}$. Thus, the Laplace transformation of the new equation (3.I.6) becomes

$$\bar{T} = -\frac{\rho_f\,\Delta H}{\lambda_f}\left(\frac{\alpha_f}{p}\right)^{1/2}\int e^{-(p/\alpha_f)^{1/2}(x-\xi)}\,d\xi + \frac{Q}{\lambda_f}\left(\frac{\alpha_f}{p^3}\right)^{1/2} e^{-(p/\alpha_f)^{1/2}x}, \tag{3.I.12}$$

where $\bar{Q} = Q/p$.

The inverse transform of equation (3.I.12) is given by

$$T = -\frac{\rho_f\,\Delta H}{\lambda_f}\left(\frac{\alpha_f}{\pi}\right)^{1/2}\int \frac{e^{-(x-\xi)^2/4\alpha_f t}}{t^{1/2}}\,d\xi + 2\left(\frac{\alpha_f t}{\pi}\right)^{1/2}\frac{Q}{\lambda_f}e^{-x^2/4\alpha_f t} - \frac{xQ}{\lambda_f}\operatorname{erfc}\frac{x}{2(\alpha_f t)^{1/2}}$$

$$= -\frac{2\rho_f\,\Delta H}{\lambda_f}\left(\frac{\alpha_f t}{\pi}\right)^{1/2}\dot{X} + 2(\alpha_f t)^{1/2}\frac{Q}{\lambda_f}\operatorname{ierfc}\frac{X}{2(\alpha_f t)^{1/2}} \tag{3.I.13}$$

The interface depth $X(t)$, although a function of time, is here not considered proportional to the square root of time as for the previous case. Instead, let it be determined by a power series expansion: $X(t) = C_1 t + C_2 t^2 + C_3 t^3 + \cdots$ where $C_1, C_2, C_3 \cdots$ are constants to be determined. Using the method of expansion for $X(t-\tau)$ and $\dot{X}(t-\tau)$, where τ is the initial time when $T = 0$, suggested by Evans *et al.*, the integrand of equation (3.I.13) is approximated by

$$C_1 t^{1/2} - \tfrac{2}{3}C_2 t^{3/2} + \tfrac{3}{5}C_3 t^{5/2} + \cdots \tag{3.I.14}$$

and the error function of equation (3.1.13) is approximated by

$$2t^{1/2} + \frac{C_1^2}{2\alpha_f}t^{3/2} - \frac{C_1 C_2}{\alpha_f}t^{5/2} + \cdots. \tag{3.I.15}$$

Comparing the coefficients of $t^{1/2}$ and $t^{3/2}$ in the above two expressions for $T=0$ at $\tau \to 0$ initially in equation (3.I.13) and assigning the appropriate constants, namely $-(2\rho_f \,\Delta H/\lambda_f)(\alpha_f/\pi)^{1/2}$ and $2\alpha_f^{1/2}Q/\lambda_f$, yields

$$C_1 = \frac{Q}{\rho_f \,\Delta H}$$

$$C_2 = \frac{3}{8\alpha_f}\left(\frac{Q}{\rho_f \,\Delta H}\right)^3, \text{ etc.} \tag{3.I.16}$$

Thus the interface depth is given by

$$X(t) = \frac{Q}{\rho_f \,\Delta \mathrm{H}}t - \frac{3}{8\alpha_f}\left(\frac{Q}{\rho_f \,\Delta H}\right)^3 t^2 + \cdots. \tag{3.I.17}$$

In another theoretical approach, Luikov (1963) deals with drying with a moving boundary by expressing the heat conduction equation (3.I.1) in terms of the velocity of the evaporation front.

4 Vapour Transfer

4.1 Gas diffusion in porous materials

Freeze-dried products are not unlike other porous materials in that they consist mainly of large and small particles or fibres, mixed with some larger ones. The pores which run through are the result of passages created in the packing of the smaller particles, and consist of irregular networks, some linked or cross-linked and others which are blind. Porous materials occur naturally, but they can also be made by consolidating fine particles or fibres under pressure or by freezing liquids and other substances and then freeze-drying them.

When the diffusion coefficient or the permeability of a gas flowing in a porous material is measured these parameters can then be described in terms of its pore geometry, i.e. structural characteristics such as pore size, porosity and tortuosity factor. The form and shape of the porous structure are often too complex for theoretical treatment and must be considered in terms of a simple geometrical system such as a bed of spheres or a bed of cubes; however, the simplest and most realistic equivalent structure is that of parallel cylindrical capillaries.

4.2 Cylindrical capillary model

4.2.1 SIMPLE THEORY

Assume a cylindrical model of a porous material as shown in Fig. 4.1 with the pore space having an equivalent radius r, a porosity ε equal to the ratio of the pore volume to the intrinsic volume of the sample, and a tortuosity factor $(L_e/L)^2$ taken as the square of the ratio of the effective flow path L_e to the length of the sample L. Consider n

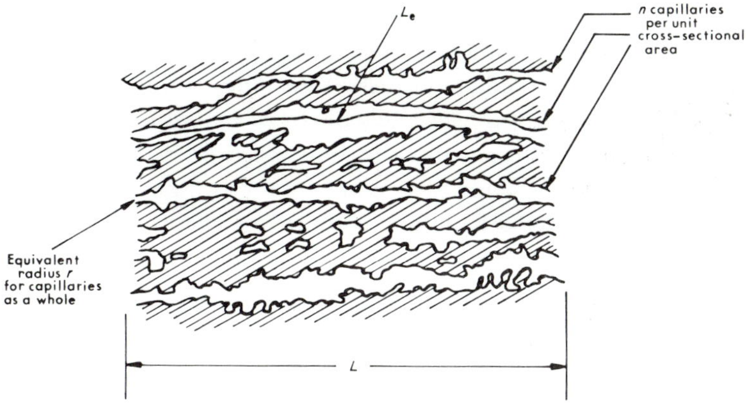

FIG. 4.1. An illustration from an enlargement of a photograph of a freeze-dried meat sample showing the porous capillary structure.

capillaries per unit area of cross-section, each with a radius r; the porosity ε is then given by

$$\varepsilon = \frac{n\pi r^2 L_e}{L}. \tag{4.1}$$

From Fick's law, the rate of diffusion of the gas equals $-D\,\mathrm{d}p'/\mathrm{d}x$, where D is the diffusion coefficient and $\mathrm{d}p'/\mathrm{d}x$ is the partial pressure gradient, which, for the porous sample, becomes $-D_e\,\Delta p'/L$ where D_e is the effective diffusion coefficient. Alternatively, the rate of diffusion equals $-n\pi r^2 D\,\Delta p'/L_e$. Eliminating $\Delta p'$ from the two expressions and substituting for ε from equation (4.1) the effective diffusion coefficient D_e becomes

$$D_e = \frac{\varepsilon}{(L_e/L)^2} D. \tag{4.2}$$

The ratio $\varepsilon/(L_e/L)^2$ is known as the diffusion ratio. In the porous material D_e varies inversely with the absolute pressure P and directly with the absolute temperature T to the 1·75 power; therefore, equation (4.2) corrected for pressure and temperature becomes

$$D_e = \frac{\varepsilon}{(L_e/L)^2} D \frac{101\,308}{P} \left(\frac{T}{273}\right)^{1.75}. \tag{4.3}$$

D_e increases with decreasing pressure until the mean free path λ of the gas molecules at intermediate pressures becomes comparable with the pore diameter, so that wall collisions begin to vie with intermolecullar collisions in the gas stream itself in limiting flow by diffusion. This condition is generally known as transition flow. Beyond this point Knudsen or molecular diffusion predominates at low pressures or in fine capillaries, as the mean free path becomes greater and the wall collisions almost completely govern the flow. The diffusion coefficient is then independent of pressure.

Turning now to gas flow in a single tube or capillary, the Knudsen diffusion coefficient D_K is expressed by

$$D_K = \frac{2}{3} r \bar{v}_m \frac{2-f}{f}, \qquad r/\lambda \to 0, \tag{4.4}$$

where the mean molecular velocity is given by $\bar{v}_m = (8RT/\pi M_g)^{1/2}$ and f = fraction of molecules undergoing diffuse reflection at the walls ($f = 1$ normally), also R = gas constant, T = absolute temperature, M_g = molecular weight of the gas.

At high pressures, the diffusion coefficient D as given by kinetic theory is

$$D = \frac{\bar{v}_m \lambda}{3}, \qquad r/\lambda \to \infty. \tag{4.5}$$

And at intermediate pressures when transition flow occurs, Bosanquet (1944) expressed the effective diffusion coefficient for diffusion in a tube by means of an interpolation formula for a random walk in which the successive steps of the individual molecules are terminated by collisions either with other molecules or with the tube wall. Since the diffusion coefficients are inversely proportional to their resistances and are additive, he simply wrote

$$\frac{1}{D_e} = \frac{1}{D} + \frac{1}{D_K}. \tag{4.6}$$

From which the transposition from equations (4.4) and (4.5) gives

$$D_e = \frac{2r}{\lambda + 2r} D, \qquad f = 1. \tag{4.7}$$

The diffusion ratio for a single tube is therefore $2r/(\lambda + 2r)$. On the other hand, if the interpolation formula is applied to a porous material, considered as a parallel set of capillaries with D_e at higher pressures given by equation (4.3) and Knudsen's diffusion coefficient (intercept value, see equation 4.13) given by

$$D'_K = \frac{2\delta_1 \varepsilon r \bar{v}_m}{3k''(L_e/L)^2}$$

then the transition diffusion coefficient so obtained is of doubtful value. Experience shows that the interpolation formula cannot be used with freeze-dried products.

4.2.2 COLLISION THEORY

The problem has been considered by Mellor and Lovett (1968), who developed an equation for the mechanics of molecular collisions in a circular capillary and applied it to porous freeze-dried products. The equation for the transition flow curve in terms of the permeability coefficient K against the pressure P approaches asymptotically a viscous-slip flow straight line in the medium pressure range. The slope and the asymptote intercept on the K-axis yield the equivalent pore radius r and the diffusion ratio $\varepsilon/(L_e/L)^2$ of the porous material. These quantities can be deduced by developing an equation for the asymptote in terms of an equivalent cylindrical model for the porous material along similar lines to those proposed by Barrer (1963).

If diffusion is considered over a wide pressure range, and the diffusion coefficient calculated in terms of Knudsen's diffusion coefficient, first, the self-diffusion equation (4.2) becomes

$$D_e = \frac{\varepsilon}{(L_e/L)^2} \frac{D_K}{1 + 2r/\lambda},$$

second, the transition equation (4.7) applied to the porous material becomes

$$D_e = \frac{\pi}{4}\frac{\varepsilon}{(L_e/L)^2}\frac{2(r/\lambda)D_K}{1+2r/\lambda}$$

and Poiseuille's equation for viscous flow

$$D_e = \frac{3\pi}{64}\frac{\varepsilon}{(L_e/L)^2}\frac{rD_K}{\lambda}.$$

The resultant coefficient K is given by

$$K = \frac{\varepsilon}{(L_e/L)^2}D_K\left(\frac{3\pi}{64}\frac{r}{\lambda}+\frac{\pi}{4}\frac{2r/\lambda}{1+2r/\lambda}+\frac{1}{2r/\lambda}\right). \tag{4.8}$$

Substituting $D_K = \frac{2}{3}r\bar{v}_m$ (equation 4.4) in equation (4.8) and the relationships:

$$\bar{P} = \tfrac{1}{3}\rho_g\bar{v}_m^2, \qquad \mu_g = \tfrac{1}{3}\rho_g\bar{v}_m\lambda$$

in the first term of equation (4.8) yields

$$K = \frac{\varepsilon}{(L_e/L)^2}\left(\frac{\pi}{32}\frac{r^2\bar{P}}{\mu_g}+\frac{2}{3}\frac{\pi}{4}r\bar{v}_m\frac{2r/\lambda}{1+2r/\lambda}+\frac{2}{3}r\bar{v}_m\frac{1}{1+2r/\lambda}\right), \tag{4.9}$$

where ρ_g = gas density and μ_g = viscosity. As the pressure is high where the transition curve approaches the asymptote, so $2r/\lambda \to \infty$, and equation (4.9) reduces to

$$K = \frac{\varepsilon}{(L_e/L)^2}\left(\frac{\pi}{32}\frac{r^2\bar{P}}{\mu_g}+\frac{2}{3}\frac{\pi}{4}r\bar{v}_m\right). \tag{4.10}$$

To account for the flow in the porous material in terms of the cylindrical model equation (4.10) embodies two structural constants k' and k'', where $k' = 2$ or $2{\cdot}5$ and $k'' = 1$. For non-circular capillaries $\pi/32$ becomes $1/4k'$ in the first term of equation (4.10) and a roughness factor $\delta_1 = \pi/4$ (smooth pores) or $\delta_1 = 3\pi/16$ (rough pores) in the second term. Therefore

$$K = \frac{\varepsilon r^2\bar{P}}{4k'(L_e/L)^2\mu_g}+\frac{2\delta_1\varepsilon r\bar{v}_m}{3(L_e/L)^2k''}. \tag{4.11}$$

A factor of 8 instead of 4 in the first term seems appropriate for freeze-dried products and has been confirmed by experimental results

(Table 4.1 and Fig. 4.2). As such, equation (4.11) applies to the asymptote and its slope $\Delta K/\Delta P$ is given by

$$\frac{\Delta K}{\Delta P}=\frac{\varepsilon r^2}{8k'(L_e/L)^2\mu_g}, \tag{4.12}$$

and the intercept D'_K is given by

$$D'_K=\frac{2\delta_1\varepsilon r\bar{v}_m}{3k''(L_e/L)^2}. \tag{4.13}$$

TABLE 4.1
Structural data for freeze-dried products that have been obtained through the measurement of permeability (after Mellor, 1966b)

Product	Slope $\times 10^5$ ($m^2\,s\,kg^{-1}$)	Intercept $\times 10^4$ ($m^2\,s^{-1}$)	Equivalent pore radius r (μ)	Diffusion ratio $\varepsilon/(L_e/L)^2$	Porosity ε	Tortuosity factor $(L_e/L)^2$
Beef	1·5	21	58	0·18	0·75	4·2
Cooked meat product	2·7	22	77	0·16	0·58	3·7
Apple	4.3	32	87	0·20	0·87	4·4
Strawberry	5·9	43	88	0·26	0·85	3·2
Onion	1·1	10	96	0·05		
Tomato concentrated	5·8	30	126	0·13	0·81	6·3
Sugar cane juice	9·0	49	119	0·21	0·95	4·5

By transposing equations (4.12) and (4.13) the equivalent pore radius can be found from

$$r=\frac{16\delta_1\bar{v}_m\mu_g k'\,\Delta K}{3k''D'_K\,\Delta P}, \tag{4.14}$$

and the diffusion ratio $\varepsilon/(L_e/L)^2$ can be obtained from either the slope or the intercept.

4.2.3 EXPERIMENTAL RESULTS

During freeze-drying the rate of water removal depends on the rate at which water vapour can pass through the dry product, and may be calculated according to the theory outlined if the appropriate coefficients of diffusion or of permeability are known. There are difficulties in measuring the diffusion coefficient directly, especially if

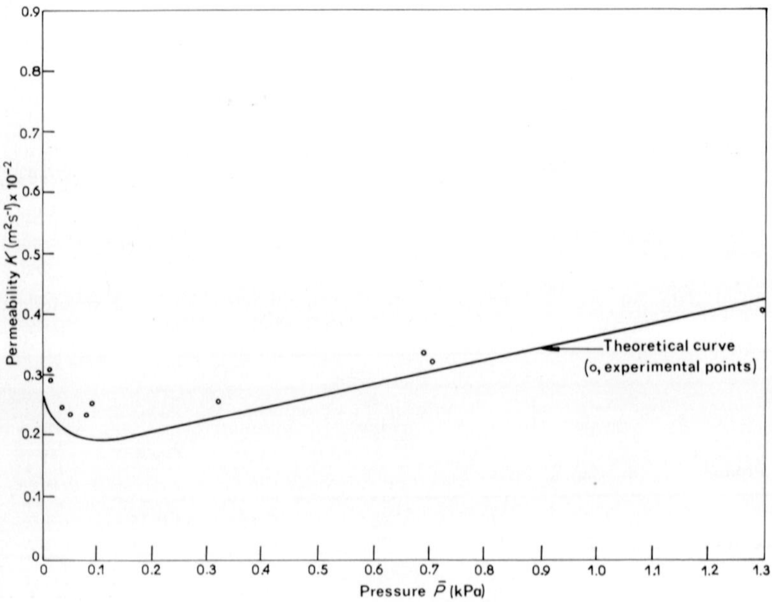

FIG. 4.2. Permeability–pressure relationships for freeze-dried beef.

sorption of water vapour occurs, and so it is usually derived from the diffusion coefficient for free water vapour in accordance with equations (4.2) and (4.3). Harper (1962) made direct measurements of the water vapour diffusivity for beef, apple and peach with an uncertainty of 10–15 per cent.

In contrast, the permeability coefficient can be obtained directly by measuring the amount of an inert gas flowing through the product under an absolute pressure gradient. Correlation of permeability and mean absolute pressure provides basic information about the porosity and tortuosity factor of the material in order to obtain a diffusion ratio as explained in equation (4.2) and in equation (4.15) to follow. Also from equation (4.11) the experimental value of the permeability K provides unique information on the structural characteristics of the product that are not given by the measurement of diffusivity alone. And, in fact, the effective diffusivity D_e appropriate to the particular product and gas used in freeze-drying can be derived from equation

(4.3) knowing the structural parameter obtained, viz. $\varepsilon/(L_e/L)^2$. Some typical parameter values are given in Table 4.1.

It follows that the equation for the asymptote involves molecular collisions between successive wall collisions, whereas under the conditions where the capillary diameter equals the mean free path (transition flow) a considerable number of molecules do not behave in this way. Instead, if the fraction of the molecules that, on the average, do collide with other molecules between two wall collisions in a viscous-slip flow term and the fraction that do not collide in an additional molecular flow term are considered an equation, similar to one developed by Pollard and Present (1948), can be derived on the basis of collision theory (Appendix 4.I, equation (4.I.9)) for the permeability in terms of the mean free path, giving

$$K = \frac{\varepsilon}{(L_e/L)^2} D_K \left[\left(\frac{3}{32k'} \frac{2r}{\lambda} + \delta_1 \right)(1 - e^{-2r/\lambda}) + e^{-2r/\lambda} \right]. \qquad (4.15)$$

In the broadest sense the term

$$\frac{\varepsilon}{(L_e/L)^2} \left[\left(\frac{3}{32k'} \frac{2r}{\lambda} + \delta_1 \right)(1 - e^{-2r/\lambda}) + e^{-2r/\lambda} \right]$$

may be considered as a diffusion ratio for the extended pressure range and it is seen that it approaches $\varepsilon/(L_e/L)^2$ at low pressures since the value of $e^{-2r/\lambda}$ approaches 1. Equation (4.15) has been found to fit the experimental results as shown in Fig. 4.3 for a model consisting of a bundle of 61 wires 0.6 mm diameter by 12 mm long, and for some freeze-dried products (Fig. 4.2). It is seen (Figs 4.2 and 4.3) that the theoretical curves are a good fit to the experimental results over the pressure ranges (transition to viscous flow regimes), and they are the ones of interest in relation to freeze-drying. Below a pressure of 66·7 Pa it would be better to fit Knudsen's diffusion coefficient directly as given by equation (4.4). In any event, Knudsen's regime is only of interest to freeze-drying when extremely low temperatures and high vacua are used, as in drying histological specimens.

4.3 Momentum transfer theory

In another approach to determining an effective diffusion coefficient for transition flow in porous materials, Dyer and Sunderland (1968a) considered momentum transfer of a binary gas mixture, and compared it with the simple kinetic theory relationship between thermal conductivity and self-diffusion proposed by Harper and Chichester (1963) to explain improvements in the rate of freeze-drying. They considered intermolecular and wall collisions of unlike molecules in

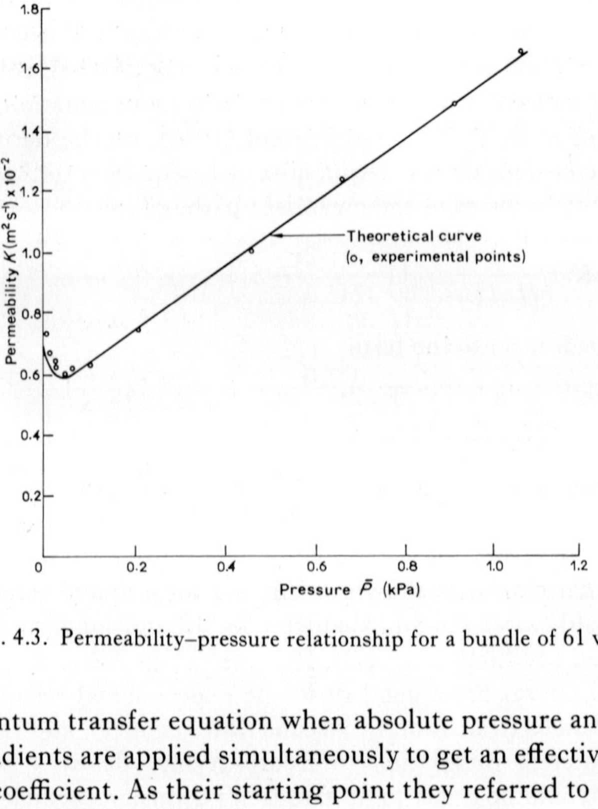

FIG. 4.3. Permeability–pressure relationship for a bundle of 61 wires.

the momentum transfer equation when absolute pressure and concentration gradients are applied simultaneously to get an effective mutual-diffusion coefficient. As their starting point they referred to Chapman and Cowling (1970) for the equation of motion of one of the constituent gases, originally the fundamental interdiffusion equation of Maxwell and, in the presence of an external force, this is given by

$$\frac{F_2}{A_2} - dp_2 - \frac{RT}{n_g M_2 D_{12}} n_1 n_2 (u_2 - u_1)\, dx = 0, \qquad (4.16)$$

where F_2/A_2 = external force per unit area (or total pressure) on gas component 2, dp_2 = partial pressure gradient in gas component 2, D_{12} = mutual-diffusion coefficient, n_1, n_2 = molecular densities of gas components, u_1, u_2 = macroscopic velocities of gas components, n_g = molecular density of gas mixture.

The third term represents molecular transfer due to molecular collisions either for a uniform pressure or for an average pressure taken over a small absolute pressure gradient. For the latter, $F_2/A_2 = p_2\,\mathrm{d}P/P$ can be deduced from Boyle's law.

Equation (4.16) requires an additional wall collision term if it is to apply to transition flow. Now, the momentum exchange with the wall of a capillary is $M_2u_2\delta'$ and from kinetic theory the number of colliding molecules per unit area is $n_2\bar{v}_m\,\mathrm{d}x/2r$; thus the pressure effect resulting from the wall collisions is $\delta' M_2 u_2 \bar{v}_m n_2\,\mathrm{d}x/2r$. Dyer and Sunderland state that the final form of the differential equation is valid for combined bulk and diffusional flow in the transition regime under a total pressure gradient, and they go on to reduce the equation by taking

$$p_2\frac{\mathrm{d}P}{P} - \mathrm{d}p_2 = -P\,\mathrm{d}\left(\frac{p_2}{P}\right) = -P\,\mathrm{d}y_2. \tag{4.17}$$

Substitution from equation (4.17) and $\bar{v}_m = (8RT/\pi M_2)^{1/2}$ in the final form of equation (4.16) yields

$$-\frac{P\,\mathrm{d}y_2}{RT} = N_2\left(\frac{8\delta'}{3\pi D_K} + \frac{n_1 - n_2}{n_g D_{12}}\frac{N_2}{N_1}\right)\mathrm{d}x. \tag{4.18}$$

where N_1, N_2 = molar flow rates of the gas components.

Applying Dalton's law of partial pressures $y_1 + y_2 = 1$, molar flux factor $a = 1 + N_2/N_1$, the definition of z ($z = x/L$), and integrating between the limits $y_2 = y_{2,0}$ at $z = 0$, $y_2 = y_{2,L}$ at $z = L$, equation (4.18) reduces to

$$N_2 = \frac{PD_e}{RTLa}\ln\frac{1 - y_{2,L}}{1 - y_{2,0}}. \tag{4.19}$$

The term $8\delta' D_{12}/3\pi D_K$ has been dropped for convenience since Dyer and Sunderland define an effective diffusion coefficient D_e to replace D_{12}.

In order to determine D_e by an iterative procedure, Dyer and Sunderland further deduced an auxiliary equation for $y_{2,L}$ from equation (4.18) by taking $y_1 + y_2 = 1$, $\mathrm{d}x = L\,\mathrm{d}z$, $PD_{12}/RT = c$, where c is a constant for the thermal conductivity–diffusion coefficient relationship, and expanding P in a Taylor series thus $P = P_0 + (\Delta P/L)x$. On substitution, these relationships yield

$$\frac{dy_2}{dz} - b_1 y_2 = -\left(b_1 + \frac{b_2}{P_0 + \Delta P z}\right) \tag{4.20}$$

which, on integrating between the limits $y_2 = y_{2,0}$ at $z = 0$, $y_2 = y_{2,L}$ at $z = L$ gives

$$y_{2,L} = 1 + \frac{b_2}{\Delta P} \frac{E_i\,[b_1(1+P_0/\Delta P)]}{\exp\,[b_1(1+P_0/\Delta P)]} + \frac{y_{2,0} - 1 - (b_2/\Delta P)\exp\,[-b_1(P_0/\Delta P)]\,E_i\,(b_1 P_0/\Delta P)}{\exp\,(b_1)} \tag{4.21}$$

where $E_i\,[b_1(1+P_0/\Delta P)]$ and $E_i\,(b_1P_0/\Delta P)$ are exponential integrals, and the dimensionless coefficients are given by

$$b_1 = \frac{(D_{12})_e}{(D_{K,2})_e} \ln \frac{1-y_{2,L}}{1-y_{2,0}}, \qquad b_2 = \frac{8\delta' R T b_1 c_1}{3\pi a D_{K,2}}. \tag{4.22}$$

Thus, by assuming a value of D_e they were able to calculate N_2 from equation (4.19), calculate $y_{2,L}$ from equation (4.21) and use them to compare by trial and error these with the experimental ones at the test-chamber exit to get the correct value of D_e. To determine the value of b_1, they used the relationship between true and effective diffusion coefficient (Scott and Dullien, 1962), namely $(D_{12}/D_{K,2})_{effec} = (D_{12}/D_{K,2})_{true}$ to obtain $(D_{K,2})_e$ from a single experimental measurement of $(D_{12})_e$ and the true values, and they took $\delta' = 1$.

4.3.1 EXPERIMENTAL PROCEDURE AND RESULTS

The experimental apparatus adopted by Dyer and Sunderland (1966) for the measurement of D_e is shown in Fig. 4.4. Several samples of freeze-dried beef were tested in turn, with an air–vapour mixture flowing through them. During pump down with valves 2 and 4 open, solonoid valve 1 remained open so no pressure gradient existed between the vacuum and test chambers until the desired pressure had been reached; then valve 2 was closed. The needle valve 3 was opened when required for air admission. Valve 1 was then closed to vaporize the ice, thereby providing the right air–vapour mixture, and throttle valve 4 adjusted to maintain a suitable back pressure.

Equilibrium conditions were sufficient when the pressure gauges, vapour concentration by electrical hygrometer and temperatures by thermocouples above and below the sample gave steady readings over

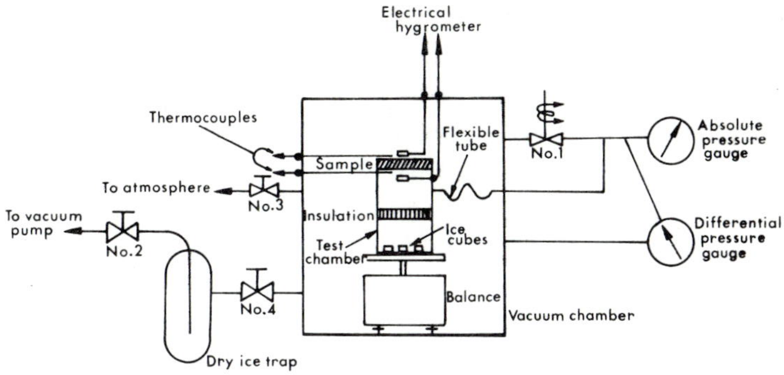

FIG. 4.4. Schematic diagram of diffusivity apparatus (after Dyer and Sunderland, 1966).

a period of 2 hours. The change in weight was recorded on the balance every 30 min during this period.

The data obtained, Table 4.2, were used to calculate the effective diffusion coefficient D_e from equation (4.19). Dyer and Sunderland's momentum transfer theory is not unlike Ede's theory of diffusion in the vapour space (Appendix 2.I) since the equations for the flow rates are similar. The difference lies in assumptions about the distribution of pressures; the first leads to a complicated vapour-pressure function

TABLE 4.2
Effective diffusion coefficients for freeze-dried beef (after Dyer and Sunderland, 1966)

Mean pressure (Pa)	$y_{w,0}$	$y_{w,L}$	Effective diffusion coefficient—transition flow D_e: Experimental ($m^2 s^{-1}$)	Theoretical (equation 4.19) ($m^2 s^{-1}$)	Theoretical (Harper and Chichester, 1963) ($m^2 s^{-1}$)
192·0	0·805	0·585	0·0014	0·0016	0·0029
210·6	0·77	0·565	0·0018	0·0017	0·0026
218·6	0·856	0·57	0·0011	0·0014	0·0026
299·9	0·813	0·52	0·0009	0·0009	0·0019
303·9	0·683	0·55	0·0016	0·0015	0·0019

and the second to a simple exponential function for the air pressure. The data in Table 4.2 are not sufficiently conclusive evidence that the effective diffusion coefficient D_e could be determined along with the flow rate and vapour pressures by an iterative procedure. Take, for instance, mean values for D_e for the pressure range listed; from Harper and Chichester's data $\bar{D}_e = 0{\cdot}0024\ \mathrm{m^2\ s^{-1}}$ and from the curve for freeze-dried beef as shown in Fig. 4.2 $D_e = 0{\cdot}002\ \mathrm{m^2\ s^{-1}}$. On the other hand, from Dyer and Sunderland's theoretical and experimental data the values are 0·0014 and 0·0016 $\mathrm{m^2\ s^{-1}}$ respectively; these values are about 40 per cent below the generally accepted value given above for freeze-dried beef. Mellor and Greenfield (1972) have found similar low values, varying by a factor, namely the cross-section $\varepsilon A/(L_e/L)$ (see equation 4.32). Thus it would appear that insufficient care has been taken with the experimental procedure (Fig. 4.4) to ensure that the ice cubes are in contact with the dried beef sample and the condenser is in an unrestricted vapour space in order to simulate actual freeze-drying conditions. Moreover, it seems doubtful whether diffusion coefficients for bulk and diffusional flow, and a total pressure gradient, are the correct analytical assumptions for freeze-drying.

No substantial evidence appears in the literature for a total pressure gradient across the dry layer–vapour space, except in so far as the vapour pressure for the sublimation can be greater than the total pressure. It would appear that more experimental evidence is needed to test the above assumptions, especially evidence based on actual or closely simulated freeze-drying procedures.

4.4 Measurement of porosity and permeability

4.4.1 POROSITY

Both liquid and gas porosimetry have been used to measure the porosity of freeze-dried materials. Spiess *et al.* (1969) used mercury porosimetry on model porous structures and on freeze-dried egg white and potato starch. When mercury is used as the penetrant the pore size distribution may be determined from the applied pressure of mercury according to Washburn's relation:

$$r = \frac{2\sigma_M \cos\theta}{P}, \tag{4.23}$$

where r = pore radius, σ_M = surface tension of mercury, θ = angle of contact of mercury with the material, P = pressure.

Pressures $10^3\ \mathrm{kg\ m^{-2}}$ tend to deform most freeze-dried biological materials. However, Triebes and King (1966) obtained good results with pressures 10^3–10^4 times greater using mercury porosimetry; they found a porosity of 0·68–0·72, due to pores 200 μm diameter and less, for two freeze-dried turkey samples. The results agreed well with a value of 0·70 derived independently from measurements of the thermal conductivity of the gas. As Blond (1972) has shown for freeze-dried orange juice (Fig. 4.5), low mercury pressures within the permitted range give a satisfactory pore-size distribution of between 5 and 200 μm radius.

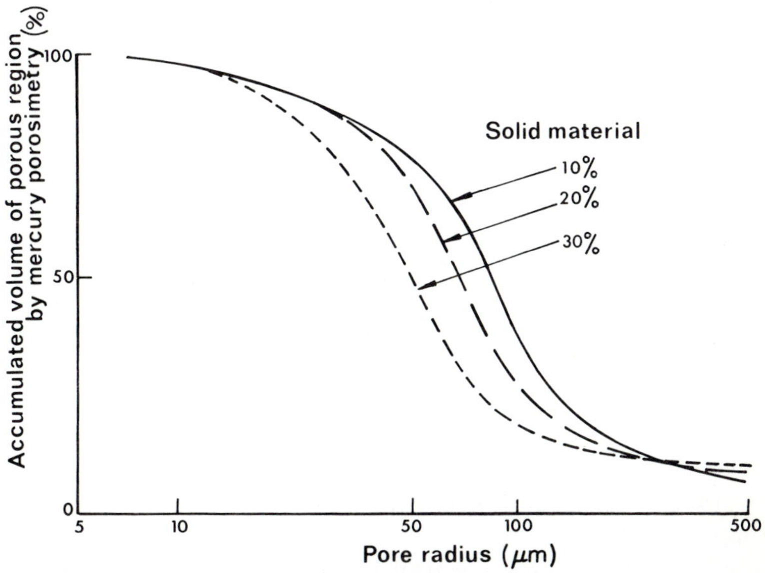

FIG. 4.5. Variation of volume of the porous region with pore radius for freeze-dried orange juice.

Other methods of pore-size determination are: (1) independently from permeability measurements (equation 4.14) or (2) directly by microscopic examination.

Harper (1962) determined the porosity from measurements of air pressure and volume discharged to atmosphere from a cell containing the sample and from the empty cell. Using air again, Mellor (1966b,

1971) measured the porosity by means of a simple U-tube manometer (Fig. 4.6), based on Boyle's law. A measured volume of the sample is enclosed in a glass vessel, the volume of which may be altered by

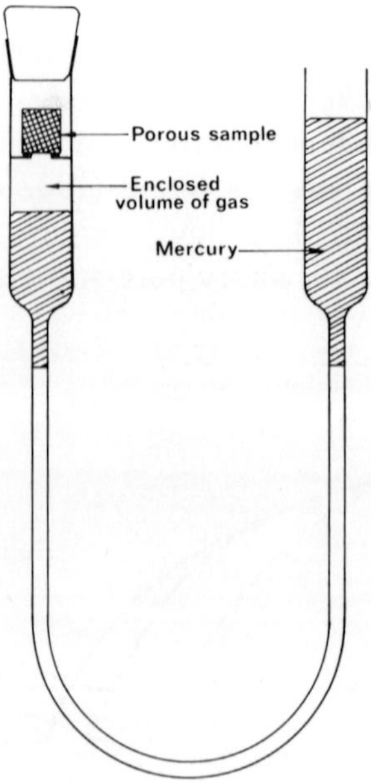

FIG. 4.6. Porosity apparatus.

movement of a mercury piston. The same mercury column serves as a means for simultaneously measuring both the volume of the vessel and the pressure within it.

Since by definition the porosity ε equals the ratio of the volume of the voids to the intrinsic volume V, the volume occupied by the solid matrix comprising the sample is equal to $(1-\varepsilon)V$. If V_1 and V_2 are the volumes above the mercury at pressures P_1 and P_2, then under these conditions application of Boyle's law leads to the relation $P_1V_1-(1-\varepsilon)P_1V=P_2V_2-(1-\varepsilon)P_2V$ and the porosity is accordingly given by

$$\varepsilon=\frac{P_1(V_1-V)-P_2(V_2-V)}{(P_2-P_1)V}. \tag{4.24}$$

4.4.2 PERMEABILITY BY STEADY-STATE FLOW

Mellor's (1971) equipment for measuring permeability consisted essentially of a glass tube in which a plug of freeze-dried product of known thickness and cross-sectional area was inserted (Fig. 4.7). A

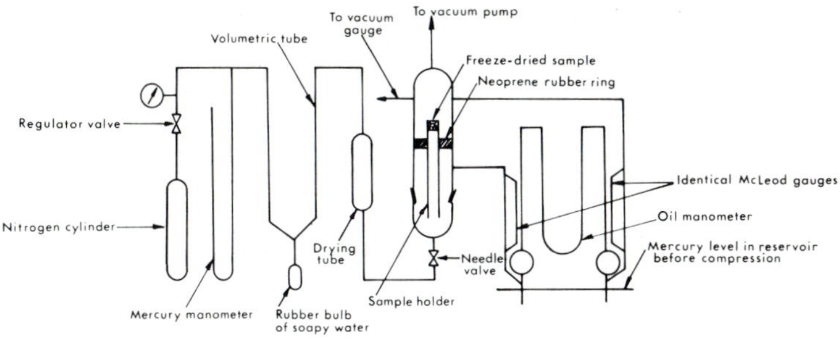

FIG. 4.7. Permeability apparatus.

needle valve, a sensitive flowmeter comprising a soap film moving in a volumetric tube, an absolute mercury manometer and a nitrogen cylinder and regulator were connected in series at one end of this tube. A vacuum gauge and a two-stage vacuum pump were connected to the other end and a differential oil manometer was connected across the sample tube. To increase the sensitivity of the oil manometer, two identical McLeod gauges were inserted, one in each arm of the tubes leading from the sample tube to the oil manometer. When the mercury was raised simultaneously in the two McLeod gauges the difference in the levels of oil was increased by a factor equal to the compression ratio. A compression ratio of 60 was found to be suitable.

A measurement of permeability is made by adjusting the regulator to obtain atmospheric pressure in the flowmeter, at the same time adjusting the needle valve to cause the soap film to move over a measured volume in a given time. Both the vacuum gauge and the oil manometer are then read, after compressing the gas in each arm by the McLeod gauges. The permeability coefficient K for different mean pressures $\bar{P}$ is then calculated using the formula

$$K = \frac{V_1 P_1 l}{A \, \Delta P t}, \tag{4.25}$$

where V_1/t = flowmeter reading, P_1 = absolute pressure, l = sample thickness, ΔP = pressure difference.

When K is plotted against $\bar{P}$ a transition curve is obtained which approaches asymptotically a viscous-slip flow straight line in the medium pressure range. The equations for the slope and the asymptote intercept on the K-axis were derived (equations 4.12 and 4.13); they yield equation (4.14) for the equivalent pore radius r, and the diffusion ratio $\varepsilon(L_e/L)^{-2}$ could also be found from them. The transition curve has been found to give a minimum with a change in slope in the transitional slip-flow region, where the mean free path of the gas molecules is about equal to the average pore diameter. Theoretically it should rise at extremely low pressures so as to cut the K-axis at a value equal to the effective Knudsen diffusion coefficient $\varepsilon(L_e/L)^{-2}D_K$. These properties of the curve have been investigated by Mellor and Lovett (1968).

Typical plots of the permeability data for several freeze-dried foods are in Figs 4.8 and 4.9.

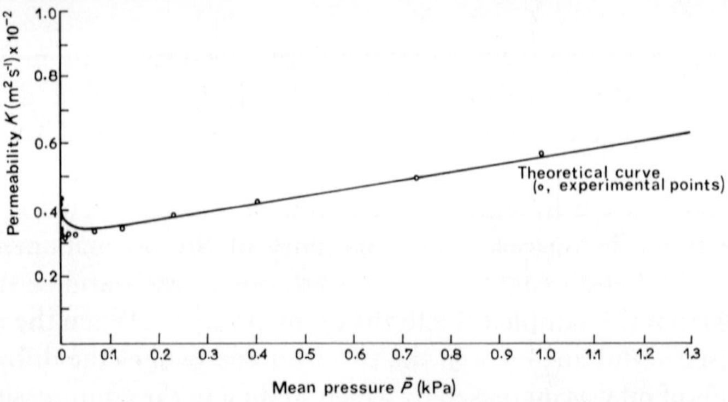

FIG. 4.8. Permeability–pressure relationships for freeze-dried prawns.

It is difficult to measure the small pressure drop across the sample at low absolute pressures. Harper (1962) used the motion of a stainless steel diaphragm detectable to 0·133 Pa but found it unstable; nevertheless, he demonstrated the existence of a minimum in the permeability–pressure curves for freeze-dried beef and peach. On the other hand, the compression system applied to the arms of the oil manometer (Fig. 4.7) was found to be stable with an accuracy of ±0.27 Pa.

Other types of transition flow are represented by the permeability–pressure curves in Fig. 4.9. The curves for freeze-dried carrot and strawberry have practically no curvature down to the limit of pressure measurement (0·67 Pa) in the permeability apparatus. Pore size for the

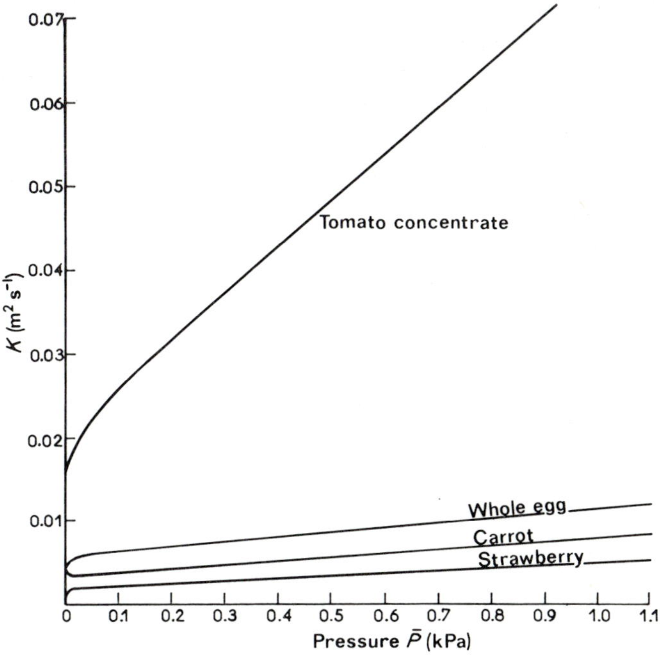

FIG. 4.9. Permeability–pressure plots showing linear and concave downward curves.

two products is about 60 μm radius. The minimum condition where the capillary diameter equals the mean free path is no criterion on which to judge the anomalous transition flow behaviour either in these products or in the freeze-dried tomato concentrate and whole egg having a calculated pore radii of about 200 μm; these curves are concave downward at low pressures. Pore size in the latter products has been found to be highly dependent on the rate of freezing before freeze-drying.

The departure from classical transition flow behaviour as suggested by the tomato-concentrate curve might be due to the molecules that are undergoing wall collisions interacting with adsorbed molecules on the surfaces of the pores instead of with a solid wall. Hirsch (1961) considers that the tangential momentum imparted to the adsorbed layer causes it to slide in the direction of gas flow and the motion eventually transforms the random diffusive flow at low pressures into ordered molecular flow. This could not occur in porous materials having straight capillaries because there would be fewer adsorbed molecules, and molecules on paths within the capillaries are likely to have a finite probability of passing through without a single collision. The latter effect might explain the linearity of the curves for carrot and strawberry.

Regarding the effect of freezing on pore size, Blond (1972) has shown (Fig. 4.10) that the difference between slow and rapid freezing

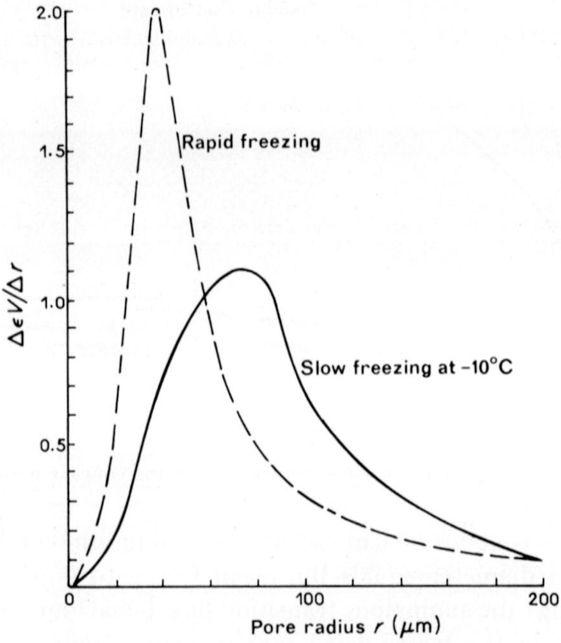

FIG. 4.10. Distribution of pore radii in freeze-dried 10 per cent Dextran.

of 10 per cent Dextran solution produces different distributions of pore size measured by mercury porosimetry, and discloses pore sizes corresponding in size to the ice crystals from which they evolved through freeze-drying. In contrast, Guex (1969) reports the creation of a microporosity in this material by the passage of the vapour during freeze-drying. This could be pertinent to the method for the preparation of adsorbents and catalysts by "Structural Freeze-Drying".

4.4.3 PERMEABILITY BY GAS EXPANSION

Measurement of permeability by the steady-state method is usually long and tedious because of the slow approach to equilibrium. Instead, a method has been developed by Mellor and Greenfield (1972) which allows rapid determination of the permeability under conditions which are similar to those experienced in cyclic-pressure freeze-drying.

Consider an enclosed volume V, into which gas may expand through a plug of freeze-dried material of cross-sectional area A and length L. The pressure in this volume at any time t is denoted by P and the external pressure, which is assumed constant, by P_a. Any change in the pressure may be expressed by an energy balance between the expansion and gas flow as

$$V\frac{\mathrm{d}P}{\mathrm{d}t}=\frac{KA}{L}(P_a-P), \tag{4.26}$$

where K = permeability. Integration of equation (4.26) and substitution of the initial condition $P=P_0$ in the enclosed volume when $t=0$ gives

$$P_a-P=(P_a-P_0)\exp\left(-\frac{KAt}{VL}\right). \tag{4.27}$$

Experimentally, the pressure in a previously evacuated cell (Fig. 4.11) is recorded when gas is expanded into the cell through the porous plug. The external gas pressure is maintained by holding the gas in a large ballast tank. The experimentally observed pressure change can be represented by the relationship

$$P_a-P=(P_a-P_0)\exp\left(-\frac{2{\cdot}303t}{t_0'}\right). \tag{4.28}$$

A plot of $\log(P_a-P)$ against t yields a straight line with the slope $-2{\cdot}303/t_0'$; and the permeability, comparing equations (4.27) and (4.28), is given by

$$K=\frac{2{\cdot}303\,VL}{At_0'}. \tag{4.29}$$

The average pressure difference across the plug during gas flow for a time t is given by the integral of equation (4.28) between the limits zero and t_c, and is

$$\overline{\Delta P} = \frac{P_a - \bar{P}}{t_c} \int_0^{t_c} \exp\left(-\frac{2{\cdot}303t}{t'_0}\right) dt. \tag{4.30}$$

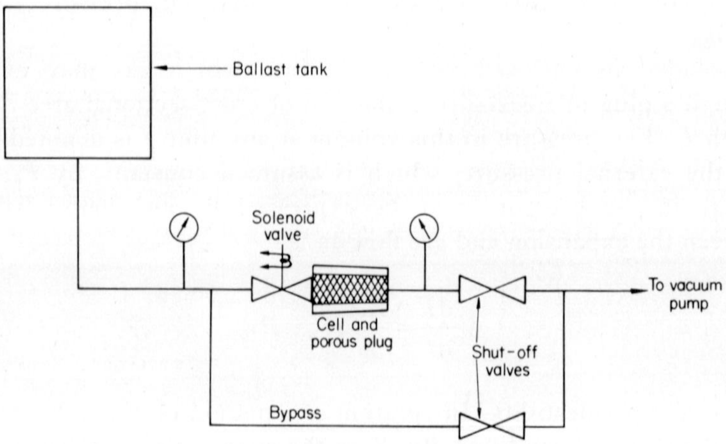

FIG. 4.11. Gas expansion apparatus.

Hence the mean pressure, at which the permeability is determined, is given by

$$\bar{P} \simeq P_a - \overline{\Delta P}. \tag{4.31}$$

In a further development, Mellor and Middlehurst (1971) have applied a correction factor to the equation for the permeability to account for the fact that the effective area of flow under sudden expansion is not necessarily equal to the total cross-sectional area because porous materials consist of direct as well as blind pores and the expanding gas would be expected to flow through the direct pores and not spread throughout the porous solid. They employed a cylindrical model for their analysis for which the porosity $\varepsilon = n\pi r^2 L_e/LA$ and, therefore, the cross-sectional area of the open pathways $n\pi r^2 = \varepsilon A/(L_e/L)$ replaces A in equation (4.29), giving

$$K = \frac{2{\cdot}303\, VL}{At'_0\varepsilon/(L_e/L)}. \tag{4.32}$$

Permeabilities measured by the gas expansion method for freeze-dried beef, fish and a bundle of wires as a model are like those shown in Figs 4.2, 4.3 and 4.8. Note the comparison with the values measured by the steady-state method should be made at the same mean pressures. There is excellent agreement between the gas expansion and steady-state results for the bundle of wires, beef and fish samples, where the flow paths are considered to approximate to a cylindrical model. However, the agreement is only fair for the apple and carrot samples which have higher porosities and homogeneous non-cylindrical pore structures. For these, the cylindrical model overestimates the gas expansion permeabilities by 14–30 per cent.

Greenfield (1972) suggested sealing the sides of the porous plug carefully to the cell with a strong adhesive to prevent the gas bypassing it, followed by degassing under vacuum before measurements were made. As the pressure can be made to rise in the cell a thousandfold in less than 10 seconds, it should be recorded to obtain the best results. Greenfield estimated the permeability also by derivative and integral methods, besides the semi-log method of equation (4.28). It appears that the new method has some further possibilities:

1. Testing the pressure rise in cyclic-pressure freeze-drying to see whether the sublimation stage is completed.
2. As a rapid qualitative test of the texture of samples, since an almost instantaneous measure of the permeability could give data on structure upon which the texture depends.

4.5 Desorption in secondary drying

In secondary drying, moisture desorption takes place in the drying product to produce a low moisture content. It possibly takes place also in primary drying as the remaining ice sublimes. Desorption requires a high temperature and good vacuum, but heating to obtain desorption is slow and inefficient because the vacuum conditions produce an insulating effect in the porous dry product. In terms of the normal dehydration of hygroscopic materials, this stage corresponds to the third falling-rate period and is due to the partial water-vapour pressure in the drying product being everywhere below saturation. The third falling-rate period is therefore important in secondary drying. It would be interesting if the equilibrium rate for moisture desorption in freeze-dried materials and the dependent heat and vapour transfer interactions were known during this stage.

4.5.1 THEORY OF DESORPTION

Most authors have derived the theory of desorption based on Fick's equation (4.33):

$$\frac{\partial m}{\partial t}=\frac{\partial}{\partial x}\left(D_e\frac{\partial m}{\partial x}\right). \tag{4.33}$$

Karel (1974) for one, produced an oversimplified derivation of this equation by assuming at the beginning of secondary drying a uniform moisture content throughout a slab:

$$\frac{dm}{dt}=\frac{K_v}{\rho}\frac{\partial^2 p}{\partial x^2}. \tag{4.34}$$

He also assumed a linear isotherm with a constant of proportionality γ between the vapour pressure p and the moisture content m : with these assumptions equation (4.34) becomes

$$\frac{dm}{dt}=\frac{K_v\gamma}{\rho}\frac{\partial^2 m}{\partial x^2}=D_e\frac{\partial^2 m}{\partial x^2} \tag{4.35}$$

where D_e is the effective diffusivity of the vapour in the dry product and K_v the permeability.

4.5.2 VARIABLE DIFFUSIVITY

Where there is a non-uniform moisture content and thus D_e is a variable diffusivity depending on certain process parameters, equation (4.35) usually takes the form of equation (4.33). By assuming: (1) mass transfer only in the vapour phase, (2) no shrinkage, and (3) heat of sorption constant, King (1968) derived equation (4.33) and showed that the variable diffusivity D_e could be rationalized (see Appendix 4.II) and defined by

$$D_e=\frac{D'P}{P-p'_w}\cdot\frac{Mp^\circ_w}{\rho_d RT}\left(\frac{\partial r'}{\partial m}\right)_T\frac{\alpha'}{1+\alpha'} \tag{4.36}$$

$$\alpha'=\frac{\lambda R^2T^3}{(\Delta H)^2D'r'p^\circ_w}, \tag{4.37}$$

where $D' = \varepsilon (L_e/L)^{-2} D$, diffusivity of the vapour in porous product, D = diffusivity of vapour in air, M = molecular weight of water, p_w° = saturated vapour pressure of water, ρ_d = density of dry porous solid, P = total pressure, p'_w = partial pressure of water, $r' = p'_w/p_w^\circ$ vapour pressure ratio (relative humidity), T = temperature, α' = constant defined by equation (4.37), λ = thermal conductivity of drying solid, ΔH = heat of sorption, $\varepsilon (L_e/L)^{-2}$ = ratio: porosity to tortuosity factor (diffusion ratio).

Thus the left-hand side of the complete equation (4.33) describes the condensed phase and the right-hand side the moisture transfer in the vapour phase with a driving force determined by the moisture gradient. The term $(Mp_w^\circ/\rho_d RT)(\partial r'/\partial m)_T$ expresses the ratio of moisture densities in the two phases, and $\alpha'/(1+\alpha')$ determines whether heat or vapour transfer is controlling for the reverse of this. Such is the trend where there exist high heats of sorption, high humidities or high temperatures. Similar theories derived by Krischer (1956) and van Arsdel (1963) are exactly the same except for the ratio $\alpha'/(1+\alpha')$.

In order to evaluate D_e when it is appreciably dependent on concentration and temperature (i.e. $D_e \propto m$ and T), it is necessary to use equation (4.38) (Appendix 4.II, equation (4.II.11)).

$$T - T_1 = -\int_{m_1}^{m} \frac{RT^2(\partial r'/\partial m)_T}{r'\Delta H}\left(\frac{1}{1+\alpha'}\right) \mathrm{d}m \tag{4.38}$$

since the derivative $(\partial r'/\partial m)_T$ can be determined progressively either numerically or analytically for all points (m, T) away from some fixed point (m_1, T_1)—for example, at the surface.

For air drying, Jason (1958) reported a constant value of $D_e = 0{\cdot}032\ \mathrm{m^2\,s^{-1}}$ in the range $(0{\cdot}1 < m < 1)$ at atmospheric pressure and 30°C. This value is largely fortuitous in terms of equation (4.36); and from his sorption isotherm where $(\partial r'/\partial m)_T \simeq 2$.

On the other hand, for freeze-drying Saravacos and Charm (1962) reported an average value of D_e, obtained from equation (4.40), which was lower by a factor of 4 because a half-thickness instead of a whole thickness of potato slabs was considered; its corrected value is $0{\cdot}16\ \mathrm{m^2\,s^{-1}}$ $(0{\cdot}1 < m < 1)$.

From Görling's (1958) sorption isotherm it is noted also that $(\partial r'/\partial m)_T \simeq 2$.

The most sensitive term in equation (4.36) is p_w°; hence it is not surprising that Jason, and Saravacos and Charm reported that D_e increases exponentially with temperature.

4.5.3 CONTROLLING FACTORS

There are three possible restrictions imposed on moisture desorption: internal or external resistances and resistances in combination due to either the nature of the product sample itself or to an impervious skin (pellicle) around the sample resisting the outward flow of moisture. The two effects often combine, and the latter effect may have resulted either from the method of preparation (e.g. cutting, slicing and pre-freezing) or from incipient melting during drying. The criterion as to whether internal resistance is controlling is when the half-equilibrium time $t_{1/2}$ increases as the second power of the sample size L, i.e. ($t_{1/2} \propto L^2$), and it is obtained from

$$D_e = 0{\cdot}049\frac{L^2}{t_{1/2}} \tag{4.39}$$

as a simplified solution to

$$\frac{m - m_0}{m_e - m_0} = 1 - \frac{8}{\pi^2}\sum_{\nu=0}^{\infty}\frac{1}{(2\nu+1)^2}\exp\left[-(2\nu+1)^2\frac{\pi^2 D_e t}{L^2}\right], \tag{4.40}$$

where equation (4.40) is the solution of (4.33) for a slab having half-thickness L and both sides exposed to a moisture gradient with the following boundary conditions:

$$\begin{aligned} m &= m_0, \quad t = 0, \quad \text{all } x \\ m &= m_s = m_e, \quad t > 0, \quad x = \text{surface} \\ m &= \text{finite}, \quad t > 0, \quad \text{all } x \end{aligned} \tag{4.41}$$

where m = moisture content (dry basis) and subscripts 0, e and s denote the initial, equilibrium with $p_{w,e}$ and T_e and surface values respectively.

Now, Saravacos and Stinchfield (1965) reported that some of the sorption rates vary as the sample thickness to the first power ($t_{1/2} \propto L$) and not the second as expected, indicating that mass transfer is controlling through the external resistance. They found also that the diffusivities varied as the moisture content of the sample. Under such conditions, m and p'_w may be considered uniform within the sample and so the equilibrium rate is given by

$$\frac{dm}{dt}=\frac{AMk_c^*}{\rho V}(m_e-m), \tag{4.42}$$

where A = surface area, V volume of sample, and k_c^* = effective mass transfer coefficient. This has a solution given by

$$\frac{m-m_0}{m_e-m_0}=1-\exp\left(-\frac{k_c^* Mt}{\rho L}\right) \tag{4.43}$$

with the boundary condition given by the first relation—equations (4.41).

Where internal and external resistances are controlling, the boundary condition, as given by the second relation (equation 4.41), must be changed, giving

$$\begin{aligned}-\frac{D'}{RT}\left(\frac{\partial p'_w}{\partial x}\right)_s &= \frac{k_c}{RT}(p_{w,e}-p_{w,s})\\ -\lambda\left(\frac{\partial T}{\partial x}\right)_s &= h(T_e-T_s)\end{aligned} \tag{4.44}$$

$p_{w,s}$ (equilibrium with T_s and m_s) at $t>0$, where h = external heat-transfer coefficient, k_c = external mass-transfer coefficient and subscripts e and s denote equilibrium and surface values respectively. Equations (4.44) may be linearized (see Appendix 4.II) if the overall changes in m and p'_w are small, giving

$$k_c^*(m_e-m_s)=-D_e\left(\frac{\partial m}{\partial x}\right)_s, \qquad t>0, \tag{4.45}$$

where m_e is in equilibrium with T_e and $p_{w,e}$

$$\begin{aligned}k_c^* &= \frac{M}{\rho_d}\left(\frac{\partial r'}{\partial m}\right)_T \frac{p_w^\circ}{RT}\,\frac{\beta'}{1+\beta'}k_c\\ \beta' &= \frac{hR^2T^3}{(\Delta H)^2 k_c r' p_w^\circ}.\end{aligned} \tag{4.46}$$

The constant β' is analogous to α' in determining whether mass or heat transfer is controlling (relationship 4.38).

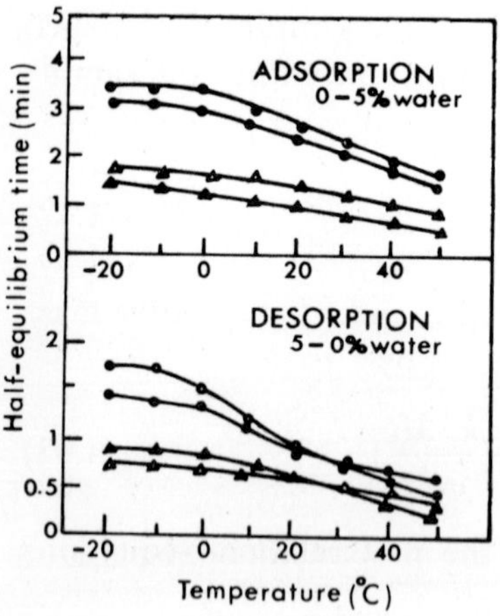

FIG. 4.12. Effect of temperature on the rates of water sorption of freeze-dried foods in vacuo. ○, Beef; ●, potato, △, peach; ▲, starch. (After Saravacos and Stinchfield, 1965.)

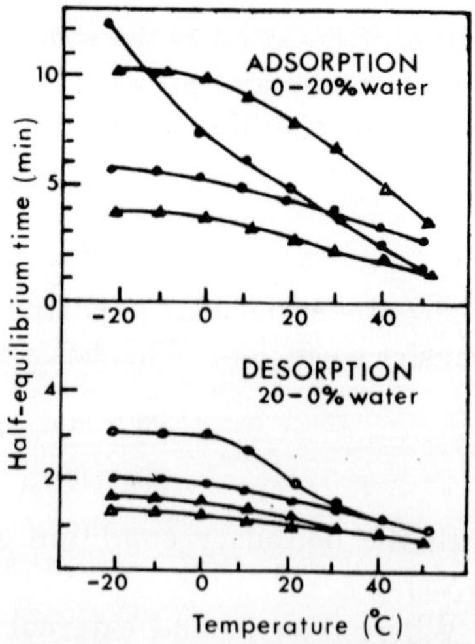

FIG. 4.13. Effect of temperature on the rates of water sorption of freeze-dried foods in vacuo. ○, Beef; ●, potato; △, peach; ▲, starch. (After Saravacos and Stinchfield, 1965.)

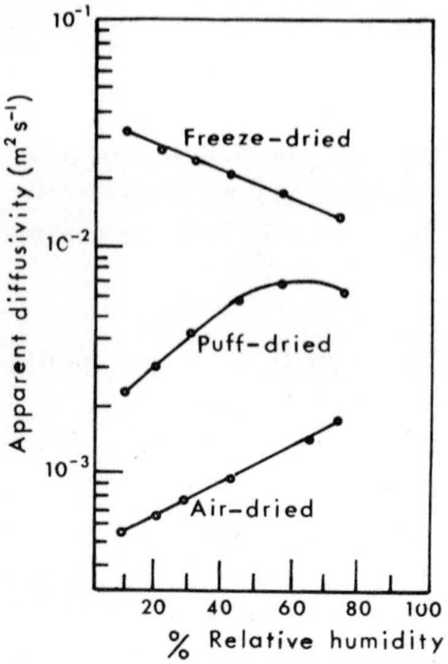

FIG. 4.14. Apparent diffusivities of water vapour in dehydrated potato (temperature 30°C, vacuum pressure 13·3 Pa). (After Saravacoos, 1967.)

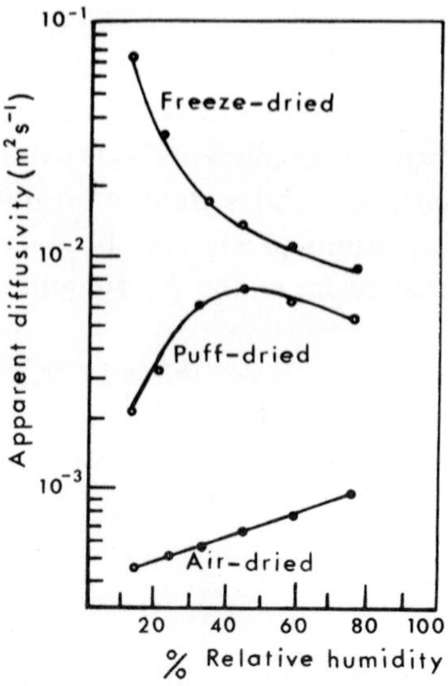

FIG. 4.15. Apparent diffusivities of water vapour in dehydrated apple (temperature 30°C, vacuum pressure 13·3 Pa). (After Saravacos, 1967.)

Saravacos and Stinchfield (1965) and Saravacos (1965, 1967) found that adsorption and desorption rates for freeze-dried foods increased with decreasing moisture content. Equation (4.46) shows that k_c^* does indeed follow this trend when external heat transfer is controlling, i.e. when β' is small (low pressures). They found also that the desorption rate, in going from $\bar{m}$ to m_0, was faster than the adsorption rate in the reverse direction, and this is interpreted as being a large change in m which invalidates the linearization leading to equation (4.45). Typical curves of half-equilibrium time, which is inversely proportional to sorption rate, are presented in Figs 4.12 and 4.13.

In comparing data for freeze-dried apple and potato against those for similar puff-dried and air-dried samples, as shown in Figs 4.14 and 4.15, Saravacos (1967) found that the former have considerably higher adsorption and desorption rates at 13·3 Pa, which appear to indicate that internal resistances are controlling in the case of puff-dried and air-dried samples and external resistance in the case of freeze-dried samples, implying a porosity effect as a result of the shrinkage with air drying and changes in texture with puff drying.

Appendix 4.I Theory for molecular collisions in a porous material

Consider both intermolecular and wall collisions inside a cylindrical tube to represent the collision phenomena of molecules flowing in a single capillary of a porous material, as depicted in Fig. 4.16. After impact with an element of area of the wall dA' at P the molecules are reflected and some take the path PQ to an element of area dA in the cross-section of the tube at Q; thereafter intermolecular collisions occur. Let the polar coordinates of Q be a, θ, ϕ in a half cylinder ($0<\theta<\pi/2$, $0<\phi<\pi/2$) and the solid angles $d\omega$ and $d\omega'$ and let the molecular path make an angle α with the normal at P.

From kinetic theory the probability of a molecular path of given length a is $e^{-a/\lambda}$ where λ is the mean free path, and by Knudsen's cosine law the fraction of molecules reaching $\mathrm{d}A$ in unit time is

$$\frac{\mathrm{d}\nu}{\nu}=\mathrm{e}^{-a/\lambda}\frac{\mathrm{d}\omega' \cos\alpha \;\mathrm{d}A'}{\pi}. \tag{4.I.1}$$

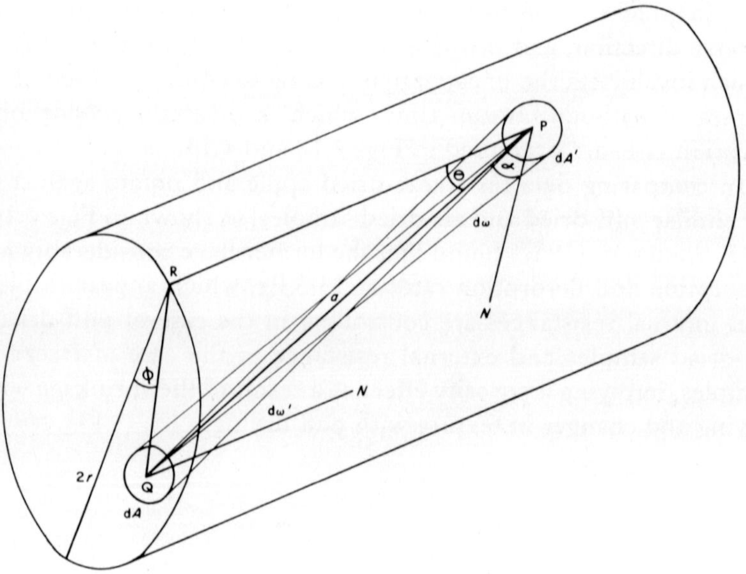

FIG. 4.16. Section of a cylindrical capillary showing the path of molecules PQ after impact with the wall.

Since there are no intermolecular collisions between P and Q it can be assumed that the elements of area on spheres, having radii r and centres at these points, are equal, that is, $a^2 = \mathrm{d}\omega = a^2\,\mathrm{d}\omega'$. However, these areas are given by

$$\begin{aligned} a^2\,\mathrm{d}\omega &= \mathrm{d}A' \cos\alpha,\\ a^2\,\mathrm{d}\omega' &= \mathrm{d}A \cos\theta. \end{aligned} \tag{4.I.2}$$

Therefore

$$\mathrm{d}A' \cos\alpha = \mathrm{d}A \cos\theta. \tag{4.I.3}$$

$\mathrm{QR} = a \sin\theta$, and, if the angles take increments $\mathrm{d}\theta$ and $\mathrm{d}\phi$, then

$$a^2\,\mathrm{d}\omega' \simeq a \sin\theta \;\mathrm{d}\phi \,.\, a\,\mathrm{d}\theta. \tag{4.I.4}$$

And for direct paths it can be assumed that

$$a \sin \theta \simeq 2r \cos \phi. \tag{4.I.5}$$

Substitution from equations (4.I.3), (4.I.4) and (4.I.5) in (4.I.1) yields

$$\frac{d\nu}{\nu\, dA} = \exp(-2r \cos \phi/\lambda \sin \theta) \frac{\sin \theta \cos \theta \, d\theta \, d\phi}{\pi}, \tag{4.I.6}$$

where $d\nu/\nu\, dA$ is the fraction of the molecules reaching dA in a half cylindrical space. For the whole space the total fraction $F(r/\lambda)$ reaching the cross-section of the tube after impact with the walls is given by

$$F(r/\lambda) = \frac{2}{\pi} \int_0^{\pi/2} \int_0^{\pi/2} \exp(-2r \cos \phi/\lambda \sin \theta) \sin \theta \cos \theta \, d\theta \, d\phi. \tag{4.I.7}$$

And the total fraction undergoing intermolecular collisions is $1 - F(r/\lambda)$.

Substituting $x = \operatorname{cosec} \theta$ and $q = (2r/\lambda) \cos \phi$, where $\cos \phi = 1$, would be a suitable constant to assume for the purpose of integration in equation (4.I.7), and integrating by parts gives

$$\begin{aligned} F(r/\lambda) &= \int_1^{\infty} \frac{e^{-qx}}{x^3} \, dx \\ &= \left[-\frac{e^{-qx}}{2x^2} + q \frac{e^{-qx}}{2x} \right]_1^{\infty} + \frac{q^2}{2} \int_1^{\infty} \frac{e^{-qx}}{x} \, dx \\ &= e^{-q}(1-q) + \frac{q^2}{2} E_i(q), \end{aligned} \tag{4.I.8}$$

where the exponential integral

$$E_i(q) = -\gamma_1 - \ln q + q - \frac{q^2}{2 \cdot 2!} + \frac{q^3}{3 \cdot 3!} - \cdots$$

and $\gamma_1 = 0{\cdot}5772$ (Euler's constant).

It is usual to take the approximation $F(r/\lambda) \simeq e^{-q} = e^{-2r/\lambda}$ as the graph of $e^{-2r/\lambda}$ against $F(r/\lambda)$ (Fig. 4.17) shows that the curve passes through the points 0 and 1, and is thus not much in error at the extreme ends of its range. Such an approximation could therefore usefully represent the Knudsen minimum as well as the asymptotic parts of the curve. On the other hand, Scott and Dullien (1962) obtained $F(r/\lambda) = e^{-\sinh^{-1}(2r/\lambda)}$ for a long capillary. This equation gives a large error for the asymptote but a good fit for the minimum; it would appear that it would only fit the minimum in transition flow perfectly when r/λ is small, i.e. when $e^{-\sinh^{-1}(2r/\lambda)} \to e^{-2r/\lambda}$. Dyer and

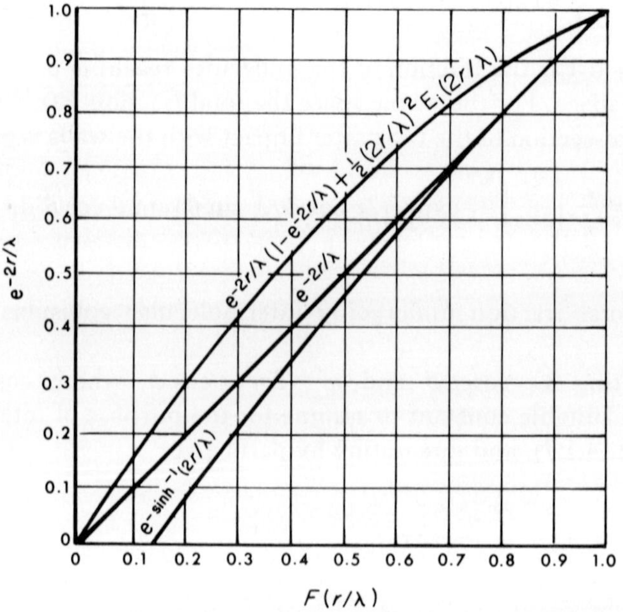

FIG. 4.17. Approximations to the fraction of the molecules undergoing wall collisions in a tube.

Sunderland (1968a), using Scott and Dullien's result, claimed to have obtained a good fit for the permeability for transition flow in porous freeze-dried products, although they did not substantiate this by examples.

The Poiseuille and transition flow terms in equation (4.15) (Section 4.2) represent the permeabilities when intermolecular collisions occur and the Knudsen term when wall collisions occur; these terms can be associated with the fractions $1 - e^{-2r/\lambda}$ and $e^{-2r/\lambda}$ to yield a complete permeability equation given by

$$K = \frac{\varepsilon}{(L_e/L)^2} D_K \left[\left(\frac{3}{32k'} \frac{2r}{\lambda} + \delta_1 \right) (1 - e^{-2r/\lambda}) + e^{-2r/\lambda} \right]. \qquad (4.I.9)$$

In this equation the factors in the term $(3/32k')(2r/\lambda)+\delta_1$ are considered to be the appropriate ones for freeze-dried products.

Appendix 4.II Theory of desorption

Variable diffusivity

King (1968) considered that desorption in a porous dry semi-infinite solid involves moisture diffusion in the vapour phase and possibly surface diffusion across moisture adsorption sites as well. The interaction of desorption with heat transfer must also be considered, because of the large heat of sorption and the likelihood of heat transfer controlling. King's theory is best understood with reference to the model of moisture desorption shown in Fig. 4.18.

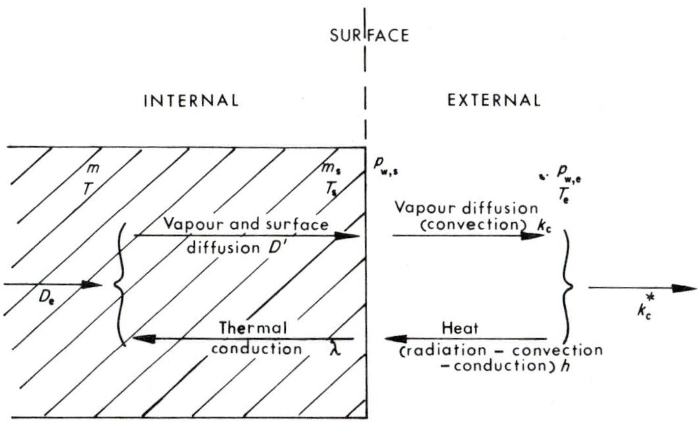

FIG. 4.18. Model of moisture desorption: heat and vapour transfer interactions.

Firstly, assume a vapour phase that is in equilibrium with the sorbed moisture, a constant heat of sorption which is large compared with sensible heat and no shrinkage. Next, by analogy with mutual diffusion in a gas layer, the vapour diffusion rate G in a gas-filled porous solid is given by

$$G=-\frac{D'}{RT}\frac{P}{P-p'_w}\left(\frac{\partial p'_w}{\partial x}\right). \tag{4.II.1}$$

And the equilibrating vapour pressure gradient, resulting from changes in moisture content and in temperature, may be written as

$$\frac{\partial p'_{\mathrm{w}}}{\partial x}=\left(\frac{\partial p'_{\mathrm{w}}}{\partial T}\right)_m\frac{\partial T}{\partial x}+\left(\frac{\partial p'_{\mathrm{w}}}{\partial m}\right)_T\frac{\partial m}{\partial x}. \tag{4.II.2}$$

In considering the sorption isotherm, one of these derivatives can be written in terms of the heat of sorption ΔH (really the Clausius–Clapeyron equation) and the other in terms of the change in relative humidity with moisture content $(\partial r'/\partial m)_T$; thus

$$\left(\frac{\partial p'_{\mathrm{w}}}{\partial T}\right)_m=\frac{\Delta H}{RT^2}r'p^{\circ}_{\mathrm{w}}$$
$$\left(\frac{\partial p'_{\mathrm{w}}}{\partial m}\right)_T=\left(\frac{\partial r'}{\partial m}\right)_T p^{\circ}_{\mathrm{w}} \tag{4.II.3}$$

Combining equations (4.II.1), (4.II.2) and (4.II.3) yields

$$G=-\frac{D'}{RT}\frac{Pp^{\circ}_{\mathrm{w}}}{P-p'_{\mathrm{w}}}\left[\frac{\Delta H}{RT^2}r'\frac{\partial T}{\partial x}+\left(\frac{\partial r'}{\partial m}\right)_T\frac{\partial m}{\partial x}\right]. \tag{4.II.4}$$

Now, for a heat of sorption greater than the sensible heat and the amount of desorbed moisture greater than the amount of vapour in the pores, the heat-mass flux may be written as

$$Q=-G\,\Delta H. \tag{4.II.5}$$

Combining equations (4.II.4), (4.II.5) and $Q=-\lambda(\partial T/\partial x)$ to eliminate Q and $\partial T/\partial x$ yields

$$G=-\frac{D'Pp^{\circ}_{\mathrm{w}}}{RT(P-p'_{\mathrm{w}})}\left(\frac{\partial r'}{\partial m}\right)_T\frac{\alpha'}{1+\alpha'}\left(\frac{\partial m}{\partial x}\right), \tag{4.II.6}$$

where

$$\alpha'=\frac{\lambda R^2T^3}{(\Delta H)^2D'r'p^{\circ}_{\mathrm{w}}}. \tag{4.II.7}$$

And $p'_{\mathrm{w}}<P$.

Now a differential mass balance gives

$$\frac{\rho_{\mathrm{d}}}{M}\left(\frac{\partial m}{\partial t}\right)=-\frac{\partial G}{\partial x}. \tag{4.II.8}$$

Hence, substitution from equations (4.II.6) and (4.II.8) gives

$$\frac{\partial m}{\partial t}=\frac{\partial}{\partial x}\left(D_e\frac{\partial m}{\partial x}\right), \tag{4.II.9}$$

where

$$D_e=\frac{D'P}{P-p'_w}\frac{Mp^\circ_w}{\rho_d RT}\left(\frac{\partial r'}{\partial m}\right)_T\frac{\alpha'}{1+\alpha'}. \tag{4.II.10}$$

Alternatively, substitution from equations (4.II.4), (4.II.5) and $Q=-\lambda_d(\partial T/\partial x)$ to eliminate G, Q and x yields

$$\frac{dT}{dm}=-\frac{RT^2(\partial r'/\partial m)_T}{r'\,\Delta H}\frac{1}{1+\alpha''}, \tag{4.II.11}$$

which is an important relation from the standpoint of evaluating D_e through the derivative $(\partial r'/\partial m)_T$ of equation (4.II.10).

Internal and external resistances

The importance of the internal–external resistances to heat and/or mass transfer will now be considered. Referring to Fig. 4.18, the rate of mass transfer from the surroundings to the surface, neglecting sensible heat, is given by

$$k_c^*(m_e-m_s)\frac{\rho_d}{M}=\frac{k_c}{RT}(p_{w,e}-p_{w,s})$$

$$=\frac{h}{\Delta H}(T_s-T_e). \tag{4.II.12}$$

Now, equation (4.II.13) is the usual expression for a linear sorption isotherm

$$p_{w,e}-p_{w,s}=\left(\frac{\partial p'_w}{\partial m}\right)(m_e-m_s)+\left(\frac{\partial p'_w}{\partial T}\right)(T_e-T_s), \tag{4.II.13}$$

which is valid only for small intervals, here: $(p_{w,e}-p_{w,s})$ and (T_e-T_s). Substitution from equations (4.II.3) and (4.II.13) yields

$$p_{w,e}-p_{w,s}=\left(\frac{\partial r'}{\partial m}\right)_T p^\circ_w(m_e-m_s)+\frac{\Delta H}{RT^2}r'p^\circ_w(T_e-T_s). \tag{4.II.14}$$

Dividing equation (4.II.12) by (4.II.14) and rearranging yields

$$k_c^* = \frac{M}{\rho_d}\left(\frac{\partial r'}{\partial m}\right)_T \frac{p_w^\circ}{RT}\frac{\beta'}{1+\beta'}k_c, \tag{4.II.15}$$

where

$$\beta' = \frac{hR^2T^3}{(\Delta H)^2 k_c r' p_w^\circ}. \tag{4.II.16}$$

Nomenclature

D' diffusivity of vapour in porous product
D_e effective diffusivity of vapour in porous product
ΔH heat of sorption
k_c external mass transfer coefficient
k_c^* effective mass transfer coefficient
m moisture content
M molecular weight
G vapour flux
p vapour pressure
P absolute pressure
Q heat flux
r' vapour pressure ratio (relative humidity)
R gas constant
t time
T temperature
x distance
α' defined by equation (4.II.7)
β' defined by equation (4.II.16)
ρ_d density porous product
λ thermal conductivity

Subscripts

e equilibrium value
s surface value

Superscripts

° saturated vapour
′ partial

5 Drying Rate

5.1 Introduction

In many theoretical models of freeze-drying the equations for the interaction of the heat and vapour transfer processes lead to the drying time, temperature profile and position of the ice interface. Some quasi-steady state models have analytical solutions which are satisfactory even though they do not involve the sublimation process. It is in fact rarely a limiting factor requiring consideration in the analytical theory. In their study of the general problem, Dyer and Sunderland (1969) listed the following seven possible conditions upon which the model would depend, including variable thermal properties and variable heat sources.

1. Arbitrary initial temperatures.
2. Variable heating by radiation and convection.
3. Local and distributed heat sources and sinks.
4. Variable interface temperature.
5. Convective heat transfer in the porous dry layer.
6. Dry layer assumed to be an absorber as well as an emitter for thermal radiation.
7. Variable thermal properties.

They conceded that it would be impossible to achieve an analytical solution and that even a numerical solution would be very complex unless many of these conditions were either simplified or completely ignored.

For the most part, the analytical solutions rely on heat transfer by conduction through the material and vapour transfer by diffusion through the porous dry layer, ignoring flow resistances at the interface and in the vapour space, as well as convection and turbulence of the vapour stream. Assuming freeze-drying with a uniform retreating ice front, either in slabs or in spheres of frozen material, the following two kinds of analytical approach have been adopted:

1. The heat and vapour flow rates are expressed in terms of overall heat and mass transfer coefficients.
2. Transport equations are expressed in terms of the conservation of mass, momentum and energy as well as equations of state.

It appears that the pattern for the first approach was set by Harper and Tappel (1957) with the theory for a slab; later this approach was used by Ginnette *et al.* (1958) for spherical pieces and by others, including Sandall *et al.* (1967) and Clark and King (1968). On the other hand, Dyer and Sunderland (1968a, 1969) favoured the second approach. However, they have also published general solutions for heat transfer for freeze-drying in materials in the form of slabs, cylinders and spheres. According to Krischer (1956) the drying times for such shapes are related as follows:

$$t_{sph} : t_{cyl} : t_{slab} = 1/3 : 1/2 : 1.$$

As a unifying step, it seems important to examine not only the heat transfer but also the vapour transfer for the three geometrical shapes, on the lines adopted by Dyer and Sunderland (1971), since many products nowadays are freeze-dried from small spherical and cylindrical pieces and, to a lesser extent, from slabs. Slabs are used for liquid products and for sliced foodstuffs freeze dried in trays, while cylindrical plugs and shells are used for pharmaceutical and medical products freeze dried in ampoules or vials, and spherical shapes are used for granulated products which can be handled in quantity and are free flowing and also for a few natural products, e.g. peas. And it is possible to relate the various theories that have been proposed for these geometrical shapes and covered by category (1) on the basis of a common theory.

5.2 Heat transfer

It is assumed that heat conduction is controlling for this theoretical approach to freeze-drying. The method of Dyer and Sunderland

(1971) assumes (1) quasi-steady state heat transfer such that the sensible heating is negligible compared with the change in latent heat at an interface R_i and at a temperature T_i; and (2) that the product surface R_s is already at a safe temperature T_s. Then, according to Carslaw and Jaeger (1959), the heat conduction is given by

$$\frac{d}{dr}\left(r^{\nu}\frac{dT}{dr}\right)=0, \tag{5.1}$$

where $\nu=0$, 1 and 2 for a slab, cylinder and sphere respectively. And the boundary conditions are:

$$\begin{aligned} T &= T_i, \qquad r = R_i, \\ T &= T_s, \qquad r = R_s, \\ Q &= \lambda\left(\frac{dT}{dr}\right)_{r=R_i}, \end{aligned} \tag{5.2}$$

where Q is the heat flux in the direction of r and λ is the thermal conductivity.

On integrating equation (5.1) twice and substituting the first and second boundary conditions, one integration constant is eliminated and the other is given by

$$\begin{aligned} c_1 &= (1-\nu)\frac{T_s-T_i}{R_s^{1-\nu}-R^{1-\nu}}, \qquad \nu=0,2, \\ c_1 &= \frac{T_s-T_i}{\ln R_s/R_i}, \qquad \nu=1. \end{aligned} \tag{5.3}$$

Using the third boundary condition, equations (5.3) become:

$$\begin{aligned} Q &= \frac{(1-\nu)\lambda(T_s-T_i)}{R_i^{\nu}(R_s^{1-\nu}-R_i^{1-\nu})}, \qquad \nu=0,2, \\ Q &= \frac{\lambda(T_s-T_i)}{R_i \ln R_s/R_i}, \qquad \nu=1. \end{aligned} \tag{5.4}$$

5.2.1 SPHERICAL PIECES

To determine the drying rates of small pieces of food heated by conduction through the dry layer, Ginnette *et al.* (1958) assumed for their analysis spherical shapes and were able to relate the temperature difference to the drying rate. For spheres, $n = 2$ and so equation (5.4) becomes

$$Q = \frac{\lambda R_s(T_s - T_i)}{R_i(R_s - R_i)}, \tag{5.5}$$

where Q is the heat flux across the interfacial area $= 4\pi R_i^2$ and where the surface heat-transfer boundary condition

$$Q' = h(T_s' - T_s), \tag{5.6}$$

T_s' being the temperature in the boundary layer adjacent to the spherical piece and Q' the heat flux across the external surface area $4\pi R_s^2$.

Ginnette *et al.* expressed the heat flux in terms of the drying rate by taking $Q = (R_s^3/R_i^2)K(\mathrm{d}\mu/\mathrm{d}t)$ where K is an arbitrary constant and μ is the weight fraction of ice sublimed at time t, that is $R_i/R_s = (1-\mu)^{1/3}$, thus, substituting in equations (5.5) and (5.6) and adding yields

$$\frac{T_s' - T_i}{\mathrm{d}\mu/\mathrm{d}t} = \frac{K_2 R_s}{h} + \frac{K_2' R_s^2}{\lambda}\left[\frac{1-(1-\mu)^{1/3}}{(1-\mu)^{1/3}}\right], \tag{5.7}$$

where K_2, K_2' are the adjusted constants.

Equation (5.7) predicts that if the temperature difference divided by the drying rate is plotted against $[1-(1-\mu)^{1/3}]/(1-\mu)^{1/3}$ a linear relationship should be obtained. Correlation with the experimental results of Ginnette *et al.* was adequate, but deteriorated as the drying run progressed.

In a different approach, Steinbach and Rothley (1973) derived the heat conduction equation for spherical pieces by considering additional shape factors to Krischer's, already considered. Their tests on 40 and 60 mm diameter spheres of mashed potato showed that the additional factors were hyperbolic functions and that the factor 1/3 was considerably higher. This was considered to be due to shielding of the radiant heat to the spheres in the experimental equipment.

Other researchers have considered heat and vapour transfer to a bed of spherical granules and particles (Oetjen and Eilenberg, 1969, and Kessler, 1975).

Clark and King (1968) used a similar analysis to relate the temperature difference $T_s' - T_s$ to the rate of change of water content expressed as dW/dt, where $W = 1 - \mu$ is the fraction of the initial water content remaining. They used the normal interface boundary condition, namely

$$Q = -\frac{\Delta H_m}{M_w V}\frac{dR_i}{dt} = -\frac{\Delta H_m R_s}{3M_w V}\left(\frac{R_s}{R_i}\right)^{2/3}\frac{dW}{dt}, \tag{5.8}$$

where ΔH_m = molar latent heat of sublimation, M_w = molecular weight of water, V = volume of the material occupied by unit weight of water initially (specific volume).

Combining the heat fluxes, equations (5.5) and (5.6), for the interfacial and external spherical surface areas respectively, equations (5.8) yield

$$\frac{T_s' - T_i}{-dW/dt} = \frac{\Delta H_m R_s}{3M_w V}\left(\frac{R_s}{R_i}\right)^{2/3}\left[\frac{1}{h} + \frac{R_i^2}{\lambda R_s}\left(\frac{1}{W^{1/3}} - 1\right)\right]. \tag{5.9}$$

Clark and King have derived a similar equation without the term $(R_s/R_i)^{2/3}$ which they apparently ignore—possibly this term could be embodied in the specific volume V.

5.2.2 CYLINDRICAL PIECES

An analogous equation for a cylindrical piece can be obtained by adding equations (5.4) and (5.6) and by substituting the appropriate interface boundary condition, namely

$$Q = -\frac{\Delta H_m}{M_w V}\frac{dR_i}{dt} = -\frac{\Delta H_m R_s^2}{2M_w V R_i}\frac{dW}{dt}. \tag{5.10}$$

This yields

$$\frac{T_s' - T_i}{-dW/dt} = \frac{\Delta H_m R_s^2}{2M_w V R_i}\left[\frac{1}{h} + \frac{1}{\lambda}\ln W^{-1/2}\right]. \tag{5.11}$$

King (1970) gave this same equation except that in his case the ratio R_s^2/R_i was replaced by R_s through adjustments to V.

5.2.3 SLABS

The drying rate for slabs can be treated similarly. With $\nu = 0$, the heat flux, equation (5.4), becomes

$$Q = \frac{\lambda(T_s - T_i)}{R_s - R_i}. \tag{5.12}$$

The corresponding boundary surface condition for this case is given by

$$Q = h(T'_s - T_s) = KR_s\frac{d\mu}{dt}, \tag{5.13}$$

where K is an arbitrary constant and μ is the weight fraction of the ice sublimed, that is $(R_s - R_i)/R_s = \mu$.

Combining equations (5.12) and (5.13) in a similar way leads to

$$\frac{T'_s - T_i}{d\mu/dt} = KR_s\left[\frac{1}{h} + \frac{\mu R_s}{\lambda}\right]. \tag{5.14}$$

However, Sandall *et al.* (1967) and King (1970) used in their derivation the normal interface boundary condition. Putting $\mu = R_i/R_s$, in this case for $1-\mu$, yields

$$Q = -\frac{\Delta H_m}{M_w V}\frac{dR_i}{dt} = -\frac{\Delta H_m R_s}{M_w V}\frac{d\mu}{dt}. \tag{5.15}$$

Therefore, equation (5.14) leads finally to

$$\frac{T'_s - T_i}{-dW/dt} = \frac{\Delta H_m R_s^2(1-W)}{4\lambda M_w V} + \frac{\Delta H_m R_s}{2hM_w V} \tag{5.16}$$

for freeze-drying from both sides of a thin slab.

Bralsford (1967) gave a similar equation; it can be derived from equation (5.16) by neglecting the term containing the boundary surface condition and by taking $1 - W = 2X/R_s$, to give

$$X^2 = \frac{2\lambda(T'_s - T_i)M_w V}{\Delta H_m}t, \tag{5.17}$$

where X is the interface position at time t. However, this is of the same form as $X^2 = 4\alpha\gamma^2 t$ obtained by Mellor (1962b). Bralsford found that a graphical evaluation of experimental results for freeze-dried beef, according to this equation, demonstrated a moving boundary for the ice interface for this product.

King (1970) assumed $T'_s - T_i$ constant in order to integrate equation (5.16), giving

$$1 - W = \frac{8\lambda M_w V}{\Delta H_m L^2}(T'_s - T_i)\frac{t}{1 - W} - \frac{4\lambda}{hL}. \tag{5.18}$$

Neglecting the second term on the right-hand side of this equation yields an equation which Gaffney and Stephenson (1968) derived, namely

$$\frac{T'_s - T_i}{(1 - W)^2} = \frac{\Delta H_m L^2}{8\lambda M_w V}\frac{1}{t}. \tag{5.19}$$

Their data for the freeze-drying of cornstarch solutions at different vacuum pressures, plotted according to equation (5.19), were found to be linear; they were therefore able to deduce the thermal conductivity, which increased with increasing pressure.

Other researchers tested the model of Sandall *et al.* (1967) by plotting $\mathrm{d}W/\mathrm{d}t$ against $(T'_s - T_i)/(1 - W)$ and found a linear relationship over a portion of the drying curves. For instance such a relationship was found by Lusk *et al.* (1964) in their work on freeze-drying of salmon, haddock and perch; their curves for slabs of haddock (Figs 5.1, 5.2 and 5.3) are reasonably linear when the slabs are fairly thin and the drying rate is fast. Massey and Sunderland (1967) have tested the same model for the freeze-drying of beef.

5.3 Vapour transfer

A similar derivation for quasi-steady state vapour transfer in slabs, cylinders and spheres can be inferred from Section 5.2. The appropriate equation for the vapour-pressure variation in the dry layer becomes

$$\frac{\mathrm{d}}{\mathrm{d}r}\left(r^{\nu}\frac{\mathrm{d}p}{\mathrm{d}r}\right) = 0, \tag{5.20}$$

where $\nu = 0$, 1 and 2 for a slab, cylinder and sphere respectively. The internal boundary conditions are:

$$\begin{aligned} p &= p_i, \qquad r = R_i, \\ p &= p_0, \qquad r = R_s, \\ G &= -K_e\left(\frac{\partial p}{\partial r}\right)_{r=R_i}, \end{aligned} \tag{5.21}$$

where G is the drying rate and K_e the effective permeability.

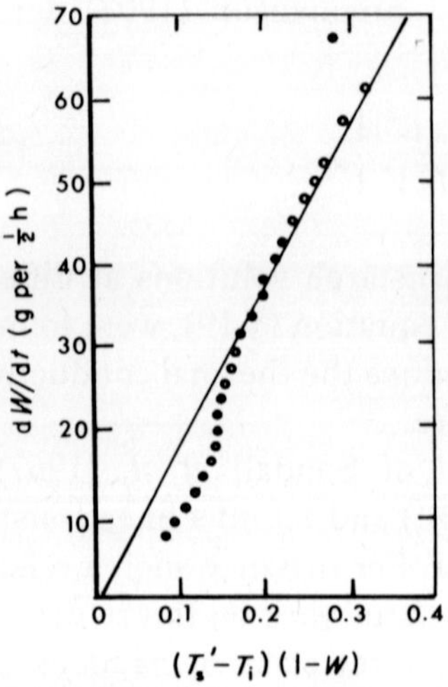

FIG. 5.1. Experimental plots for a 12·7 mm slab of haddock.

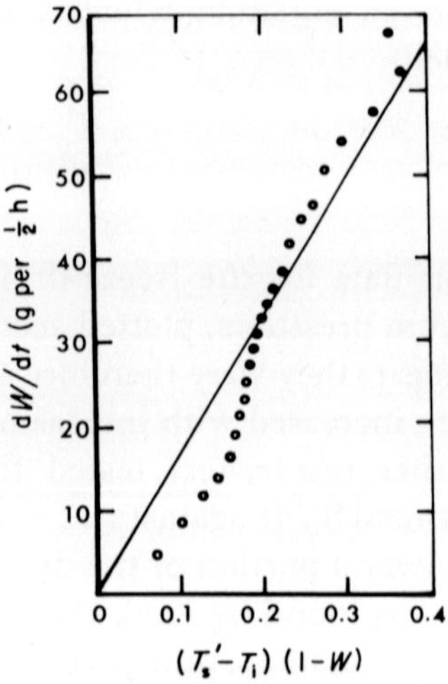

FIG. 5.2. Experimental plots for a 19·1 mm slab of haddock.

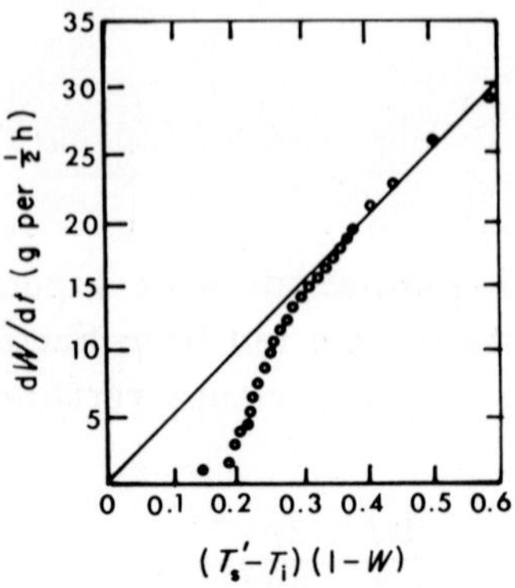

FIG. 5.3. Experimental plots for a 25·4 mm slab of haddock.

With these conditions the drying rate is given by the following set of equations:

$$G=\frac{(1-\nu)K_e(p_i-p_0)}{R_i^{\nu}(R_s^{1-\nu}-R_i^{1-\nu})}, \qquad \nu=0,2$$
$$G=\frac{K_e(p_i-p_0)}{R_i \ln R_s/R_i}, \qquad \nu=1. \tag{5.22}$$

5.3.1 SPHERES

For spheres, $\nu=2$ and so equation (5.22) becomes

$$G=\frac{K_eR_s(p_i-p_0)}{R_i(R_s-R_i)}$$
$$=K_d\frac{p_i-p_0}{R_s-R_i}, \tag{5.23}$$

where G is the internal drying rate over the interfacial area $4\pi R_i^2$, and K_e varies as the ratio of the radii of the interface and surface in the dry spherical shell; that is, $K_e=K_dR_i/R_s$ (K_d is the permeability of the dry material).

$G_m \propto G$ is also the external migration rate of vapour over the surface area $4\pi R_s^2$, and is given by

$$G_m=\frac{K'(p_0-p_c)}{L} \tag{5.24}$$

where $K'=DM/RT$ (D = diffusivity of water vapour), p_c is the pressure in the vapour space or of the condenser if it is in close proximity to the product and L is the distance through which the vapour flows.

Ginnette *et al.* (1958) expressed G in terms of the weight fraction μ of ice sublimed at time t, where $R_i/R_s=(1-\mu)^{1/3}$, as

$$G=-\rho\varepsilon\left(\frac{dr}{dt}\right)_{r=R_i}$$
$$=\frac{\rho\varepsilon}{3}\frac{R_s^3}{R_i^2}\frac{d\mu}{dt}. \tag{5.25}$$

Adding equations (5.23) and (5.24) and substituting for G from equation (5.25) yields

$$\frac{p_i - p_c}{G'_m} = \frac{L}{4\pi R_s^2 K'} + \frac{1}{4\pi R_i^2 K_d}(R_s - R_i),$$
$$\frac{p_i - p_c}{d\mu/dt} = \frac{K_2 R_s}{3} + \frac{K'_2 R_s^2}{3}\left[\frac{1-(1-\mu)^{1/3}}{(1-\mu)^{2/3}}\right], \qquad (5.26)$$

where K_2 and K'_2 are adjusted constants and $G'_m = 4\pi R_s^2 G_m$.

Plots of experimental results of Ginnette *et al.* for the freeze-drying of carrot, according to equation (5.26), gave a better linear relationship than equation (5.7) for heat transfer, as shown in Fig. 5.4. Substitution of the constants a and b for the intercept and slope respectively for this line in the integral of equation (5.26) gave:

$$\text{Minimum drying rate} = \frac{a + 3b/2}{p_i - p_c}$$

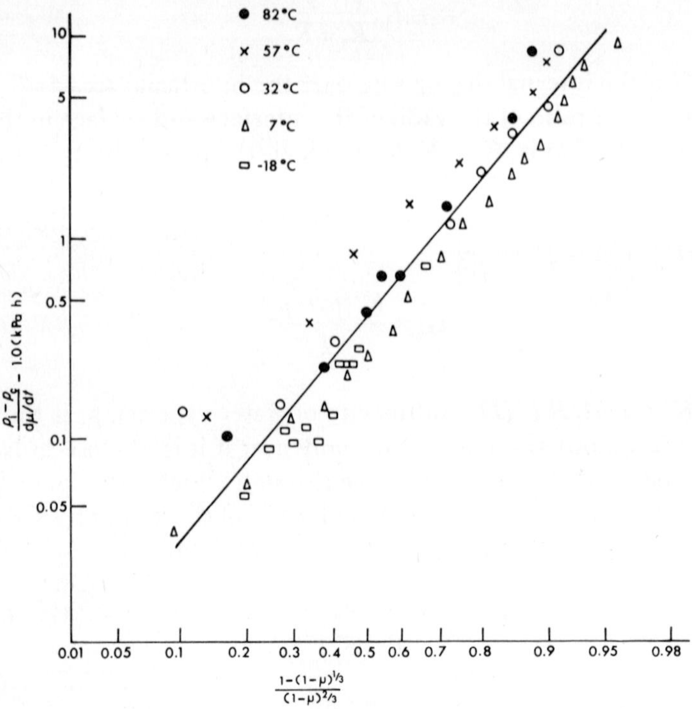

FIG. 5.4. Relationship of vapour pressure to carrot-drying rate (after Ginnette *et al.* 1958).

5.3.2 CYLINDERS

King (1970) refers to an analogous equation for cylinders which can be written in terms of his notation by considering a similar approach to that adopted in the theory for slabs, which is given below.

5.3.3 SLABS

For slabs, $\nu = 0$ and the derived equation for the internal drying rate is the same as for spheres, as is the equation for the external drying rate. However, equation (5.25) changes since $(R_s - R_i)/R_s = \mu$; therefore

$$G_m = \rho\varepsilon R_s \frac{d\mu}{dt}. \tag{5.27}$$

Adding equations (5.22) and (5.24) and substituting for G_m from equation (5.27) yields

$$\begin{aligned} \frac{p_i - p_c}{G_m} &= \frac{L}{K'} + \frac{1}{K}(R_s - R_i), \\ \frac{p_i - p_c}{d\mu/dt} &= K_1 L R_s + K_2 R_s^2 \mu, \end{aligned} \tag{5.28}$$

where K_1, K_2 are adjusted constants.

Using the notation of Sandall *et al.* (1967), equation (5.28) becomes

$$\frac{p_i - p_c}{-dW/dt} = \frac{LR_s}{2k_c M_w V} + \frac{RTR_s^2}{4D'M_w V}(1 - W). \tag{5.29}$$

This is for freeze-drying from both sides of a thin slab, k_c being the external mass transfer coefficient and D' the diffusivity of the material.

Bralsford (1967) has shown by graphical evaluation of the experimental data for freeze-dried beef that the process can be described in terms of the sublimation rate $\dot{W}X$, as the constant of proportionality between the interface position and the square root of the drying time; thus, in effect, proving the existence of a moving boundary for the ice interface. The same conclusion could have been reached by neglecting the first term on the right-hand side of equation (5.29) and substituting for $1 - W = 2X/R_s$, where X equals the interface position after time t, to yield

$$X^2 = \frac{2D'M_w V(p_i - p_c)}{RT} t$$
$$= \gamma^2 t \tag{5.30}$$

Thus, $\gamma^2 = 2D'M_w V(p_i - p_c)/RT$ which is of the same form as the constant of proportionality used by Mellor (1966b).

Equally, equation (5.30), expressed in terms of the sublimation rate constant $\dot{W}X$ as in Bralsford's derivation, simplifies to

$$X^2 = \dot{W}XR_s t. \tag{5.31}$$

5.4 Heat and vapour transport

5.4.1 BACK-FACE HEATING

A heat and vapour transport theory has been developed by Dyer and Sunderland (1968a) for a uniformly retreating ice interface with heat conduction through the dried layer occurring at the same time as back-face heating across a vapour seal and across the frozen layer. This is in line with the approach defined in category (2) of Section 5.1.

Drying time can be significantly reduced by transferring part of the radiant heat to the ice interface through the frozen layer, provided that close and intimate contact exists between the tray receiving the heat and the material to be dried. It is a problem of ensuring continuous contact so that a layer of dry material does not form between it and the frozen layer. A higher rate of heating through the frozen layer (back-face heating) can be expected since the thermal conductivity of the frozen layer is about 45 times greater than that of the dried layer, and so the combined heating effects offer some advantages for freeze-drying.

Dyer and Sunderland (1968a, 1968b) used a transport theory approach, based on the conservation of energy, momentum and mass, together with the ideal gas equation, to obtain a set of algebraic and differential equations which they were able to solve by trial and error; and they assumed an infinitesimally small interface thickness between the dry and frozen layers which moves slowly, so that temporal effects can be ignored compared with spatial effects. In addition, for the frozen layer, they assumed a constant interface temperature and a back-face sealed against vapour flow (Fig. 5.5).

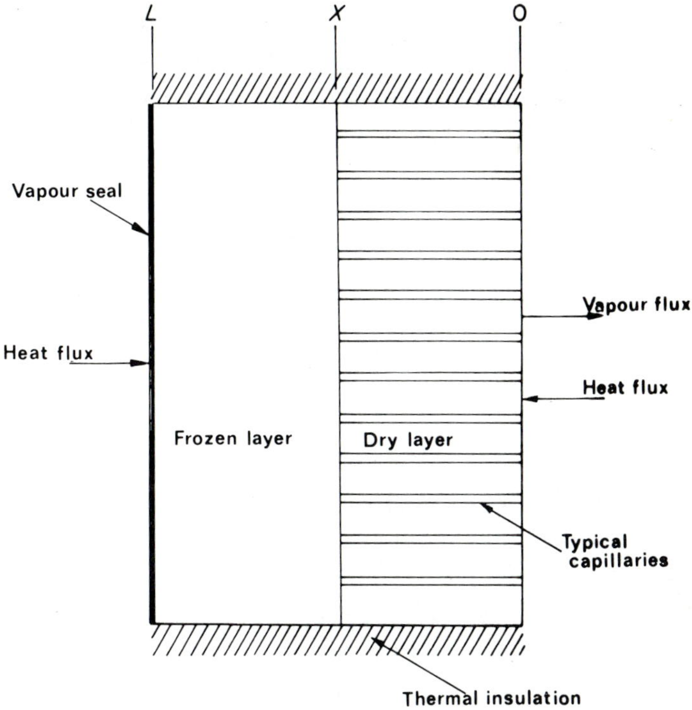

FIG. 5.5. Drying model.

5.4.2 ENERGY EQUATION

By assuming the gas temperature to be in equilibrium with the dried layer, Hardin (1965) and Dyer (1966) derived the differential equation for the temperature distribution in the dry layer, namely

$$\frac{d^2\theta_d}{dx^2}+\frac{Gc}{\lambda_d}\frac{d\theta_d}{dx}=0, \tag{5.32}$$

where $\theta_d=(T_d-T_s)/(T_i-T_s)$ is the normalized temperature of the dry layer, G = vapour flux, c = specific heat of the vapour–air mixture,

λ_d = effective thermal conductivity of the dry layer.

The boundary conditions are:

$$\begin{aligned} \theta_d &= 0, \qquad x = 0, \\ \theta_d &= 1, \qquad x = X. \end{aligned} \tag{5.33}$$

The solution of equation (5.32), using the boundary conditions given in (5.33), yields the temperature in the dried layer for steady-state heat conduction:

$$\theta_d = \frac{1 - \exp(-Gcx/\lambda_d)}{1 - \exp(-GcX/\lambda_d)}. \tag{5.34}$$

The differential equation for the temperature distribution in the frozen layer is given simply by

$$\frac{d^2\theta_f}{dx^2} = 0, \tag{5.35}$$

where $\theta_f = (T_f - T_L)/(T_i - T_L)$ is the normalized temperature in this layer. The boundary conditions are:

$$\begin{aligned} \theta_f &= 1, \quad T_f = T_i, \quad x = X, \\ \theta_f &= 0, \quad T_f = T_L, \quad x = L. \end{aligned} \tag{5.36}$$

Solving equation (5.35), using the boundary conditions, yields

$$\theta_f = \frac{x - L}{X - L}. \tag{5.37}$$

At the interface the gradients for these two temperature distributions are linked by equation (5.38)

$$\lambda_f(T_i - T_L)\frac{d\theta_f}{dx}\bigg|_{x=X} + \lambda_d(T_s - T_i)\frac{d\theta_d}{dx}\bigg|_{x=0} - Gc(T_s - T_i) = G\,\Delta H, \tag{5.38}$$

where ΔH is the heat of sublimation. Here the left-hand side of the equation is the heat gain by conduction and the increase in enthalpy of the vapour, and the right-hand side the heat loss due to sublimation.

Taking the derivatives of equations (5.34) and (5.37) with respect to x and substituting in equation (5.38) leads to

$$\frac{\lambda_f(T_i - T_L)}{X - L} + \frac{\lambda_d(T_s - T_i)(Gc/\lambda_d)}{1 - \exp(-GcX/\lambda_d)} = G[\Delta H + c(T_s - T_i)]. \tag{5.39}$$

Now, for freeze-drying the argument in the exponential $GcX/\lambda_d \ll 1$, so that $1 - \exp(-GcX/\lambda_d) \simeq GcX/\lambda_d$ leads to the reduction

$$GX = \frac{X\lambda_f(T_i - T_L)/(X - L) - \lambda_d(T_i - T_s)}{\Delta H + c(T_s - T_i)} \tag{5.40}$$

5.4.3 MOMENTUM EQUATION

As a consequence of diffusion of water vapour in the porous dry material, Dyer and Sunderland (1966) expressed the momentum balance (see also Section 4.3) in terms of the effective diffusion coefficient D_e for transition flow, writing

$$G' = \frac{\rho D_e}{X} \ln \frac{1 - N_s}{1 - N_i}, \tag{5.41}$$

where G' = drying rate, taken as positive in the negative x-direction, D_e = effective diffusivity in the dry material, N_s = mole fraction of water vapour at the surface, N_i = mole fraction of water vapour at the interface. From the definition of mole fractions:

$$N_s = \frac{p_s}{P}, \qquad N_i = \frac{p_i'}{p_i^\circ}, \tag{5.42}$$

where p_s, p_i' = partial pressures of water vapour at the surface and interface respectively, and P, p_i° = absolute pressure and vapour pressure of sublimation at the interface respectively.

Dyer and Sunderland calculated D_e by trial and error from the following equation:

$$N_i = 1 + \frac{b_3}{\Delta p} \frac{\ln(P+\Delta p) + \left[b_4\left(1+\frac{P}{\Delta p}\right)\right] + \frac{1}{2\cdot 2!}\left[b_4\left(1+\frac{P}{\Delta p}\right)\right]^2 + \frac{1}{3\cdot 3!}\left[b_4\left(1+\frac{P}{\Delta p}\right)\right]^3 + \cdots}{\exp\left[b_4\left(1+\frac{P}{\Delta p}\right)\right]}$$

$$+ \frac{N_s - 1 - \frac{b_3}{\Delta p}\exp\left(-\frac{Pb_4}{\Delta p}\right)\left\{\ln P + \frac{b_4 P}{\Delta p} + \frac{1}{2\cdot 2!}\left(\frac{Pb_4}{\Delta p}\right)^2 + \frac{1}{3\cdot 3!}\left(\frac{Pb_4}{\Delta p}\right)^3 + \cdots\right\}}{\exp(b_4)} \tag{5.43}$$

where

$$b_3 \equiv [8\delta'\rho(D_t)_e/3\pi(D_{K,w})_e] \ln \frac{1-N_s}{1-N_i}, \tag{5.44}$$

$$b_4 \equiv [D_t/(D_{a,w})_e] \ln \frac{1-N_s}{1-N_i}. \tag{5.45}$$

The reader should refer to Dyer and Sunderland (1966) for the derivation of equation (5.43).

5.4.4 MASS TRANSFER

Dyer and Sunderland (1968a) considered for steady flow the vapour flux to be constant and the residual gas flow flux to be zero throughout the entire dry layer.

Now, the driving force $\Delta P = p_i^\circ - P$, but $G' = \rho K_v \Delta P/\mu X$, assuming the vapour behaves according to d'Arcy's law for the viscous regime in the dry layer. Thus $p_i^\circ = P + G'\mu X/\rho K_v$ and so

$$N_i = \frac{p_i'}{P + G'\mu X/\rho K_v}, \tag{5.46}$$

where K_v = viscous permeability and μ = viscosity of the vapour. Substitution from equations (5.41), (5.42) and (5.46) yields

$$G' = \frac{\rho D_e}{X} \ln \frac{1 - p_s/P}{1 - p_i'(P + G'\mu X/\rho K_v)^{-1}} \tag{5.47}$$

A further substitution from equations (5.40) and (5.47) $G \equiv G'$ leads finally to

$$\frac{\lambda_f X(T_i - T_L)/(X - L) - \lambda_d(T_i - T_s)}{\Delta H + c(T_s - T_i)}$$
$$= \rho D_e \ln \frac{1 - p_s/P}{1 - p_i'\{P + \mu[\lambda_f X(T_i - T_L)/(X - L) - \lambda_d(T_i - T_s)]/\rho K_v[\Delta H + c(T_s - T_i)]\}^{-1}} \tag{5.48}$$

Dyer and Sunderland considered also in their analysis the variation of partial pressure with temperature for the interface, that is, $p_i' = f(T_i)$.

5.4.5 TRIAL AND ERROR SOLUTION

They found a solution by incrementing the interface position ΔX in equations (5.40) and (5.48) and using a trial and error procedure to determine T_i starting from an assumed value. They determined also values of the product: (flow rate) × (interface depth), and of $G'X$ and the diffusion coefficient D_e.

Dyer and Sunderland (1968a) found time increments from $G' \simeq (\Delta X/\Delta t)\rho\varepsilon$, which, by addition, gave the drying time at a given interface position, and eventually the complete drying time for the slab. They obtained also a simplified solution with the temperature of the frozen layer equal to the temperature of the interface, i.e. $T_L = T_i$, whence it follows from equation (5.48) that

$$\frac{\lambda_d(T_s - T_i)}{\Delta H + c(T_s - T_i)}$$
$$= \rho D_e \ln \frac{1 - p_s/P}{1 - p_i'\{P + \mu\lambda_d(T_s - T_i)/\rho K_v[\Delta H + c(T_s - T_i)]\}^{-1}} \tag{5.49}$$

This is ideally the case for an insulated back-face or for parallel drying from both surfaces. With $T_L = T_i$ substituted in equation (5.40) the value of X is obtained and is constant. Integrating $G' = \rho_i(dX/dt)$ with $t = 0$, $X = 0$ gives the drying time for this case, namely

$$t = \frac{\rho_i X}{G'}\bigg|_{X=L}. \tag{5.50}$$

5.4.6 RESULTS FOR BOVINE MUSCLE

Dyer (1966) and Dyer and Sunderland (1968a) gave several example calculations based on their theory for freeze-dried bovine muscle. One of these, taking $P = 267$ Pa, $p_s = 133$ Pa, $\lambda_d = 0.021$ W m^{-1} K^{-1} and $L = 3{\cdot}8$ cm, gives $D_e = 0.012$ m^2 s^{-1}, and shows a marked effect of back-face heating on drying time (Fig. 5.6).

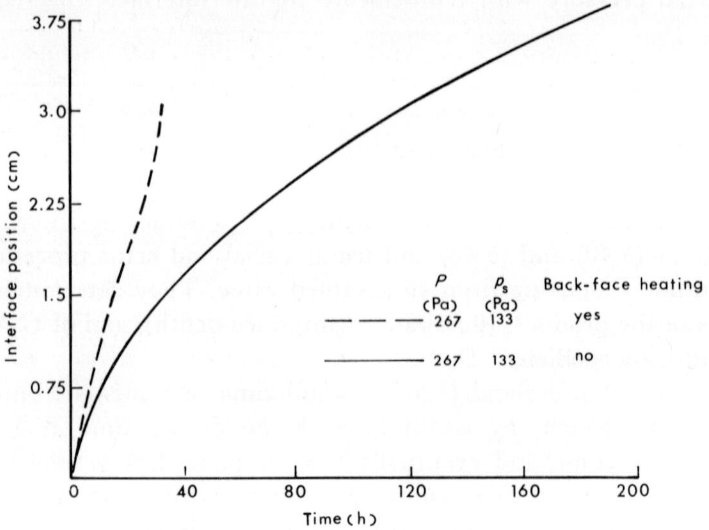

FIG. 5.6. Effect of back-face heating on drying time.

On the basis of their trial and error calculations these authors claimed that (1) back-face heating has a marked effect on drying time and the highest possible temperature at the back-face should be used consistent with avoiding melting; (2) drying time is not significantly reduced by decreasing the water vapour pressure at the drying surface; and (3) drying time decreases with decreasing absolute pressure below 133 Pa. However, it is doubtful whether the graphical results obtained by these authors could be used to assist freeze-drying in practice because the effects really only show up in the extended interface and time scales.

It was noted in Section 5.3 that the relationship between the vapour flow rate and the rate at which the interface advances is given by $G' = \rho_i\,(dX/dt)$, and in Section 5.4 that $G' = \rho K_v\,\Delta P/\mu X$ for viscous flow in the dried layer with a permeability K_v. Therefore, from these two relationships: $dX/dt = \rho K_v\,\Delta P/\rho_i \mu X$ which, after integration, where the interface position increases from zero to X in time t, gives

$$X^2 = \frac{2\rho K_v\,\Delta P}{\rho_i \mu} t. \tag{5.51}$$

Thus, Dyer and Sunderland deduced as a direct consequence of their theory that the interface position varies directly with the square root of time; that is, $X = \gamma t^{1/2}$ where $\gamma^2 = 2\rho K_v\,\Delta P/\rho_i \mu$. On the other hand, Mellor (1966b) showed that $\gamma^2 = 2DM\,\Delta P/\rho_i RT$ where D is an overall diffusion coefficient. In conclusion, therefore, their theory reverts to a moving boundary problem.

Earlier Dyer and Sunderland (1967) showed that the problem could be solved analytically if a constant temperature for the interface and constant parameters were assumed. Their solution is in fact very similar to the one in Section 3.3 for heat transfer with a constant surface temperature.

5.5 Shorter solutions for back-face heating

Another conclusion to be drawn from the foregoing analysis is that the convection heat transfer term $G'c(T_s - T_i)$ is quite small and therefore equation (5.38) becomes

$$\lambda_f(T_i - T_L)\left.\frac{d\theta_f}{dx}\right|_{x=X} + \lambda_d(T_s - T_i)\left.\frac{d\theta_d}{dx}\right|_{x=0} = G'\,\Delta H. \tag{5.52}$$

5.5.1 CONSTANT INTERFACE TEMPERATURE

To determine the drying time, Dyer and Sunderland (1968b) assumed a constant interface temperature, quasi-steady state temperature conditions and constant parameters.

Equations (5.34) and (5.37) remain the same, except that the zero on the x-axis (Fig. 5.5) is changed over to the back-face and this gives

$$\theta_d = \frac{1-\exp\left[-G'cL/\lambda_d(1-x/L)\right]}{1-\exp\left[-G'cL/\lambda_d(1-X/L)\right]}, \tag{5.53}$$

$$\theta_f = \frac{x}{X}. \tag{5.54}$$

Kan and de Winter (1968) proposed a similar equation to (5.53) for the temperature in the dry layer but with a different argument for the exponentials—a heat-flux term instead of a vapour-flux term.

Substitution then from equations (5.52), (5.53) and (5.54) leads to

$$\frac{\lambda_f(T_i - T_L)}{X} + \frac{\lambda_d(T_s - T_i)G'c/\lambda_d}{\exp(G'cX/\lambda_d) - \exp(G'cL/\lambda_d)} = G'\,\Delta H. \tag{5.55}$$

Dividing through by $\lambda_f G'c(T_i - T_L)/\lambda_d$ yields

$$\frac{\lambda_d}{G'cX} - \frac{A}{\exp(G'cX/\lambda_d) - \exp(G'cL/\lambda_d)} - B = 0, \tag{5.56}$$

where

$$A = \frac{\lambda_d}{\lambda_f}\frac{T_i - T_s}{T_i - T_L},$$

$$B = \frac{\lambda_d}{\lambda_f}\frac{\Delta H}{c(T_L - T_i)},$$

and the expansion of the exponentials for small arguments gives

$$1 - \exp\left(\frac{G'cL}{\lambda_d}\right)\left(1 - \frac{X}{L}\right) \simeq -\frac{G'cL}{\lambda_d} + \frac{G'cX}{\lambda_d}$$

which is approximately equal to the denominator in the second term of equation (5.56), and this leads to

$$\frac{1}{W'} - \frac{Ay}{1-\exp\,[W'(1-y)]} - By = 0, \qquad (5.57)$$

where $W' = G'cL/\lambda_d$ and $y = X/L$.

Equation (5.57) was solved for W', and hence for G' as a function of B by the Newton–Raphson method. The dimensionless drying time $\bar{\tau}$, as defined by Dyer and Sunderland, was determined from

$$\bar{\tau} = \frac{\tau\lambda_d}{\Delta x \rho \varepsilon c L}. \qquad (5.58)$$

Equation (5.58) results from $G' = \rho\varepsilon\ \Delta x/\tau$ and $W' = G'cL/\lambda_d$ where τ is the actual drying time for an incremental change in interface position Δx.

In another approach, Dyer and Sunderland assumed the interface position to vary directly with the square root of time, as had been observed to be the case by a number of researchers (see Section 3.3); that is, $X = 2\gamma(\alpha t)^{1/2}$. Once again, they assumed a constant interface temperature equal to the frozen layer temperature and, taking the equations for the conservation of energy, they developed equations for the temperature profile and for the constant γ, thus allowing the drying time to be determined. Agreement with experiment was satisfactory as long as there was negligible heat transfer through the frozen layer, that is, the correlation became weaker as the drying progressed.

5.5.2 VARIABLE INTERFACE TEMPERATURE

In their analysis Dyer and Sunderland (1969) took into account the fact that a variable interface temperature often occurs in practice, by assuming a linear change from the initial temperature T_0 to the back-face temperature T_L for the slab model—a method which has been employed also by Hill (1967).

Assuming the initial temperature T_0 equals the equilibrium temperature of the frozen layer in the model (Fig. 5.5), that is, $T_0 = T_f$ at the operating pressure, then the addendum of equation (5.37), namely

$$1+\frac{T_{\mathrm{L}}-T_{\mathrm{f}}}{T_{\mathrm{i}}-T_{\mathrm{L}}}=1+\frac{L-x}{X-L}$$

leads to

$$\frac{T_{\mathrm{i}}-T_0}{T_{\mathrm{i}}-T_{\mathrm{L}}}=\frac{X}{X-L}, \quad T_{\mathrm{f}}=T_0, \quad x=0. \tag{5.59}$$

Substitution in the original equation (5.37) leads to

$$\frac{T_{\mathrm{L}}-T_0}{T_{\mathrm{i}}-T_{\mathrm{L}}}=\frac{L}{X-L}, \quad T_{\mathrm{f}}=T_0, \quad x=0. \tag{5.60}$$

Substitution from the two expressions for the ratio $(T_{\mathrm{i}}-T_0)/(T_{\mathrm{L}}-T_0)$ to obtain the interface temperature leads to

$$T_{\mathrm{i}}=T_0+(T_{\mathrm{L}}-T_0)\frac{X}{L}. \tag{5.61}$$

Thus, the temperature profiles T_{d} and T_{f} in the original equations (5.34) and (5.37) become

$$T_{\mathrm{d}}=T_{\mathrm{s}}+\left[T_0-T_{\mathrm{s}}+(T_{\mathrm{L}}-T_0)\frac{X}{L}\right]\frac{1-\exp(-W''z)}{1-\exp(-W''y)}, \tag{5.62}$$

$$T_{\mathrm{f}}=T_{\mathrm{L}}+\left[T_0-T_{\mathrm{L}}+(T_{\mathrm{L}}-T_0)\frac{X}{L}\right]\frac{1-z}{1-y}, \tag{5.63}$$

where $z=x/L$ and $y=X/L$.

$W''/L=-(\rho_{\mathrm{i}}c/\lambda_{\mathrm{d}})(\mathrm{d}X/\mathrm{d}t)$ is the convective heat-transfer term.

Writing the energy balance equation at the interface as

$$\lambda_{\mathrm{f}}\left.\frac{\mathrm{d}T_{\mathrm{f}}}{\mathrm{d}x}\right|_{x=X^+}-\lambda_{\mathrm{d}}\left.\frac{\mathrm{d}T}{\mathrm{d}x}\right|_{x=X^-}=\rho_{\mathrm{i}}\,\Delta H\frac{\mathrm{d}X}{\mathrm{d}t}, \tag{5.64}$$

and substituting from equations (5.62) and (5.63):

$$\left[T_0-T_{\mathrm{L}}+(T_{\mathrm{L}}-T_0)\frac{X}{L}\right]\frac{\lambda_{\mathrm{f}}}{X-L}$$

$$-\left[T_0-T_{\mathrm{s}}+(T_{\mathrm{L}}-T_{\mathrm{i}})\frac{X}{L}\right]\frac{\lambda_{\mathrm{d}}\frac{W''}{L}\mathrm{e}^{-W''X/L}}{1-\mathrm{e}^{-W''X/L}}=\frac{\lambda_{\mathrm{d}}W''\,\Delta H}{cL}. \tag{5.65}$$

Dividing through by $\lambda_f(T_0 - T_L)$ and rearranging the above equation yields

$$1+\frac{W''}{e^{W''X/L}-1}\left[\frac{\lambda_d}{\lambda_f}\frac{T_0-T_s}{T_0-T_L}-\frac{\lambda_d}{\lambda_f}\frac{X}{L}\right]-\frac{\lambda_d W''\,\Delta H}{\lambda_f c(T_L-T_0)}=0. \qquad (5.66)$$

Defining

$$A=\frac{\lambda_d(T_0-T_s)}{\lambda_f(T_0-T_L)},$$

$$B=\frac{\lambda_d\,\Delta H}{\lambda_f c(T_L-T_0)},$$

$$K=\frac{\lambda_d}{\lambda_f},$$

substituting these definitions and those for y and W'' in equation (5.66) gives

$$1+\frac{W''}{e^{W''y}-1}[A-Ky]-BW''=0. \qquad (5.67)$$

Making the approximation $e^{W''y}-1=W''y$, since $W''y \ll 1$, equation (5.67) becomes

$$W''\simeq\frac{1+(A-Ky)/y}{B}. \qquad (5.68)$$

Dyer and Sunderland (1969) defined

$$\bar{\tau}=\frac{L}{\Delta x}\int_0^1\frac{dy}{W''}$$

and obtained the following approximation for the drying time

$$\bar{\tau}\simeq\frac{L}{\Delta x}\frac{B}{1-K}\left[1+\frac{A}{1-K}\ln\frac{1-K+A}{A}\right]. \qquad (5.69)$$

They noted the close comparison between the analysis which proved to be accurate within 2 per cent in equation (5.57) and that leading to equation (5.67), since $1-e^{W'(1-y)}\simeq -W'$, and concluded that the linearly variable interface temperature only involves the introduction of one additional parameter K into the equation and the replacement of T_0 by T_i.

5.5.3 COMPARISON OF DRYING TIME SOLUTIONS

The constants should have the following ranges: $0<A<1$, $0<B<1000$ and $0{\cdot}033<K<0{\cdot}1$ for calculating $\bar{\tau}$. The results (Fig. 5.7) show

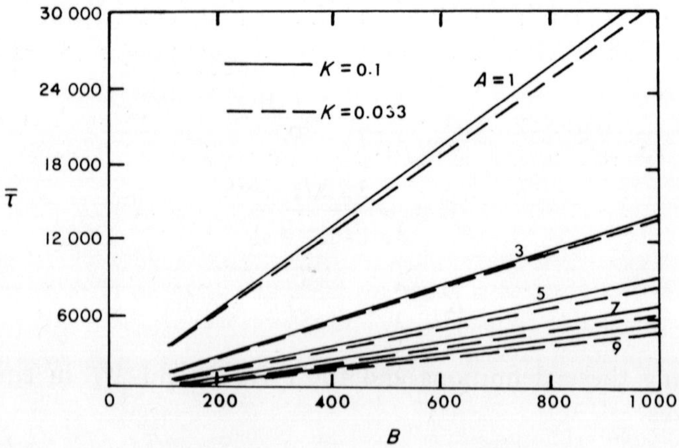

FIG. 5.7. Dimensionless drying time as a function of B for various values of A and K (after Dyer and Sunderland, 1969).

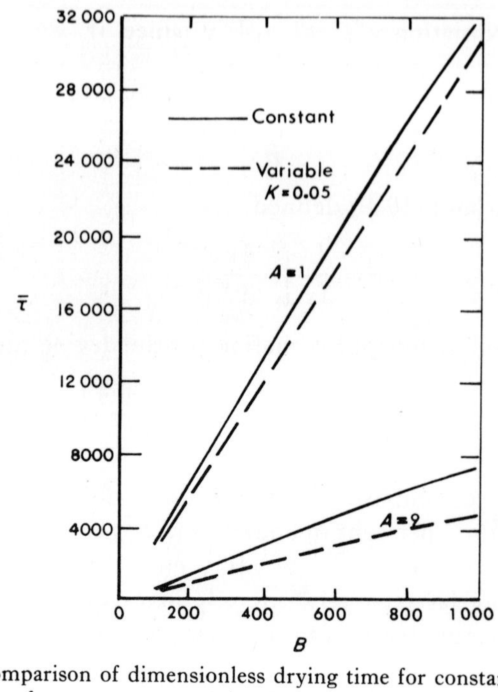

FIG. 5.8. Comparison of dimensionless drying time for constant and variable interface temperatures (after Dyer and Sunderland, 1969).

that the drying rates tend to improve as K increases, due to the high thermal conductivity of the frozen layer.

Calculations, based on equations (5.57) and (5.67), are compared in Fig. 5.8 and show somewhat longer drying times for variable compared with steady interface temperatures, probably because of reductions in the temperature gradients for heat conduction as drying progressed.

Figure 5.9 is a comparison between three analytical methods for dealing with the back-face heating problem. Two of them deal with

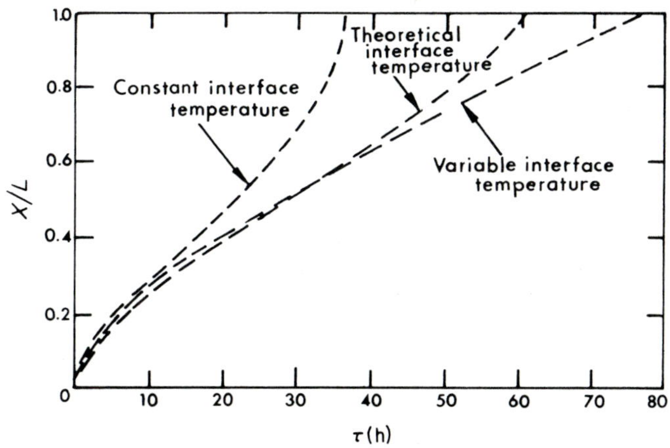

FIG. 5.9. Comparison of drying times for different interface temperature variations against interface position (after Dyer and Sunderland, 1969).

assumptions about constant and linearly variable interface temperatures and a third about theoretical interface temperatures which have been estimated according to the experimental results of Dyer (1966). The first assumption gives a grossly overestimated drying time and the second a better result.

5.6 Constant- and falling-rate drying periods in the course of freeze-drying

5.6.1 ANALOGY WITH AIR-DRYING

It is questionable whether the time for freeze-drying can be divided into constant- and falling-rate periods by analogy with air-drying of wet materials. In air-drying a constant-rate period indicates that excess liquid moisture, rising to the surface, evaporates freely; and a falling-rate period indicates that moisture loss from the receding wet portion of the material is gradually reduced. However, with freeze-drying, liquid moisture is immobilized by freezing before the start of freeze-drying, and so it might be expected that a constant-rate period would cease to exist after the surface ice had sublimed. Dyer and Sunderland (1969) are adamant that no clear division of the time for freeze-drying into two periods can be justified or assumed; they point out that the analytical problem is similar to heat conduction with a moving boundary, or resembles heat conduction with ablation. However, it is hard to see the connection with ablation, unless the dry layer is continuously removed during freeze-drying. Greaves (1962) has been able to achieve this, but only on a laboratory scale.

On the other hand, Burke and Decareau (1961, 1964) found two such periods when freeze-drying 10-mm cubes of chicken meat, and

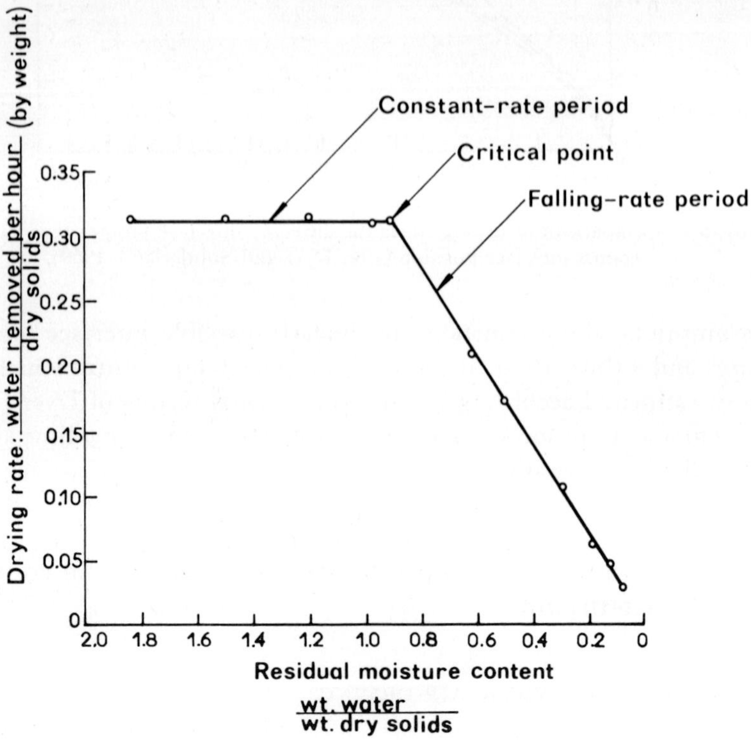

FIG. 5.10. Sherwood plot of drying-rate data (after Burke and Decareau, 1961).

attempted an explanation in terms of heat transfer by convection as well as conduction. Their experimental results (Fig. 5.10) look convincing enough, with the critical point being reached after 3 h in a 12-h drying run, until the effect of the temperature on drying rate is examined. No doubt their drying rate–temperature curves are linear in the range 82°–102°C; however, the paucity of experimental points below this range provides no grounds for believing in a constant-rate period for freeze-drying. In the range of high temperatures evaporation of melted liquid reaching the surface of the product cannot be entirely ruled out. Further, it has been suggested that evidence of convection playing a part in the mode of heat transfer is provided by the fact that the curve is not concave upwards.

5.6.2 COMMENCEMENT OF FREEZE-DRYING

Measurements by Mellor (1966b) have shown that when sublimation commences from a frozen product the water vapour pressure in the vicinity of the dry part of the product rapidly approaches the external vacuum pressure. However, as the vapour pressure at the ice interface is meanwhile at its saturation value, a large vapour-pressure difference must ensue, as shown in Fig. 5.11, when the external vacuum pressure is reduced to below the water vapour pressure within the product. Consequently, there is instantaneously a sudden outward hydrodynamic flow of water vapour.

The measurements plotted in Fig. 5.12 also make it clear that, following the commencement of sublimation, quasi-equilibrium is reached within a reasonably short time and the saturation vapour pressure at the ice interface then closely approaches the vacuum pressure and the vapour pressure at the surface is not much lower than that within the product. Consequently, the mechanism of vapour transfer is by normal diffusion through the porous product.

According to Mellor (1966b), the constant flow rate for surface sublimation, or for sublimation just under the surface, is given by

$$G = \frac{KM\,\Delta p}{RT\bar{L}}, \tag{5.70}$$

where K is the permeability coefficient of water vapour in the porous product and is obtained from the corresponding permeability coefficient K_g using nitrogen. Since K varies inversely with the square root of the molecular weight of the gas, $K = K_g(M_g/M)^{1/2}$. Its pressure variation can be read from a graph of K_g against $\bar{P}$ and can be corrected for temperature by multiplying by $(T/273)^{1 \cdot 75}$.

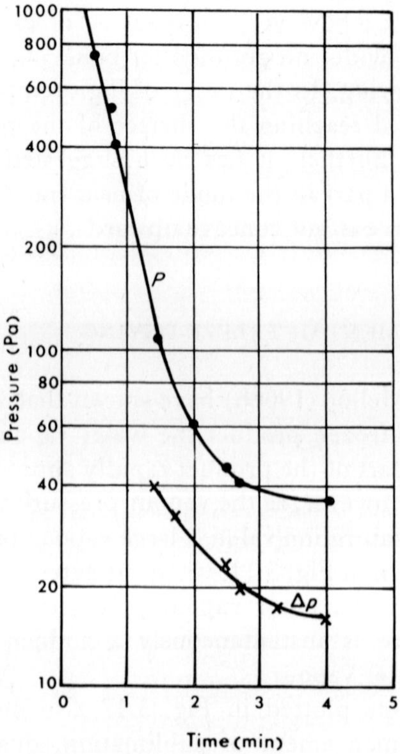

FIG. 5.11. Variation of vapour-pressure difference Δp with sudden change in vacuum pressure P.

From Appendix 2.II the falling flow rate G is given by

$$G = \frac{DM\ \Delta p}{RT\gamma t^{1/2}}, \tag{5.71}$$

where D is the diffusion coefficient for water vapour in the porous product and the constant γ is determined from $\gamma^2 = 2DM\ \Delta p/\rho_i RT$.

These equations have been used to calculate the vapour-pressure difference ΔP for the two periods following the commencement of freeze-drying. For example, using data on freeze-drying of orange juice (Lambert and Marshall, 1962), a constant-rate period of 1·4 h was estimated (Table 5.1). During this phase the sublimation vapour pressure was 57 Pa and the vacuum pressure was 32 Pa but, after 2 h in the falling-rate stage, the values were 31 and 29 Pa respectively.

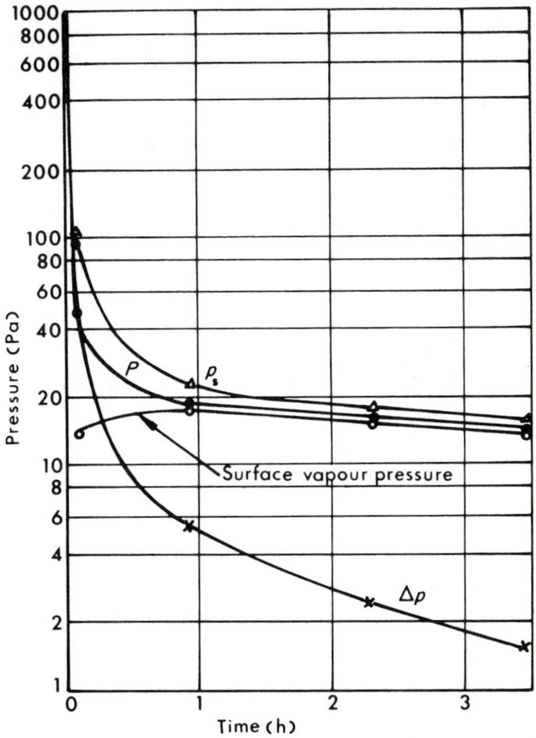

FIG. 5.12. Variation of vapour pressure following the commencement of sublimation.

Similarly, in the freeze-drying of cooked meat (Mellor, 1966b), the sublimation vapour pressure after 15 min was 105 Pa with the vacuum pressure at 27 Pa, but after 2·5 h these values had fallen to 19 and 17 Pa respectively. In general it has been observed that the falling-rate period during sublimation can represent between seven and nine-tenths of the time for sublimation. This time is generally taken to be about two-thirds of the total drying time, and thus nine-tenths is an obvious conclusion to be drawn from Table 5.1. Substantial amounts of sublimation will have taken place before the commencement of the falling-rate period, but this does not mean that sublimation then

TABLE 5.1

Calculated and observed values of vapour-pressure difference and associated data for freeze-dried products

Product	Time (h)	Vacuum pressure P (Pa)	Sublimation Temperature (°C)	Sublimation p_s (Pa)	Condenser Temperature (°C)	Condenser p_c (Pa)	Vapour-pressure difference Δp Calculated (Pa)	Vapour-pressure difference Δp Observed (Pa)	Remarks
Orange juice (16-h drying)	0·5[a] 1·4[a]	32·0	−26	57·3	−40 (end of constant-rate period)	13·3	42·7	44·0	$K = 60 \times 10^{-4}$ $A = 0{\cdot}0422$ $\Delta p/\bar{L} = 22{\cdot}8$ kPa m^{-1}
	2·0	29·3	−32	30·7	~−40	13·3	10·7	13·3	$D = 0{\cdot}0121$ $\gamma = 5{\cdot}2 \times 10^{-5}$
	3·0	26·7	−33	28·0	−38·5	16·0	4·0	5·3	$D = 0{\cdot}0127$ $\gamma = 3{\cdot}3 \times 10^{-5}$
Cooked meat	0·25[a]	26·7	−19·8	105·3	−54	2·7	90·6	92·0	$K = 30 \times 10^{-4}$ $A = 0{\cdot}0365$ $\Delta p/\bar{L} = 3{\cdot}8$ kPa m^{-1}
(12-h drying)	2·5	17·3	−37	18·7	−54	2·7	2·7	2·7	$D = 0{\cdot}0250$ $\gamma = 3{\cdot}2 \times 10^{-5}$
(20-h drying)	8·0	80·0	−26	57·3	−47	5·3	4·0	2·7	$D = 0{\cdot}0054$ $\gamma = 1{\cdot}8 \times 10^{-5}$

[a] Constant-rate period.

ceases; in fact, it continues slowly for many hours, until thinning sections and pockets of ice have disappeared and the final dry state is attained.

5.7 Superheated Vapour

When the water vapour moves away from the interface it is no longer in equilibrium with ice but is in contact with the interstices of the dry layer and with the heater; it therefore attains a correspondingly high temperature whilst the absolute pressure remains unchanged, i.e. it becomes unsaturated or superheated. This superheating effect in freeze-drying has occasionally been referred to by Rowe *et al.* (1963), Rowe (1964a) and Mellor (1962b) as follows.

At a fixed temperature the drying rate varies with the thickness of the dry layer and the vapour-pressure difference. With superheating it can therefore be inferred that the drying rate is not necessarily a simple function of the method of heating alone but is due to changes in the vapour-pressure gradient, both in the dry layer and in the space surrounding it. The vapour becomes superheated and its molecular motion is increased at a constant absolute pressure in the pores of the dry layer and within the confines of the heater plates such as expanded metal plates used for contact heating. The superheated vapour behaves like a perfect gas and follows the well-known gas laws.

5.7.1 SUPERHEATING EFFECT IN POROUS MATERIALS

At low pressures the passage of vapour molecules through the dry layer, which is assumed to consist of fine capillaries, can be examined in terms of molecular streaming or Knudsen flow and, in the higher pressure range, in terms of Poiseuille flow.

For a single capillary of radius r and length L, the total flow rate at all pressures per unit area of cross-section is equal to the sum of the Knudsen and Poiseuille flows, namely

$$G = \frac{8r^3}{3L}\left(\frac{\pi M}{2RT}\right)^{1/2}(p_{\mathrm{up}} - p_{\mathrm{do}})\left(1 + \frac{3\pi r}{64\lambda}\right), \tag{5.72}$$

where $M/RT = \rho/p$, the ratio of the vapour density to the vapour pressure, λ = mean free path of the vapour molecules, $p_{up} - p_{do}$ = difference between the upstream and downstream vapour pressures, located at the interface and at the surface of the dry layer respectively.

When the temperatures T_{up} and T_{do} of the upstream and downstream vapours respectively are appreciably different and the partial vapour pressures p' depend on T, the ratio in equation (5.72) is

$$\frac{p_{up} - p_{do}}{T^{1/2}} \rightarrow \frac{p_{up}}{T_{up}^{1/2}} - \frac{p_{do}}{T_{do}^{1/2}}. \quad (5.73)$$

Where there is complex geometrical structure, as in freeze-dried materials, the theory is nevertheless the same and similar equations have been developed by Barrer (1948) for an idealized porous material having a combination series-parallel arrangement of capillaries. And he has experimental evidence for considering that vapours and gases pass through porous materials in this way. In fact, Harper and Tappel (1957) have found that the following equation governs the rate of flow of vapour through freeze-dried beef:

$$G = \frac{KM\bar{P}}{\mu L_d RT}(p_{up} - p_{do})\left(1 + \frac{b}{\bar{P}}\right), \quad (5.74)$$

where K is the permeability and L_d the thickness of the dry layer, μ is the viscosity and $\bar{P}$ the mean pressure of the vapours, and b a correction term depending on λ and K. From equation (5.72), if the total pressure is below 13 Pa, $\lambda > r$ (r is about 50 μm, for freeze-dried beef) and the value of the correction term $3\pi r/64\lambda$ becomes negligible, thus, $b/\bar{P}$ in equation (5.74) would likewise be negligible.

Now the absolute rate of sublimation of ice per unit area where there is no vapour flow resistance is given, by Knudsen's equation, as:

$$G_s = p_s\left(\frac{M}{2\pi RT_s}\right)^{1/2}, \quad (5.75)$$

which equals $1{\cdot}2 \times 10^{-1}$ kg m^{-2} s^{-1} at -20°C. In this equation assume without correction that the saturation vapour pressure at the sublimation temperature is identical with the upstream vapour pressure at the interface. For heating through the dry layer the temperature of the downstream vapour would be higher than the temperature of the ice due to the temperature gradient and heater, and therefore the downstream vapour would be superheated.

Writing the ratio of the rates G/G_s to give a dimensionless quantity for comparison of actual drying rates and with $\mu = (4/3)\bar{P}\lambda(M/2\pi RT)^{1/2}$, $M/RT = \rho/p$ and the application of the same considerations for the dependence of p on T as in equation (5.73), then equations (5.74) and (5.75) yield

$$\frac{G}{G_s} = \frac{3\pi K}{2\lambda L}\left(1 - \frac{\rho_{\mathrm{do}} T_{\mathrm{do}}^{1/2}}{\rho_{\mathrm{up}} T_{\mathrm{up}}^{1/2}}\right). \tag{5.76}$$

This equation is similar to the one for interphase vapour transfer of a pure substance, which has been discussed by Schrage (1953) in terms of a particular Maxwell velocity distribution function. Schrage's equation gives, however, erroneous values for G/G_s in the case of the actual drying of materials unless the mean velocity of the recondensing vapour can be evaluated.

5.7.2 DRYING RATES

A value $G = 1{\cdot}4 \times 10^{-4}\ \mathrm{kg\ m^{-2}\ s^{-1}}$ has been found by Harper and Tappel (1957) for beef drying at −20°C with heat transfer through the frozen layer and with a downstream pressure equal to the absolute pressure of 13 Pa. (In practice, vacuum systems operate at higher absolute pressures but pressures do not affect G greatly unless they equal the saturation vapour pressure, thus $G/G_s = 1{\cdot}2 \times 10^{-3}$.)

PART 2
CYCLIC-PRESSURE OPERATION

6 Physical Mechanism of Cyclic-Pressure Freeze-Drying

6.1 Introduction

6.1.1 CONSTANT-PRESSURE PROCESS—DISADVANTAGES

In the usual freeze-drying process, substances which have been frozen are dried by applying heat to them in a vacuum to sublime the ice. Heat is applied to the external surface of the drying product either by radiation or by direct contact and, in either case, it is conducted through a drying outer layer to the frozen core, where it converts the ice directly into vapour. A dry layer forms around a piece of frozen substance when it is placed in a vacuum and, because the layer is porous, diffusion of the vapour takes place through it, as shown in Fig. 6.1.

The sublimed vapour is usually removed from the system either (1) by condensation on a cold refrigerated surface at a temperature of about −40°C, or (2) by direct pumping to the atmosphere whilst the vacuum pressure is maintained constant at below 133·3 Pa, at which pressure the ice may be made to sublime at −20°C when heating by radiation.

A disadvantage of this system of freeze-drying is that as soon as the outer layer dries it reduces the heat transfer to the ice crystals lying further from the surface of the product and it also reduces the rate of escape of vapour because, under these conditions, the vapour flow is mainly by diffusion. Any attempt at this stage to increase the drying rate by applying more heat may cause overheating of the surface and consequent denaturing of the product. Also, with many products,

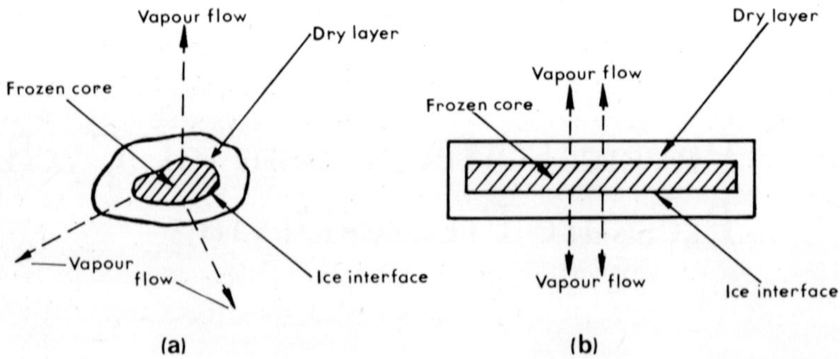

FIG. 6.1. Water vapour flow from a substance undergoing freeze-drying.

reducing the vacuum pressure to facilitate the flow of vapour is ineffectual because the heat-transfer coefficients then become too low (e.g. 0·44 W m^{-1} K^{-1} at 66·7 Pa for dried beef).

6.1.2 CYCLIC-PRESSURE PROCESS

To overcome these difficulties, a new and improved method of freeze-drying has been developed (Mellor, 1967a) in which the vacuum pressure is controlled in such a way that the initial drying rate tends to be sustained. To be more explicit, frozen substances are subjected to repeated cycles in vacuum pressure, varying between certain limits. The optimum pressures and their duration can be determined either by experiment or by calculation; they depend upon the nature of the product, the composition of the gas and the efficiency of the vacuum and refrigeration systems.

The following discussion of basic principles may lead to a better understanding of the cyclic process.

In conventional freeze-drying, it has been established by experiment that the resistance to flow of vapour through many substances can be 6–10 times greater than that of the surrounding vacuum space, which means that the drying rate is determined principally by the rate of passage of water vapour through the dry porous substance. Now at the commencement of freeze-drying, while the system is being evacuated, the absolute gas pressure falls rapidly. In this initial stage the partial water vapour pressure in the vicinity of the dry surface rapidly approaches the absolute vacuum pressure; also, at the temperature prevailing, the vapour pressure at the ice interface is close to its saturation value. There is thus a distinct difference in vapour pressure as between the ice interface and the outer surface and the flow of vapour from within the product is relatively fast, being due to hydrodynamic flow rather than diffusion. For a given heat input there is a significant flow of heat inwards across the dry layer, the thermal conductivity of which is largely determined by the presence of vapour and air (or whatever gas may be present) in the interstices within this layer.

Other experiments have shown that not long after the commencement of freeze-drying a steady-state condition is established within the dry porous layer, such that the vapour pressures at the outer surface and at the interface are both equal to the saturation value for ice at the temperature of the interface. When this state is attained, vapour passing from the interface towards the outer surface through the dry layer (which itself increases in relative thickness as drying proceeds) is driven against a temperature gradient by a relatively small vapour-pressure difference. Accordingly, vapour removal from the product is now controlled by relatively slow diffusion processes instead of by the faster hydrodynamic flow that previously prevailed. Consequently the rate of freeze-drying diminishes.

In the new cyclic process, however, as soon as the steady-state condition is attained, the vacuum pressure is deliberately increased and consequently the gas thermal conductivity, and hence the temperature of the interface, are raised. Then, in the second half-cycle, the pressure is again reduced and conditions once more favour hydrodynamic flow.

Thus with the new cyclic process a series of impulses in heat transfer is caused to occur in the dry parts of the product; these impulses are conducted inwards to supply the heat of sublimation of ice within the product each time the gas pressure outside it is allowed to rise. The air or gas pressure in the vacuum chamber is permitted to rise for a period of time just sufficient to supply the heat of sublimation to the ice, and not enough to cause overheating of the product. During this stage of the cycle, vapour transfer from the outer surface of the product may actually decrease when this is followed by a sudden decrease in air pressure in the next half-cycle.

The improvement in heat transfer is evident from Fig. 3.12 (Section 3.8) which shows the variation of thermal conductivity with air or helium pressure in freeze-dried beef and apple. The thermal conductivities of freeze-dried products have been measured by Harper and el Sahrigi (1964), Ehlers *et al.* (1962) and Harper (1962). Also, Meffert (1965) measured the thermal conductivity of partially dried vegetables in order to disentangle the effects of residual moisture upon thermal conductivity from those of occluded gas.

6.1.3 PRESSURES

An obvious requirement for efficient cyclic-pressure freeze-drying is that the pressure changes should be rapid. The change from low to high pressure can be made quickly by simply admitting gas to an evacuated chamber. The rapidity of the change from high to low pressure depends upon (1) the pumping rate, and (2) selecting for the high pressure the lowest value which is consistent with obtaining maximum difference in product thermal conductivity for minimum pressure difference. As indicated in Fig. 3.12 (Section 3.8), when using air the best pressure difference is between 26·7 Pa and 2·7 kPa and when using helium it is between 26·7 Pa and 13·3 kPa for beef and 26·7 Pa and 20 kPa for apple.

A low pressure of 26·7 Pa is necessary to ensure complete transfer of vapour during the low-pressure part of the cycle and to cause the sublimation temperature to fall again. Figure 3.12 shows that increasing the pressure range will give little gain in thermal conductivity; nevertheless, some truly labile products would be better if freeze-dried at a pressure below 13·3 Pa while for others a pressure of 133·3 Pa would be more suitable. The high pressure could lie between 13·3 and 6·7 kPa for air and between 13·3 and 26·7 kPa for high-conductivity gases such as helium, but high pressures such as this are seldom used in practice.

It is important, therefore, to select a pressure difference which will ensure that the water vapour is transferred very quickly from the product, accompanied by a fall in temperature of the product because of the heat abstracted from it by sublimation. The heat and vapour transfer effects occur alternately and more or less independently when the vacuum pressure is made to rise and fall in the manner described.

In cyclic-pressure freeze-drying, for maximum freeze-drying rates the high-pressure period should be just long enough to raise the temperature of the product to its maximum safe level and the low-pressure period just short enough to substantially complete the process of rapid vapour transfer from the product.

The optimum values for these two periods will differ according to the composition of the gas and the physical and geometric characteristics of the product being dried; they will depend also upon the range of operating pressures chosen, which in part at least will be determined by the dimensions of the chamber, its charge and the efficiency of the pumping and refrigeration systems. However, experiments and calculations have shown that for each class of product to be freeze dried a pressure cycle can be found that gives optimum performance of the freeze-drying equipment.

6.2 Vacuum pressure cycles

6.2.1 GAS THERMAL CONDUCTIVITY

To improve heat transfer by improving gas thermal conductivity, air or some other suitable gas, e.g. helium, may be admitted to the vacuum chamber when the pressure is about 26·7 Pa so as to give 100 per cent increase in pressure. However, the operating pressure of most conventional freeze-dryers is less than 133·3 Pa and is fixed.

6.2.2 OPTIMAL CYCLES—CONSTANT RATIO AND RATIO OF DRYING TIMES

The optimal periods of application of the high and low pressures have been found by comparing the heat-transfer conditions for constant-pressure and cyclic-pressure processes. The analysis for optimal cycles can clearly be developed by reference to Figs 6.2 and 6.3—the first, showing the values of the thermal conductivity for the two drying processes under consideration; and the second, the asymptotic approach to a maximum thermal conductivity in terms of the ratio of high-pressure time to low-pressure time.

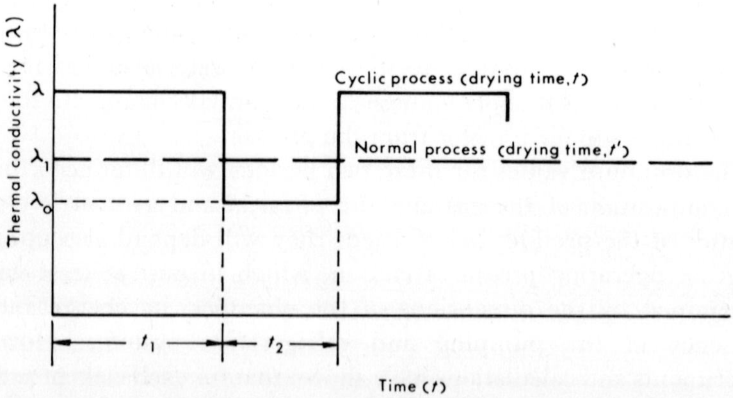

FIG. 6.2. Schematic diagram of the thermal conductivities in normal and cyclic processes.

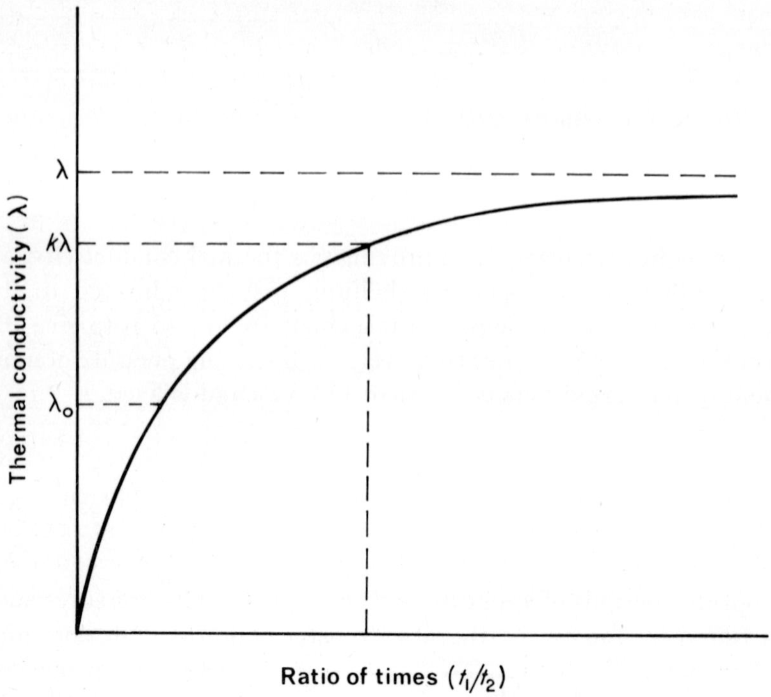

FIG. 6.3. Asymptotic approach of the thermal conductivity in the pressure cycle to the maximum in terms of the ratio of high-presure time to low-pressure time.

Consider heat transfer in the constant-pressure (normal) process for a long period t' which is equivalent to that in the cyclic-pressure process, comprising high- and low-pressure periods of t_1 and t_2 respectively, for a somewhat shorter period t. As shown in Fig. 6.2, to sublime the same mass of ice whether it is a constant- or cyclic-pressure process requires that the total quantities of heat for the separate processes must be invariant, that is

$$H_n \equiv H_{cyc.} \tag{6.1}$$

Now H_n and H_{cyc} are directly proportional to the products of the effective conductivities and drying times for the respective processes. Figure 6.3 shows that as the ratio t_1/t_2 for the high-pressure time to the low-pressure time increases, the effective thermal conductivity approaches more and more closely to its maximum value λ at the high pressure. A necessary and sufficient condition would be to allow it to approach, say, $k\lambda$ where k is defined as a constant ratio. Thus, the heat quantities H_{cyc} and H_n are proportional to the weighted mean of the thermal conductivities $k\lambda$ and λ_0 for the high- and low-pressure periods multiplied by the drying time t for the cyclic process, and by $\lambda_1 t'$ for the equivalent "normal" process. The identity (6.1) then becomes

$$\lambda_1 t' \equiv \frac{k\lambda t_1 + \lambda_0 t_2}{t_1 + t_2}\, t. \tag{6.2}$$

Simplifying, by putting $r = t/t'$ for the ratio of the cyclic to normal process drying times and $\tau = t_1 + t_2$ for the period of the cycle, equation (6.2) becomes

$$\frac{t_1}{t_2} = \frac{\lambda_1/r - \lambda_0}{k\lambda - \lambda_1/r}. \tag{6.3}$$

Now for the low-pressure period, theory and experiment (Mellor, 1966b; Mellor and Middlehurst, 1971) predict that the vapour flow rate varies with time according to the equation of hydrodynamic relaxation, namely $k = 1 - e^{-\beta_1 t}$, where k is a constant ratio which for convenience is assumed to be the same as the one above for heat transfer, and sets a limit to the flow rate if the low-pressure period is not to continue indefinitely. And β_1 is the time-constant.

On differentiating k with respect to t_1/t_2 in equation (6.3) it can be shown that the condition for a maximum is

$$r = \frac{\lambda_1}{k\lambda}. \tag{6.4}$$

As with all processes of physical transport which follow an exponential law, the vapour flow and heating rates are usually determined after their values have altered by ten times or by one-tenth of their original values, i.e. by one log cycle, or $k = 0{\cdot}9$ which on substitution in equation (6.4) gives $r = 0{\cdot}69$. This value is in general agreement with an average value of 0·65 which has been found in practice for all "cyclic" drying runs using air as the thermal gas. Therefore, the optimal drying time for "cyclic" processes should be about two-thirds the drying time of "normal" processes.

6.2.3 OPTIMAL HIGH- AND LOW-PRESSURE PERIODS

An optimal pressure cycle is necessary in order to keep the sublimation temperature below a critical level. It is therefore necessary to derive suitable expressions for the optimal high- and low-pressure periods.

Writing the right-hand side of equation (6.2) in terms of the effective thermal conductivity gives

$$\lambda_e = \frac{k\lambda t_1 + \lambda_0 t_2}{t_1 + t_2}. \tag{6.5}$$

Transposing this equation gives

$$\frac{t_1}{t_2} = \frac{\lambda_e - \lambda_0}{k\lambda - \lambda_e}. \tag{6.6}$$

Now Fig. 3.12 shows that the slope of the conductivity–pressure curve at high pressures is about double that at low pressures; that is

$$k\lambda - \lambda_e = 2(\lambda_1 - \lambda_0). \tag{6.7}$$

Further, the high- and low-pressure time limits for a cycle are at an optimum when k equals 0·9, as concluded above; that is, the logarithm of $k = 1 - e^{-\beta_1 t_2}$ for one log cycle leads to the optimal low-pressure period

$$t_2 = \frac{2 \cdot 303}{\beta_1}. \tag{6.8}$$

Substitution from equations (6.6), (6.7) and (6.8) yields the optimal high-pressure period:

$$t_1 = \frac{1 \cdot 151(\lambda_e - \lambda_0)}{\beta_1(\lambda_1 - \lambda_0)}. \tag{6.9}$$

From a comparison of equations (6.3) and (6.6) it is clear that the effective conductivity $\lambda_e = \lambda_1 / r$ where $r = 0 \cdot 69$, showing undeniably that

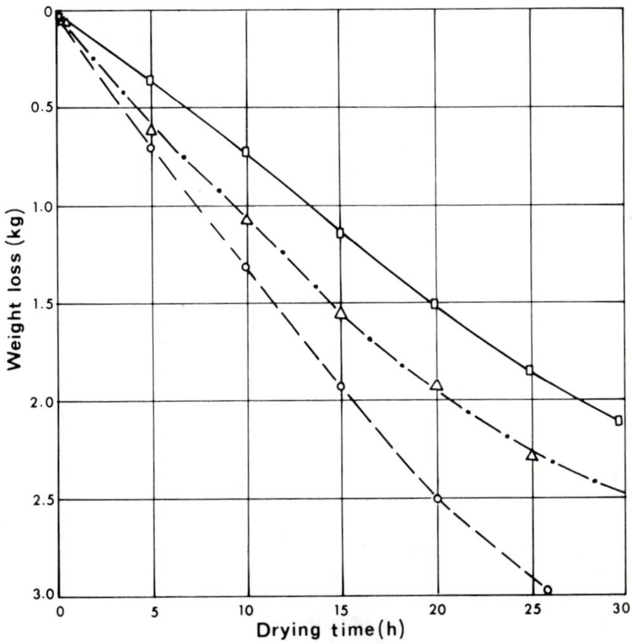

FIG. 6.4. Effect of gas pressure on freeze-drying rates for 10 per cent sucrose solutions: □, constant-pressure freeze-drying, pressure = 26·7 Pa; △, cyclic-pressure freeze-drying, air, 26·7–533 Pa; ○, cyclic-pressure freeze-drying, helium, 26·7–533 Pa. (After Greenfield, 1972.)

the optimization analysis depends implicitly on a comparison between the "normal" and "cyclic" process. This seems to be the most suitable optimization procedure so far devised.

TABLE 6.1
Derived time-constants for the cyclic process

Product	Time-constant β_1 (s^{-1})	Thermal conductivity ($W\ m^{-1}\ K^{-1}$) Effective λ_e	Low pressure λ_0	Constant pressure λ_1	Cycle time (experimental values in brackets) High-pressure time t_1 (min)	Low-pressure time t_2 (min)	Total drying time (h) Constant pressure	Cyclic pressure
Cooked meat	0·021	0·062	0·041	0·045	4·8(4·5)	1·8(1·5)	8	5
Egg	0·040	0·034	0·021	0·024	2·1(2·5)	1·0(1·7)	10	7·75
Apple	0·020	0·034	0·019	0·024	2·9(2·8)	1·9(2·3)	12·5	7·25

The typical values for the high- and low-pressure periods for a high rate of heating given in Table (6.1) show that the analysis is quite satisfactory for accelerating freeze-drying by the cyclic method. Other comparative results are shown in Fig. 6.4 for a low rate of heating.

7 Analytical Cyclic Process

7.1 Introduction

This chapter is central to the theme of Part 2: Cyclic-Pressure Operation—especially those aspects concerning the pressure and temperature effects.

Firstly, in Section 7.2 a theory for combined mechanical and condensation pumping is developed for the stage when there is a down-turn in pressure and a vapour flash occurs. The condenser then acts as a powerful pump for exhausting the vapour and air.

Next, Section 7.3 deals with the influence of pressure on temperatures in the frozen and the dry layers of the product; an iterative analysis is investigated and shows the trend in the pressure and temperature curves and in the sublimation rate.

Finally, Section 7.4 deals with a simple analytical approach to the temperature wave in the ice core, attempting to elucidate some of the basic parameters involved. Also a note (Section 7.5) on time-constants for an optimal cycle is given.

For a new process such as this the theoretical approaches are of course complicated, yet, as can be seen from Chapters 6 and 8 dealing with "cyclic" systems, these approaches show how the process can be readily applied in practice by utilizing quite simple parameters. They are the most logical theoretical concepts so far devised.

7.2 Condensation pumping—flash flow

Following the vapour flash, the pumping rate is enhanced by the rapidity with which the vapour is condensed on the refrigerated coil, so that the vapour is removed by a combined mechanical and condensation pumping effect. This is obvious from Fig. 7.1, which shows a typical pressure cycle in the vapour space where the rate of pressure reduction after 80 s is about twice that before.

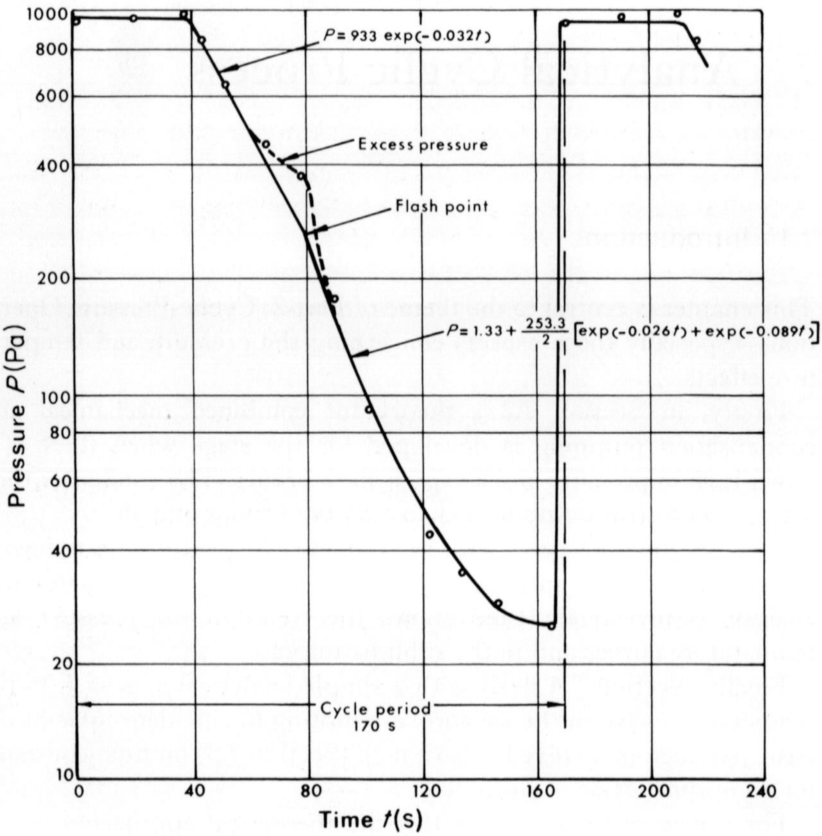

FIG. 7.1. Typical pressure cycle.

A pressure equation for the condensing vapour is derived (Appendix 7.I) by considering the rate of change in energy balance for the low-pressure part of the cycle, and this gives

$$x\frac{\partial^2 P}{\partial x \partial t} + v\frac{\partial P}{\partial x} + \frac{\partial P}{\partial t} = 0. \tag{7.1}$$

Equation (7.1) has a solution given by

$$P = P_0 \exp(-vt/x), \tag{7.2}$$

where P = pressure in vapour space, P_0 = initial pressure in vapour space, x = distance, t = time of transit, v = vapour velocity.

In the first place the pressure is reduced in the chamber by a mechanical pump according to the relation

$$P = P_0 \exp(-s_p t/V_{ch}), \tag{7.3}$$

where s_p is the displacement of the pump and V_{ch} the volume of the chamber. At the same time as the vapour flash is taking place, condensation pumping assists mechanical pumping according to relation (7.2), already derived.

Equations (7.2) and (7.3) may be combined with p_c equal to the final pressure in such a way that when the resultant equation is evaluated at $t = 0$, the "flash" pressure is given by $P_F = p_c + P_0$, and this yields

$$P = p_c + \frac{P_0}{2}[\exp(-s_p t/V_{ch}) + \exp(-vt/x)]. \tag{7.4}$$

An ordinary mean pressure for mechanical and condensation pumping may be taken since the two pressures are the same at P_F. p_c is the saturated vapour pressure of ice at the surface of the condenser and P_0 the saturated vapour pressure of water vapour in the dried layer before the flash, assuming a 10°C rise in temperature above that of sublimation.

Velocity v can be determined from permeability by gas expansion (see Section 4.4, equation 4.29), thus

$$v = \frac{L}{t'_0} = \frac{KA}{2{\cdot}303\ V}. \tag{7.5}$$

In equation (7.5), A is the area of sublimation and V the volume of vapour sublimed during the period of condensation pumping. In practice, this period is about half the cycle period so that V, determined from the specific volume for the weight of ice sublimed in each cycle, is half; the volume V has also to be corrected for an average temperature in the dried layer of −11·5°C.

Taking, for example, data by Greenfield (1972) from his experiment 29-1 the parameters chosen are: $p_c = 1{\cdot}33$ Pa (limiting pressure for condensation at about −60°C), $P_0 = 253{\cdot}3$ Pa (saturated vapour pressure of water at about $-11{\cdot}5°C$), $A = 0{\cdot}24\ m^2$, $V = 2{\cdot}56\ m^3$, $V_{ch} = 0{\cdot}384\ m^3$, $K = 0{\cdot}0035\ m^2\ s^{-1}$, $s_p = 0{\cdot}01\ m^3\ s^{-1}$ ($0{\cdot}0123\ m^3\ s^{-1}$ initial displacement), $x = 0{\cdot}16$ cm.

For mechanical pumping, initially from a pressure of 933 Pa to a flash pressure of 253·3 Pa, the pressure curve is given by $P = 933 \exp(-0{\cdot}032t)$ and thereafter by $P = 253{\cdot}3 \exp(-0{\cdot}026t)$.

For the mechanical-condensation pumping, the pressure curve is given by $P = 1{\cdot}33 + (253{\cdot}3/2)[\exp(-0{\cdot}026t) + \exp(-0{\cdot}089t)]$.

Figure 7.1 shows the calculated pressure curve plotted against the experimental values (experiment 29-1), and it is seen that the theory is quite satisfactory in spite of the fact there is an initial excess pressure at the flash point which has not been taken into account in the present theory. Perhaps it could be repeated by another exponential that would die away rapidly.

It is observed that the combined pumping rate is at least double the mechanical pumping rate.

7.3 Iterative analysis of cyclic process

By applying an iterative analysis to the complex problem of cyclic-pressure freeze-drying some of the problems inherent in analytical solutions are avoided. Once the pressure effects on heat and vapour transfer are considered in detail, the equations that result are non-linear, and only analogue or numerical solutions are possible unless a number of simple assumptions are made. Some of these are inherent to the analyses (Sections 7.2 and 7.4).

Middlehurst and Parker (1972) have adopted an analogue solution for the sublimation rate dG/dt, by including the effect of the pressure variation on thermal conductivity and on diffusion coefficient in the porous layer (Fig. 7.2). They have shown the effects of gas expansion and finite pumping resistances on their results (Fig. 7.2(C)).

For the νth node of the dry layer the basic equation is

$$2\Delta x \rho_d c_d \frac{dT_\nu}{dt} = \frac{\lambda_d(p)}{2\Delta x}(T_{\nu-1} - T_\nu) - \frac{\lambda_d(p)}{2\Delta x}(T_\nu - T_{\nu-1}). \qquad (7.6)$$

For the heating cycle they proposed $dG/dt = 0$ as the boundary condition at the interface and, for the pumping cycle, the equations at the interface given by

$$\Delta H \frac{\mathrm{d}G}{\mathrm{d}t} = \frac{\lambda_\mathrm{d}(p)}{\Delta x}(T_\nu - T_\mathrm{i}),$$
$$\frac{\mathrm{d}G}{\mathrm{d}t} = \frac{D(p)(p_\mathrm{i} - p)}{2\nu\,\Delta x}, \quad (7.7)$$

where $\lambda_\mathrm{d}(p)$ and $D(p)$ are known functions of the pressure p. Similar terms with λ_f and $T_{\nu+1}$ may be added to the equations for back-face heating (conduction to the bulk of the solid). The solution involves taking values of λ_d and D at each point in the pressure range to find T_i

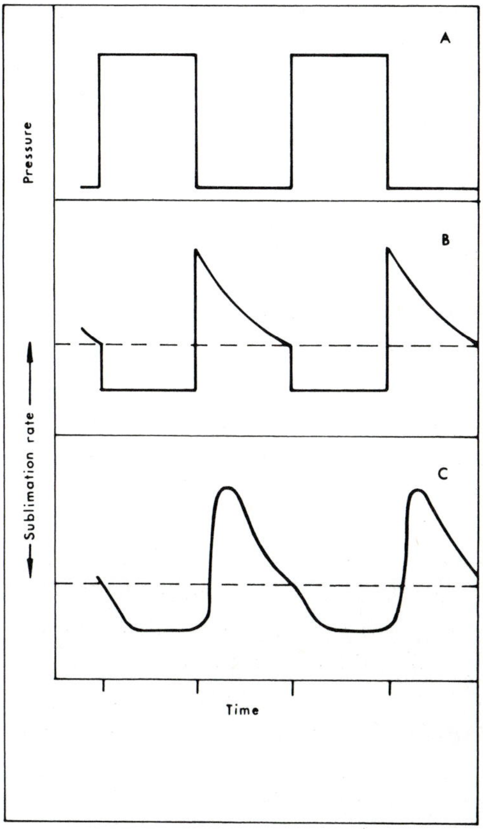

FIG. 7.2. A, square pressure wave; B, real effects—above datum line for basic equation; C, real effects—above datum line for equation with expansion and pumping terms.

and $\mathrm{d}G/\mathrm{d}t$, which gives values within 15 per cent of those measured in practice.

Greenfield (1972) sought numerical solutions to determine the pressure and temperature variations; his analysis is presented in Appendix 7.II.

His model (Fig. 7.6, Appendix 7.II) contains a slab of material of thickness $2L$, with sublimation occurring within it at an interface between the dried layer and the frozen layer. It is assumed that the heat of sublimation is conducted through the dried layer to the interface and that water vapour is removed to the surface countercurrent to the heat flow. The principal equations have been summarized by Mellor and Greenfield (1972) as follows:

Pressure variation in dried layer

a. High-pressure period:

$$\frac{\mathrm{d}G}{\mathrm{d}t}=0 \tag{7.8}$$

b. Low-pressure period:

$$\left.\frac{\mathrm{d}P}{\mathrm{d}t}\right|_{x=j}=\frac{\partial}{\partial x}\left[D_{\mathrm{K}}\frac{\partial P}{\partial x}+D_{\mathrm{v}}P\frac{\partial}{\partial x}(P+P^{*})\right]\Bigg|_{x=j}+\frac{x}{X}\frac{\mathrm{d}X}{\mathrm{d}t}\left.\frac{\partial P}{\partial x}\right|_{x=j}, \tag{7.9}$$

$$\left.\frac{\mathrm{d}G}{\mathrm{d}t}\right|_{x=j}=\left[D_{\mathrm{K}}\frac{\partial P}{\partial x}+D_{\mathrm{v}}P\frac{\partial}{\partial x}(P+P^{*})\right]\Bigg|_{x=j}. \tag{7.10}$$

Similar equations apply to a system having appreciable amounts of residual gas.

Temperature profile

a. Dried layer:

$$\left.\frac{\mathrm{d}T}{\mathrm{d}t}\right|_{x=j}=\left[\frac{1}{H}\frac{\partial}{\partial x}\left(\lambda_{\mathrm{e}}\frac{\partial T}{\partial x}\right)-cG\frac{\partial T}{\partial x}-E\right]\Bigg|_{x=j}-\frac{x}{X}\frac{\mathrm{d}X}{\mathrm{d}t}\left.\frac{\partial T}{\partial x}\right|_{x=j}, \tag{7.11}$$

b. Frozen Layer:

$$\left.\frac{\mathrm{d}T}{\mathrm{d}t}\right|_{x=j}=\frac{\lambda_\mathrm{f}}{c_\mathrm{f}\rho_\mathrm{f}}\left.\frac{\partial^2 T}{\partial x^2}\right|_{x=j}+\frac{L-x}{L-X}\frac{\mathrm{d}X}{\mathrm{d}t}\left.\frac{\partial T}{\partial x}\right|_{x=j}. \tag{7.12}$$

This equation applies for the case of back-face heating.

Interface condition

$$H_\mathrm{e}\left.\frac{\mathrm{d}T}{\mathrm{d}t}\right|_{x=X}=\left[\lambda_\mathrm{f}\frac{\partial T_\mathrm{i}}{\partial x}-\lambda\frac{\partial T}{\partial x}\right]\Bigg|_{x=X}-\bar{\rho}\,\Delta H\frac{\mathrm{d}X}{\mathrm{d}t}. \tag{7.13}$$

The above equations were solved numerically with appropriate initial and boundary conditions. Of the various solution methods used,

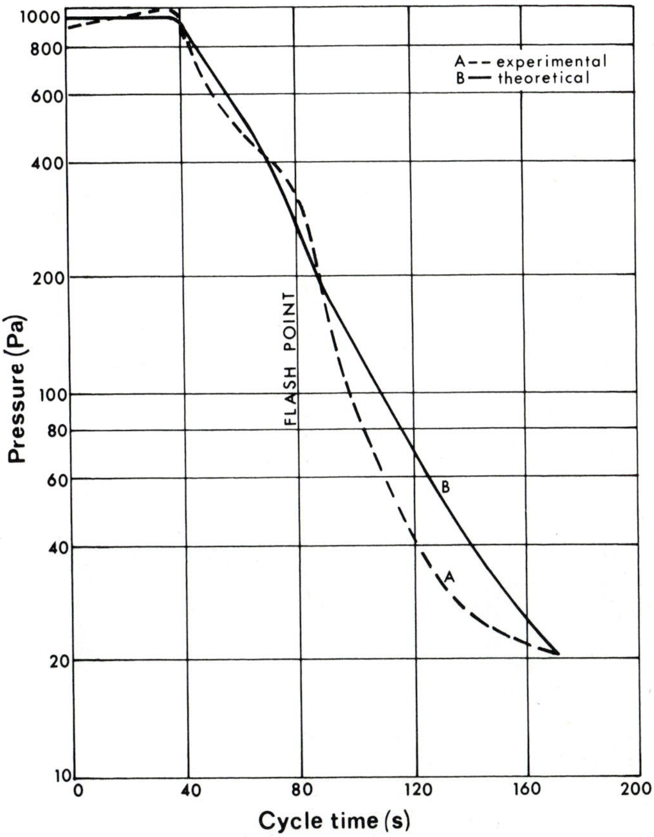

FIG. 7.3. Pressure variations in a cycle.

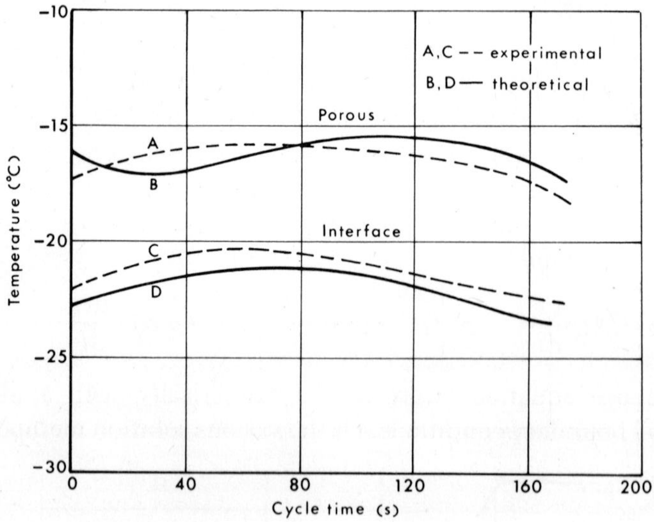

FIG. 7.4. Temperature profiles in a slab.

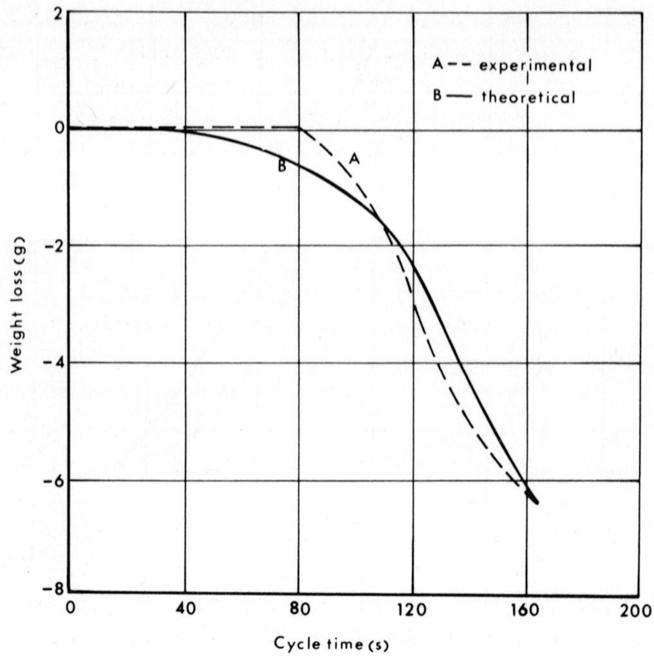

FIG. 7.5. Weight loss in a slab by sublimation.

the most successful was found to be discrete-space and discrete-time solutions by the Crank and Nicolson (1947) procedure. The principal equations can be written as matrices for which special recursive formulae were developed, so facilitating the calculations. Solutions to the various parts of the analysis (equations 7.8–7.12) were taken in turn; first, the pressure variation, next the sublimation rate and, then, the temperature profile. The value for the sublimation rate also serves to evaluate the interface condition (equation 7.13) since $G\,\Delta H = \bar{\rho}\,\Delta H(\mathrm{d}X/\mathrm{d}t)$.

A non-linear term in the radiation heating boundary condition (equation 7.II.11, Appendix 7.II) was expanded as a Taylor's series before applying another recursive formula.

Typical results for one cycle of the pressure variation, temperature profile and weight loss at the point when sublimation was taking place at a depth of 1·6 mm in a slab of 10 per cent gelatin, 13·7 mm thick, are shown in Figs 7.3–7.5. Agreement was found to be satisfactory (Greenfield, 1972). Figure 7.3 shows that a marked change in pumping rate occurs at 80 s and at about 267 Pa. This change manifests itself at about the same time in the weight loss curve (Fig. 7.5). From then on improvements in pumping rate and sublimation rate take place and are demonstrated by the results. Moreover, a pragmatic calculation of the change in enthalpy in the frozen region of the slab due to a temperature change of 2°C approximately (Fig. 7.4), shows that about one third of all heating for sublimation in a cycle is to be directly attributed to sensible heat stored before the flash.

7.4 Temperature wave: high-pressure heating and vapour flash cooling

At first sight one would expect high-pressure heating and vapour cooling to cause the amplitude of the temperature waves in the ice core and in the mantle to exceed the safe limits set for the process. Closer examination of the methods used to determine the temperature waves in the iterative analysis (Section 7.3) showed, however, that these methods were not explicitly derived and could not be prescribed; consequently it was decided to have recourse instead to an analytical procedure.

The amplitude of the temperature wave in the ice core is far more important than that of the temperature wave in the mantle. The former can be estimated with the help of some simplifying assumptions and treating the heat transfer as a "steady-state" problem over a single cycle. The total heat flux to the ice interface by conduction through the dry mantle in the high-pressure part of the cycle is then balanced by the heat removed by sublimation of ice. The heat flux at the ice interface can be represented by the terms of a Fourier series. However, the main interest is in the fundamental, that is the term with the same period as the cyclic process. That this term is dominant is confirmed by the shape of the experimentally observed temperature curves.

The heat flux at the ice interface can be represented by

$$Q(t) = Q_0 \sin(\omega t + \varepsilon'), \tag{7.14}$$

where $Q_0 = 1/\tau \oint \dot{m}\, \Delta H\, dt$, $\omega = 2\pi/\tau$, τ = period of the pressure cycle.

The solution for the temperature wave in the ice core, at a depth x, is given by Carslaw and Jaeger (1959) as

$$T = \frac{\alpha^{1/2}}{\lambda\omega^{1/2}} Q_0 \exp[-x(\omega/2\alpha)^{1/2}] \sin[\omega t + \varepsilon' - \pi/4 - x(\omega/2\alpha)^{1/2}], \tag{7.15}$$

and its amplitude is given by

$$T_{\text{amp}} = \frac{\alpha^{1/2}}{\lambda\omega^{1/2}} Q_0 \exp[-x(\omega/2\alpha)^{1/2}]. \tag{7.16}$$

In a typical example taken from Greenfield's (1972) data the sublimation rate $\dot{m} = 1{\cdot}88 \times 10^{-4}$ kg m^{-2} s^{-1} when the ice thickness is 10·5 mm and $\tau = 170$ s.

Data for ice are as follows:

$$\lambda = 2{\cdot}22 \text{ W m}^{-1} \text{ K}^{-1}$$

$$\alpha = 1{\cdot}24 \times 10^{-6} \text{ m}^2 \text{ s}^{-1}$$

$$\Delta H = 2800 \text{ kJ kg}^{-1}.$$

Thus, $Q_0 = 1{\cdot}88 \times 10^{-4} \times 2800$ kJ m^2 s^{-1}.

Upon substitution of the data in equation (7.16) the amplitude of the temperature wave at the interface ($x=0$) is found to be 1·4°C. This agrees fairly well with the measured value of 1·2°C.

The equations show that the temperature wave is attenuated with position in the ice; at a depth of 1·8 mm where it was measured the amplitude was 0·85°C. The calculated value is 1·1°C.

In the example of cyclic-pressure freeze-drying the estimated steady-state temperature gradient across a mantle, 1·6 mm thick, for the heat flux in a cycle is about 17°C, which is roughly the difference in prescribed temperatures for the surface of the mantle and for sublimation at the ice surface.

In conclusion, experience shows that freeze-drying of foods by the cyclic method is satisfactory if the surface temperatures are prescribed at around zero and the ice interface temperatures at around −20°C, at which point the temperature fluctuations are approximately ±1°C.

7.5 Time-constant: optimal pressure cycle

Assume, with each flash of vapour, that the following conditions apply:

1. The vapour pressure inside the product is greater than that outside, and lags behind when the pressure is varied.
2. The vapour has a high exit velocity and appreciable kinetic energy.
3. The vapour is accelerated along the pores of the dry product.
4. Work done on the vapour as it flows through the porous product is against viscous drag.

These conditions satisfy Bernoulli's law for fluid motion, which states that the rate of work equals the rate of change of kinetic energy plus the rate of change of viscous energy. Cancelling v in each energy term leads to a differential equation of "hydrodynamic" relaxation:

$$A\,\Delta p = m_{\mathrm{cyc}}\frac{\mathrm{d}v}{\mathrm{d}t} + \mu A'\frac{\mathrm{d}v}{\mathrm{d}r}, \tag{7.17}$$

where A is the area of the drying surface, m_{cyc} the mass of ice sublimed per cycle, v vapour velocity, μ vapour viscosity, A' internal pore surface and r equivalent pore radius.

7.5.1 TIME-CONSTANT

Separating the variables in equation (7.17) and assuming that $\mathrm{d}v/\mathrm{d}r = v/r$, the time-constant β_1 is given by

$$\beta_1 = \frac{\mu A'}{m_{cyc} r}. \tag{7.18}$$

The value of β_1 increases continuously with the advance of each cycle, because m_{cyc} and A' vary and have to be estimated in terms of the maximum number of cycles N_{max} required.

As m_{cyc} varies only slightly, a mean value,

$$\bar{m}_{cyc} = \frac{m_{tot}}{N_{max} \varepsilon LA}$$

may be taken. But the accessible internal pore surface A' available to vapour flow in the Nth cycle varies greatly and is related to N/N_{max}. Normally, $A' = 2n\pi r L_e$ for a cylindrical model of porous materials. For the case of cycling,

$$A' = \frac{2n_{cyc}\pi r L_e N}{N_{max}},$$

and as accessible internal pore surface per unit volume of the void space εLA, then

$$A' = \frac{2n_{cyc}\pi r \varepsilon L_e N}{N_{max} \varepsilon LA},$$

where the porosity $\varepsilon = n\pi r^2 L_e/LA$ and n_{cyc} is the effective number of accessible pores at the Nth cycle.

Now $L_e/L \simeq 2$ for many freeze-dried materials; this, and an additional factor 2 to account for freeze-drying slabs of materials from both sides, lead finally to equation (7.19):

$$\beta_1 = \frac{8\pi\mu\varepsilon L}{\bar{m}_{cyc}} \frac{n_{cyc} N}{N_{max}}. \tag{7.19}$$

In turn the number of continuous pores n_{cyc} becomes less important because a certain number of them, the shorter continuous ones and the smaller necking and intersecting ones, become blind to vapour flow as the depth of the porous dry material increases: that is, n is dependent on N. Assuming a parabolic distribution for n against N (parabola with origin at N_{max}, k and axis parallel to the n-axis) given by

$$(N - N_{max})^2 = 4a(n - k), \tag{7.20}$$

where $4a$ = latus rectum, $k = 0{\cdot}24\, n_1$ (number of truly accessible pores through to the Nth cycle, and n_1 the number in the first cycle).

The factor 0·24 seems appropriate since King *et al.* (1968) have shown that 69–79 per cent of all pores in freeze-dried cooked turkey are very small (~40 nm diameter). The remainder (~24 per cent) with measured diameters >175 μm would be sufficiently large to allow continuous accessibility through to the maximum number of cycles.

Now, the average water content of cooked meat is 50 per cent and the solids density is $1280\,\text{kg m}^{-3}$. Thus the estimated porosity $\varepsilon \simeq 0{\cdot}6$. Similarly, Mellor (1966b) obtained $\varepsilon = 0{\cdot}58$ and $r = 77\,\mu\text{m}$ for a cooked-meat product having an observed cylindrical pore structure.

Beginning with the first few cycles, it may be inferred from $\varepsilon = n\pi r^2 L_e/LA$ that the fractional amount of open surface for drying is 0·6. Equating this to the integrated area under the parabola defined by equation (7.20) leads to a set of tabulated values for the ratio $n_{cyc}N/N_{max}$ against N (Table 7.1). Estimates of the time-constant β_1 for three typical products which have been freeze-dried by the cyclic method are also given; they show the dependence of β_1 on N, increasing by factors of 1·5, 2·1 and 2·5 for cooked meat, apple and egg respectively over the middle range of cycling, beginning after the first ten (faster cycles) to ten before the end of cycling. Estimates from Greenfield's (1972) results give a factor of 2·75. The viscosity of water vapour $\mu = 0{\cdot}88 \times 10^{-5}\,\text{N s m}^{-2}$.

The time-constants have been used further to derive the high- and low-pressure times considered in Section 6.2; they are the values after the completion of a fifth of the maximum number of cycles. Although during the initial cycles, values were used at about half these, no real attempt was made in practice to vary the time-constants in accordance with Table 7.1.

7.5.2 OPTIMAL PRESSURE CYCLE

For an optimal cycle the duration of each low-pressure part should be so related to the duration of each high-pressure part as to ensure maximum freeze-drying efficiency.

The solution of equation (7.17) is

$$v = \frac{A\,\Delta p}{m_{cyc}\beta_1}[1 - e^{-\beta_1 t}]. \tag{7.21}$$

Since the vapour transfer varies exponentially with time, a limit must be set if the low-pressure part of the cycle is not to continue

TABLE 7.1
Freeze-dried product data and derived time-constants with $L = 15$ mm, $A = 0{\cdot}2\ m^2$

Product	N	n_{cyc}	$\frac{n_{cyc}N}{N_{max}}$	β_1
Cooked meat:	1	1000	20	0·003
$\varepsilon = 0{\cdot}58$	10	723	145	0·021
$r = 77\ \mu m$	20	508	203	0·030
$(L_e/L)^2 = 3{\cdot}37$	40	261	209	0·031
	50	231	231	0·034
Apple:	1	800	10	0·002
$\varepsilon = 0{\cdot}88$	10	665	83	0·015
$r = 87\ \mu m$	20	529	132	0·024
$(L_e/L)^2 = 4{\cdot}33$	40	338	169	0·031
	60	227	170	0·031
	80	185	185	0·034
Egg:	1	1200	12	0·003
$\varepsilon = 0{\cdot}76$	10	1015	102	0·023
$r = 99\ \mu m$	20	868	174	0·040
$(L_e/L)^2 = 1{\cdot}93$	40	609	244	0·056
	80	314	251	0·058
	100	277	277	0·064

indefinitely. A reasonable time t_2 for the latter was arrived at by allowing the flow rate to reach, say, 10 per cent greater than its steady-state value. A similar assumption was made for the heat input (Section 6.2).

Appendix 7.I Analysis of flash flow

The fundamental equations of d'Arcy and Fick for gaseous transport in transient state can hardly be expected to apply to transport accompanying sudden expansion or rarefaction of a gas, or to the pressure reduction caused by condensation pumping following a vapour flash. The extant analysis of a surge in the transport of some physical property, considered as an initial disturbance by applying the well-known fundamental equations, is not without a lot of complication. When the disturbance is, however, repetitive or consists of a succession of sudden fluctuations, exact equations for these are almost unknown or have not been fully realized.

The following analysis yields the transport equations covering a sudden fall in pressure in the vapour space. Factors taken into consideration include (1) the rate of change in the energy balance for one half-cycle in pressure, and (2) the rarefaction which occurs when the vapour is flashed off in a sublimation process.

Pressure equation for a condensing vapour

Consider a condenser lying opposite a certain volume of vapour which has suddenly flashed into the chamber through a block of porous material, causing a sudden fall in pressure in the cycle. A closed adiabatic system is assumed, at least momentarily during the flash. The rate at which the condensing vapour gives up its contraction energy and gains kinetic energy as it moves through the porous material from the sublimation front towards the condenser equals the rate of work, thus:

(Rate of contraction energy − rate of kinetic energy
= Rate of work)$_{\text{cycle}}$

$$V'\left(\frac{\partial P}{\partial t}\right)_{x=0} - V'\left(\frac{\partial P}{\partial t}\right)_{x=X} - \frac{\mathrm{d}}{\mathrm{d}t}\left(\frac{1}{2}mv^2\right) = \frac{\mathrm{d}}{\mathrm{d}t}\int_0^X (P-P')\,\mathrm{d}x$$

$$= \frac{\mathrm{d}}{\mathrm{d}t}\int_0^X P\,\mathrm{d}x - AP_0\frac{\mathrm{d}}{\mathrm{d}t}\int_0^X \mathrm{d}x. \tag{7.I.1}$$

In equation (7.I.1) the first term represents the contraction energy above a datum at $x=0$ and can be expressed as an integral having a second derivative with respect to x and t as

$$V'\int_0^X \frac{\partial^2 P}{\partial x \partial t}\,\mathrm{d}x = V'\int_0^X \frac{\partial}{\partial x}\left(\frac{\partial P}{\partial t}\right)\mathrm{d}x = V'\left(\frac{\partial P}{\partial t}\right)_{x=X} - V'\left(\frac{\partial P}{\partial t}\right)_{x=0}. \quad (7.I.2)$$

Also in equation (7.I.1) the second term can be reduced, giving

$$\frac{\mathrm{d}}{\mathrm{d}t}\left(\tfrac{1}{2}mv^2\right) = m\dot{v}v = AvP_1, \quad (7.I.3)$$

since, initially, rate of change of momentum $m\dot{v}$ equals the impressed force AP_0.

Reducing the integrals gives

$$A\frac{\mathrm{d}}{\mathrm{d}t}\int_0^X P(x,t)\,\mathrm{d}x - AP_0\frac{\mathrm{d}}{\mathrm{d}t}\int_0^X \mathrm{d}x = A\int_0^X \frac{\partial P}{\partial t}\,\mathrm{d}x - 0AP(0,t)$$
$$+AvP(X,t) - AvP_0$$
$$= A\int_0^X \frac{\partial P}{\partial t}\,\mathrm{d}x + AvP - AvP_0, \quad (7.I.4)$$

where $\dot{X}=v$.

Substituting from equations (7.I.1), (7.I.2), (7.I.3) and (7.I.4) yields

$$V'\int_0^X \frac{\partial^2 P}{\partial x \partial t}\,\mathrm{d}x + AvP + A\int_0^X \frac{\partial P}{\partial t}\,\mathrm{d}x = 0, \quad (7.I.5)$$

which, after differentiation with respect to x to remove the integrals, gives

$$V'\frac{\partial^2 P}{\partial x \partial t} + Av\frac{\partial P}{\partial x} + A\frac{\partial P}{\partial t} = 0. \quad (7.I.6)$$

Taking $x = V'/A$ as the distance in the contracting vapour, the pressure equation for the condensing vapour is given by

$$x\frac{\partial^2 P}{\partial x \partial t} + v\frac{\partial P}{\partial x} + \frac{\partial P}{\partial t} = 0. \quad (7.I.7)$$

If $\Lambda = x(\partial^2/\partial x \partial t) + v(\partial/\partial x) + \partial/\partial t$ denotes an operator then the fundamental equation for rarefaction is $\Lambda P = 0$. This equation has a unique solution given by

$$P = P_0 \, e^{-vt/x}. \tag{7.I.8}$$

Nomenclature

A	area of cross-section
m	mass of vapour produced by vapour flash
P	pressure in vapour space
P_0	initial pressure in vapour space
t	time of transit
V'	volume of vapour space
v	vapour velocity
x	distance
X	distance traversed by the vapour
Λ	$x(\partial^2/\partial x \partial t) + v(\partial/\partial x) + \partial/\partial t$, operator

Appendix 7.II Iterative analysis of flash flow

In order to elucidate the cyclic-pressure process, Greenfield (1972) derived a set of partial differential equations from basic theory for the pressure variation and another set for the temperature profile, and solved them numerically by an iterative procedure using finite differences for either continuously moving or discretely moving boundary models. Greenfield's analysis also produced a set of equations for the sublimation rate, a simplified derivation of which for one of his models is as follows.

Continuously moving boundary model

Consider, during a pressure cycle, a continuously moving boundary model with the ice interface moving across the grid lines of a network with a spacing which divides up a partly dry slab of material, and with pressure and temperature within it depending on distance and time.

The mesh spacings in the porous and frozen regions are Δx and Δx_i respectively. The porous region is assumed to be divided up into r equal increments and with the interface lying always on a mesh line. The grid spacing $\Delta x = X/r$. For q increments to the centre line, there will be $(q-r)$ increments each of width $(L-X)$ in the frozen region. Then the grid spacing here will be $\Delta x_i = (L-X)/(q-r)$.

An example, with $q = 2r = 10$ is shown in Fig. 7.6.

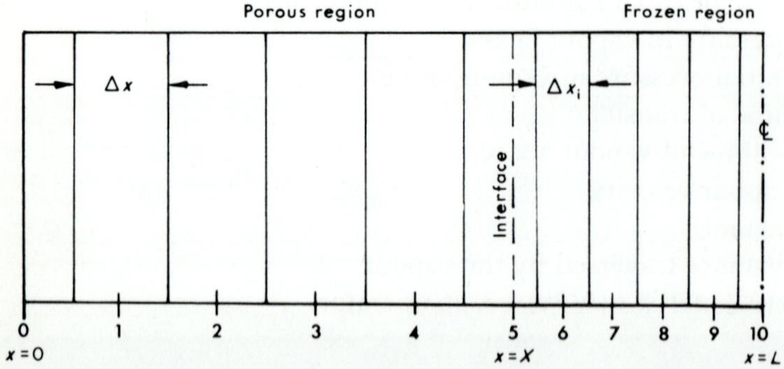

FIG. 7.6. Model of network for a continuously moving boundary model.

At grid line $x = j$, the temporal derivatives of pressure and of temperature in terms of the partial derivatives with respect to x and t are

$$\left.\frac{\mathrm{d}P}{\mathrm{d}t}\right|_{x=j} = \left.\frac{\partial P}{\partial t}\right|_{x=j} + \frac{x}{X}\frac{\mathrm{d}X}{\mathrm{d}t}\left.\frac{\partial P}{\partial x}\right|_{x=j} \tag{7.II.1}$$

$$\left.\frac{\mathrm{d}T}{\mathrm{d}t}\right|_{x=j} = \left.\frac{\partial T}{\partial t}\right|_{x=j} + \frac{x}{X}\frac{\mathrm{d}X}{\mathrm{d}t}\left.\frac{\partial T}{\partial x}\right|_{x=j} \tag{7.II.2}$$

where

$$\left.\frac{\mathrm{d}x}{\mathrm{d}t}\right|_{x=j} = \frac{x}{X}\left.\frac{\mathrm{d}X}{\mathrm{d}t}\right|_{x=j},$$

for the position of the interface.

Gas properties

It is assumed that the pressure of the mixture (vapour and residual gas) behaves ideally, that is, $P/\rho = RT/M$ and $(\partial P/\partial T)_\rho = P/T$.

The sublimation rate is given by $G=\rho(\mathrm{d}x/\mathrm{d}t)$. Since it depends on distance and time, it may be used to evaluate the large expansion in volume of the sublimed vapour. Expressed as equivalent heat units the expansion E is given by

$$E=T\left(\frac{\partial P}{\partial T}\right)_{\rho}\frac{\partial}{\partial x}\left(\frac{\mathrm{d}X}{\mathrm{d}t}\right), \tag{7.II.3}$$

Substituting the three expressions from above gives

$$\begin{aligned} E &= \frac{RP}{M}\frac{\partial}{\partial x}\left(\frac{GT}{P}\right) \\ &= \frac{R}{M}\left(G\frac{\partial T}{\partial x}+T\frac{\partial G}{\partial x}-\frac{GT}{P}\frac{\partial P}{\partial x}\right)\bigg|_{x=j} \end{aligned} \tag{7.II.4}$$

on differentiating for grid line $x=j$.

Pressure variation

The rate of change of pressure is found by applying the equation of continuity

$$\frac{M}{RT}\frac{\partial P}{\partial T}+\frac{\partial G}{\partial x}=0$$

and an expression for the vapour flow rate in a gas mixture

$$G=-\frac{M}{RT}\left[D_{\mathrm{K}}\frac{\partial P}{\partial x}+D_{\mathrm{v}}P\frac{\partial}{\partial x}(P+P^*)\right]$$

due to Hewitt and Sharratt (1964) to equation (7.II.1), to yield

$$\frac{\mathrm{d}P}{\mathrm{d}t}\bigg|_{x=j}=\frac{\partial}{\partial x}\left[D_{\mathrm{K}}\frac{\partial P}{\partial x}+D_{\mathrm{v}}P\frac{\partial}{\partial x}(P+P^*)\right]\bigg|_{x=j}+\frac{x}{X}\frac{\mathrm{d}X}{\mathrm{d}t}\frac{\partial P}{\partial x}\bigg|_{x=j}. \tag{7.II.5}$$

Boundary conditions

As there is negligible diffusion, the vapour is assumed to be saturated and in equilibrium through the dry material when the pressure is high and rising rapidly to a uniform level at each grid line:

$$P = p_s^\circ, \qquad 0 \leq x \leq X, \qquad 0 \leq t \leq t_1. \tag{7.II.6}$$

During the pumping cycle the low pressure decreases exponentially as

$$P = P_0 \, \mathrm{e}^{-C(t-t_2)}, \qquad x = 0, \qquad t_1 < t < t_2. \tag{7.II.7}$$

At the interface

$$P = p_s^\circ, \qquad x = X, \qquad t > 0. \tag{7.II.8}$$

Temperature profile

The equation for the temperature profile is found after combining the energy balance equation for the gaseous phase in the porous region:

$$\varepsilon c\rho \frac{\partial T}{\partial t}\bigg|_{x=j} = \frac{\partial}{\partial x}\left(\lambda \frac{\partial T}{\partial x}\right)\bigg|_{x=j} - cG\frac{\partial T}{\partial x}\bigg|_{x=j} - T\left(\frac{\partial P}{\partial T}\right)_\rho \frac{\partial}{\partial x}\left(\frac{\mathrm{d}X}{\mathrm{d}t}\right)\bigg|_{x=j}$$

attributed to Bird *et al.* (1960) with the conduction equation for the solid phase of the porous region:

$$(1-\varepsilon)c_\mathrm{d}\rho_\mathrm{d}\frac{\partial T}{\partial t}\bigg|_{x=j} = \frac{\partial}{\partial x}\left(\lambda_\mathrm{d}\frac{\partial T}{\partial x}\right)\bigg|_{x=j}$$

and then substituting in equation (7.II.2) to yield

$$\frac{\mathrm{d}T}{\mathrm{d}t}\bigg|_{x=j} = \frac{1}{H}\left[\frac{\partial}{\partial x}\left(\lambda_\mathrm{e}\frac{\partial T}{\partial x}\right) - cG\frac{\partial T}{\partial x} - E\right]\bigg|_{x=j} + \frac{x}{X}\frac{\mathrm{d}X}{\mathrm{d}t}\frac{\partial T}{\partial x}\bigg|_{x=j}, \tag{7.II.9}$$

where $H = \varepsilon c\rho + (1-\varepsilon)c_\mathrm{d}\rho_\mathrm{d}$, and λ_e = effective thermal conductivity.

In the frozen region the normal equation applies for back-face heating and is given by

$$\frac{\mathrm{d}T}{\mathrm{d}t}\bigg|_{x=j} = \frac{\lambda_\mathrm{f}}{c_\mathrm{f}\rho_\mathrm{f}}\frac{\partial^2 \mathrm{T_f}}{\partial x^2}\bigg|_{x=j} + \frac{L-x}{L-X}\frac{\mathrm{d}X}{\mathrm{d}t}\frac{\partial T_\mathrm{f}}{\partial x}\bigg|_{x=j}. \tag{7.II.10}$$

Boundary conditions

i. Surface:

$$T = T_\mathrm{s}, \qquad x = 0, \qquad t > 0$$

$$-\lambda_\mathrm{e}(\partial T/\partial x) = \sigma F_1(T_\mathrm{h}^4 - T_\mathrm{s}^4), \qquad x = 0, \qquad t > 0 \tag{7.II.11}$$

(radiation heating).

ii. Centre line:

$$\frac{\partial T}{\partial x} = 0. \qquad (7.II.12)$$

iii. Interface: The heat balance gives

$$H_e \frac{\partial T}{\partial t}\bigg|_{x=X} = \left[\lambda_f \frac{\partial T_f}{\partial x} - \lambda \frac{\partial T}{\partial x}\right]\bigg|_{x=X} - \bar{\rho}\,\Delta H \frac{dX}{dt}. \qquad (7.II.13)$$

iv. Other important boundary conditions are: saturated vapour pressure varying as the sublimation temperature, and the effective conductivity as the pressure according to the relation $\lambda_e = \lambda_{g,o}(1 + C/P)^{-1}$ derived by Harper (1962).

Nomenclature

c specific heat of vapour
c_f specific heat of ice phase
c_d specific heat of porous solid
$\bar{c} = \{\Delta x[\varepsilon c + (1-\varepsilon)c_d] + \Delta x_i c_f\}/\Delta x_e$
C constant
D_K Knudsen diffusion coefficient
D_v Diffusion coefficient (viscous flow with slip)
E expansion of vapour
F_1 viewing factor (radiation)
G sublimation rate
$H = \varepsilon c\rho + (1-\varepsilon)c_d\rho_d$, sensible heat quantity
$H_e = \bar{c}\bar{\rho}\Delta x_e$, effective heat quantity
ΔH latent heat of sublimation
j grid line number
L distance to centre line
M molecular weight of vapour
P pressure
p_0 initial vapour pressure
p_s° saturated vapour pressure of ice
P^* residual gas pressure
R gas constant
t time
t_1 high-pressure time

t_2 low-pressure time
T temperature in porous region
T_f temperature in frozen region
T_s surface temperature
T_h heater temperature
x depth in porous region
x_f depth in frozen region
X depth of interface
Δx spacing in porous region
Δx_i spacing in ice region
$\Delta x_e = \Delta x + \Delta x_i$, effective spacing
ε porosity
λ thermal conductivity of vapour in porous solid
$\lambda_{g,o}$ thermal conductivity of free gas
λ_e effective thermal conductivity of gas-porous solid
λ_f thermal conductivity of frozen material
λ_d thermal conductivity of porous solid
ρ vapour density
ρ_d density of porous solid
$\bar{\rho} = (\Delta x \rho_d + \Delta x_i \rho_f)/\Delta x_e$
ρ_f density of ice phase
σ Stefan's constant

8 Drying Plant and Equipment

8.1 Drying plant features and operation

Based on a patented invention by Mellor (1967b), several semi-industrial "cyclic" freeze-drying plants and as many pilot-scale units have been constructed and operated in Australia. One of these is a pilot unit (Fig. 8.1) of an advanced design in which freeze-drying can be carried out using a vacuum pumping system open to the atmosphere, or, alternatively, using a closed pumping system where helium gas is circulated at reduced pressure. Usually, one of the industrial plants has a rated capacity of 180 kg water removal per batch run of 12 h and another has 340 kg. The rated capacity of the pilot units is 15 kg per 12 h. An increase in capacity rating of 20 to 25 per cent is possible with the industrial plants and up to 45 per cent with one of the pilot units.

One of the industrial plants (Fig. 8.2) will now be described in detail.

8.1.1 DRYING CHAMBER

A drying chamber, 1·4 m diameter by 3·5 m long, is made of 6·4 mm thick mild steel and is reinforced by steel rings attached to the outside shell. Two of these rings, at the ends of the chamber, are machined flat in order to seal against O-rings in grooved flanges of dished end covers. The end cover at the back is bolted in position, whereas the one at the front forms the door to the drying chamber and is supported from an overhead rail with rollers as shown in Fig. 8.2. A guide on the bottom of the door ensures the O-ring can be seated when the door is in the closing position and prevents the O-ring from pinching in the

FIG. 8.1. A cyclic-pressure freeze-dryer in which helium is used as the heat exchange gas. (Courtesy of Commonwealth Scientific and Industrial Research Organization, Division of Food Research, North Ryde, N.S.W., Australia.)

sliding position. Four lever-action catches seal the door of the chamber ready for evacuation. The inside of the drying chamber has been coated first with zinc and then with a hard-wearing heat resistant plasticized paint, to make cleaning easier and to prevent contamination. The outside has been painted with an industrial enamel. It is not necessary to insulate the shell except around the section where the refrigerated condenser is housed. In dry climates no insulation would be required at all, as on this chamber it serves only to prevent condensate formation. Two glass inspection windows are provided in the shell of the drying chamber and one in the sliding door.

FIG. 8.2. Semi-industrial cyclic freeze-drying system. (Courtesy of Australian Armed Forces Food Science Establishment, Scottsdale, Tasmania, Australia.)

The radiant heating platens are made from roll-bonded aluminium and have been anodized to a black matt finish. They are arranged horizontally in the drying chamber to interleave between trays of frozen product situated on a trolley in the loading position, and are located side-by-side in two sets of twelve with a narrow gap between them to permit the central vertical frame of the trolley to enter the chamber. Each set of platens is connected to its own headers on racks which permit them as a unit to be readily removed from the drying chamber for service.

Glycol is the medium used to heat the platens. It is continuously recirculated (Fig. 8.3) by a centrifugal pump through a shell-and-tube heat exchanger heated by steam which can be controlled at pressures up to 690 kPa. The fluid is not pressurized but is open to atmosphere in an overhead tank where it can expand when temperatures up to

150°C occur. Temperature control of the glycol is obtained by allowing some of it to by-pass the heat exchanger via a three-way valve, which is a modulating type fitted with an electric drive and responds to a temperature-sensing resistance element attached to one of the platens.

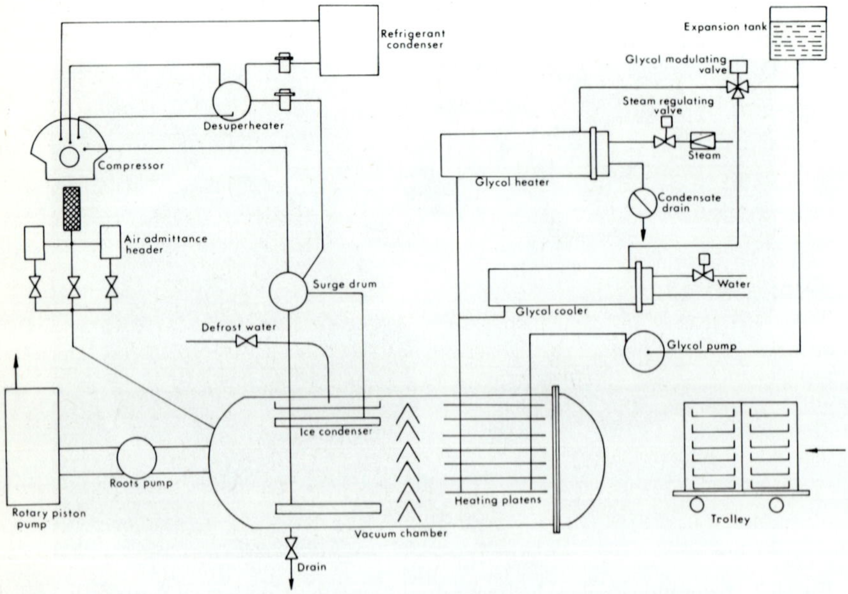

FIG. 8.3. Schematic diagram of cyclic-pressure freeze-dryer.

Platens, product trays and trolley occupy one half of the drying chamber and the other half houses the condenser, which is a flooded-type refrigerant evaporator equipped with surge drum above the drying chamber. The condenser can be defrosted by water flowing over the edges of open troughs at the top. Aluminium chevron baffles, fitted between the platens and condenser, shield the cold ends of the condenser coils from the hot platens; they are arranged so as not to restrict the flow of water vapour.

8.1.2 REFRIGERATION AND VACUUM SYSTEMS

The condenser is cooled by a compound refrigeration plant comprising a six cylinder ammonia compressor, an evaporative condenser and a flash-type interstage desuperheater. The compressor is driven by a 36 kW two-speed electric motor and is fitted with a capacity unloading mechanism. The condensed refrigerant is fed through a self-draining trap to the desuperheater. Then the cold refrigerant liquid at interstage pressure is taken from the desuperheater above a fixed level and is fed to the surge drum associated with the condenser in the end of the drying chamber.

The vacuum pumping system is connected through a large-bore pipe and flexible connector to the back-end cover of the drying chamber; it includes a Roots blower backed by an oil-sealed rotary piston pump, both driven from separate 5·4 kW electric motors. When the drying chamber is first closed after loading, it takes five to six minutes to evacuate to a pressure of 2·7 kPa. From this point the Roots blower reduces the vacuum pressure still further, down to 26·7 Pa in a few minutes, and then the pressure is repeatedly raised to 2·7 kPa by a controlled air-bleed into the drying chamber. There are air-bleeds on both the high and low pressures through suitably sized throttle valves in series with vacuum solenoid valves, which are energized one after the other to obtain the correct pressure cycle. The optimal periods for operation are predetermined for the product being dried (see Section 6.2), and for many foods these calculated periods have been verified experimentally.

8.1.3 AUTOMATIC CONTROLS

The aim in running the plant efficiently is to minimize the staff that is necessary to operate it, so that the plant can be left operating unattended except for periods of loading and unloading. Thus a programmed control system has been incorporated which, once it has been preset, will automatically control and reduce the heating and refrigeration requirements as the freeze-drying run continues towards the end.

It is necessary to reduce the temperature of the heating platens as the thickness of the dry layer about the frozen core increases to prevent the product surface temperatures from rising too high and the ice interface from going above the point of incipient melting. Therefore, to allow flexibility in maintaining low product surface and ice interface temperature levels, the control system chosen includes a set of four adjustable timers, from zero to four hours, and four adjustable potentiometers provided with temperature scales. This permits the temperature of the heating platens to be stepped down after each successive time period. A temperature-sensing element attached to one of the heating platens controls the modulating three-way valve in the hot glycol line, so that the temperatures of the heating platens are maintained at a level corresponding to the setting of the particular controlling potentiometer. When the timer associated with this particular potentiometer operates, control is passed to the next succeeding timer and its potentiometer, which is set at the lower temperature level, and so the control sequence continues through all stages until finally at the end of the freeze-drying run the hot glycol circulating pump is switched off and heating ceases. The modulating drive on the three-way glycol valve is also used to control a small step controller which, in turn, governs the operation of the compound refrigeration system through the unloading devices and speed control on the motor as the heating requirement in the drying chamber is diminished; the refrigeration requirement is correspondingly reduced.

For this freeze-dryer an automatic control system was designed and built up from a number of industrial-type control units. It has proved satisfactory in that it fails safe and is flexible for programming.

8.1.4 OPERATION

The prepared product is spread on trays in a freezing room, operated by a 5·4 kW refrigerating plant, and when frozen it is quickly transferred into the drying chamber. At the end of the drying period the dried product is removed and promptly packaged in sealed pouches; these can then be stored.

The trays, 0·8 m square, are made from aluminium, the bottoms being perforated to allow the vapour to escape from the lower as well as the upper surfaces of the product. Granular products are frozen in plain trays for freeze-drying, but liquid products are first frozen in plain trays lined with plastic sheets and then the frozen slabs are transferred to perforated trays ready for freeze-drying.

All temperatures are measured with thermocouples, some of which are inserted in the product before freezing. If experimental tests are required the trays may be weighed before being placed on the trolley.

The trolley is made from mild steel and is designed so that in the loading position in the chamber the trays are located evenly between the horizontal heating platens at about 89 mm centres. In another unit a closer separation has been obtained, at about 64 mm centres, and this gives a considerable increase in loading capacity. The trolley in the first version nominally holds 204 kg of food in 22 trays and can be loaded from both sides.

Before the drying run commences the condenser is cooled with the empty chamber closed in order to reduce its temperature in readiness for obtaining a good vacuum pressure at the beginning of the run. Now the loaded trolley can be pushed into the opened drying chamber and the door closed and secured with quick-action catches.

The vacuum pumps are started and, after a low enough vacuum pressure has been reached, the heating platens are brought up to the maximum temperature required for the particular product being freeze dried. In the early stages the heating platens can be operated at temperatures up to 55°C higher than the safe surface temperature for the product, but their temperatures must be reduced as the thickness of the dry layer around the frozen core increases, to avoid overheating.

8.2 Equipment design

8.2.1 CONDENSER

The condenser is a major vacuum-producing component of a freeze-drying installation since it is the pump for vapour removal. The design of condensing systems has not been extensively studied or developed; practical problems include the insulating effect of the growth in ice layer and the need for defrosting. Triggs (1944), appreciating these problems, designed a hollow refrigerated condenser in which the ice, as it formed, was continually scraped off and then ejected from the condenser intermittently through a vacuum lock. His condenser is not generally used today, although it seems to have been reliable. The effect of the ice layer on condenser temperature is examined theoretically in Section 3.5.

Rowe (1964b) discussed eight criteria as guidelines to the design and operation of static refrigerated condensers, as follows:

1. The temperature at the surface of the ice layer should be $-40°C$, at which point the saturated vapour pressure of ice would be approximately 13 Pa.
2. The backing pump should maintain in the condensing section of the vacuum chamber a partial air pressure equal to about one-tenth that of the saturated vapour pressure of ice on the condenser—that is, about 1·3 Pa.
3. The condensing system should comprise an array of coils or plates.
4. The longer the condensation process the denser the frost becomes, and it may even reach the density of clear ice (920 kg m^{-3}).
5. As a useful guide, the surface area of the condenser should equal that of the shelves in the drying system, although as little as half of this surface may be adequate.
6. For batch units the condenser can either be situated inside the drying chamber or in a separate external chamber; it can also form a pair of external chambers which are opened and closed alternately for defrosting purposes.
7. With most rudimentary condenser designs frosting is uneven—in particular, the frost accumulates on the front surfaces nearest the source of water vapour.
8. No overall solution has been found to the problem of frontal build-up of ice on the condenser. The problem can be avoided by condensing at a temperature as high as $-20°C$, but this temperature seems unsatisfactory for condensing in freeze-drying.

Experience with condensers for cyclic-pressure freeze-dryers indicates that once the water vapour has escaped from the product it must be removed quickly and efficiently from the vacuum chamber. Because the condensation load in cyclic-pressure freeze-drying is extremely variable, being negligible during the high-pressure time and very high during most of the low-pressure time (about 2 minutes), special care must be taken with condenser design. The important considerations are to allow a high vapour flow rate from the product down the the length of the coil system, and to ensure that the latent heat can be absorbed quickly by the refrigeration system. In a prototype industrial plant described by Mellor and Munns (1968), the coil system consists of five rows of piping, each 1·84 m long and with 230 mm separation, all connected to top and bottom headers 920 mm apart. In the space between the product and the condenser the end of each row of coils is shaded by a chevron baffle and at the far end there is a circular baffle in front of the pumping line connection. (The critical value of 230 mm for the separation between the rows was determined by comparing the conductance for the system with the actual vapour flow rate). For long channels, the conductance (Mellor and Lovett, 1968) is given by

$$F=\frac{A}{l}\left\{\frac{d_{\mathrm{H}}^2\bar{P}}{\mu_{\mathrm{g}}k_2'}+\frac{3\pi}{16}\frac{2-\delta_1}{\delta_1}\frac{d_{\mathrm{H}}\bar{v}_{\mathrm{m}}}{3k_1'}\right],\tag{8.1}$$

where F = conductivity, A = cross-sectional area of flow channel, l = channel length, d_{H} = hydraulic diameter (channel width = 2 × (area of channel)/(circumference of channel)), $\bar{v}_{\mathrm{m}}$ = mean molecular velocity, $\bar{P}$ = mean pressure, μ_{g} = gas viscosity, δ_1 = reflection coefficient of the gas molecules, k_1', k_2' = shape factors.

The constants $k_1'=0{\cdot}8$ and $k_2'=35$ are for a rectangular space between rows of coils having a breadth equal to four times the width. Taking 66·7 Pa for the operating pressure in the freeze-dryer, equation (8.1) gives $F=15{\cdot}6\times10^3\ \mathrm{m^2\,s^{-1}}$. This is five times the actual vapour flow, assuming an average ice condensation rate of 30 g s^{-1} at −20°C and taking 1100 l g^{-1} for the specific volume of the vapour.

Critical spacings for the more complicated baffle geometries were determined using the method of Levenson *et al.* (1961). In practice the design has given a working condenser on which ice builds up almost uniformly in both constant- and cyclic-pressure freeze-dryers.

The driving force for the cyclic process depends on the difference between the sublimation pressure and the falling chamber pressure in each cycle and not, as in the usual process, on vapour-pressure difference. Very low condenser temperatures are usually needed to enhance the gradient in the constant-pressure process. On the other hand, the cyclic process is far less dependent on very low condenser temperatures. Temperatures 10°C higher than with conventional freeze-dryers can be used and, in some instances, satisfactory freeze-drying has taken place when the average sublimation and condenser temperatures were about the same. Thus significant savings in running costs are possible owing to the higher condenser temperatures and to explain this at least theoretically, estimates (Table 8.1) show a

57 per cent decrease in energy for refrigeration matched by a 28 per cent increase in energy for pumping for the cyclic process. The estimated net decrease in energy works out at 40 per cent for residual gas and water vapour removal.

TABLE 8.1
Estimated energy requirements for freeze-drying processes, vacuum pumping and cooling

Type of operation	Drying time Total decrease (h)	(%)	Vacuum pumps Energy increase (kWh)	(%)	Refrigeration compressor Condenser temperature (°C)	Energy decrease (kWh)	(%)
Constant pressure	10·5	—	113	—	−45	454	—
Cyclic pressure:							
air	8·5	20	145	28	−36	193	57
helium	6·7	36					

8.2.2 VACUUM PUMPING SYSTEM

In modern freeze-drying plant the pressure in the system is usually reduced from atmospheric pressure to better than 13 Pa in 10 min. If the system is of considerable volumetric capacity a large single-stage pump is used for initial evacuation and another, smaller, one for holding the system at the required operating pressure. This is the case for bulk freeze-dryers in which the water vapour is removed by a refrigerated condenser; otherwise, for direct pumping systems where steam ejectors or rotary air-ballast pumps are used, only a single pump is used. The two pumps can be compounded by opening a valve to permit a further reduction in pressure after sublimation, if secondary drying is wanted. Lower pressures are often required in the later stages of drying, in order to reduce the residual water content of the product in a reasonable time.

A fast vacuum pumping system is needed for processes which depend on self-cooling effects, for low-temperature drying and for snap freezing of non-frothing substances. Further, it is essential to be able to reduce quickly the pressure to avoid surface melting of pre-frozen material when placed on warm shelves in the dryer.

Consider the time required to reduce the pressure in a chamber of a freeze-dryer, where its volume (free space) is about 0·7 m^3 and its internal surface area is 4·6 m^2 plus another 4·6 m^2 approximately for the condenser coil and heater platen surfaces. The surface area has to be considered because outgassing of internal surfaces below pressures of 1·3 kPa depends on it. The time t is given by

$$t = 2{\cdot}303\,\frac{V_{ch}}{s_p}\log\frac{P_a - p_c^\circ}{P - p_c^\circ}, \tag{8.2}$$

where V_{ch} = volume of vacuum chamber, s_p = displacement of the pump, P_a = atmospheric pressure, P = operating pressure, p_c° = vapour pressure equivalent to the saturated vapour pressure of ice at the temperature of the condenser.

Now $p_c^\circ < P \ll P_a$ and hence p_c° can be neglected. From equation (8.2), a pump with a displacement of 60 $m^3\,h^{-1}$ could reduce the pressure from 101 to 1·3 kPa in this chamber in about 3 min; and using a nomogram (Diels and Jaeckel, 1962) which gives the relationship between size of pump, total internal surface area and outgassing factor (= 133 μPa m s^{-1}) which is considered fairly strong since there are rough surfaces and leaks in the chamber, the time required to reduce the pressure from 1·3 kPa to 1·3 Pa would be about 7·5 min. Thus, a pump-down time of 10·5 min should be possible.

The holding pump should have a displacement of about one-sixth of that for the evacuation pump and should be capable of maintaining a pressure of 1·3 Pa in a perfectly dry vacuum chamber.

Vacuum lines should be of sufficiently large diameter and of minimum length, so as to limit resistances to gas and vapour flow. Where valves are used, these should also have a throat diameter the same as the vacuum lines. The design of vacuum lines is adequately described by Diels and Jaeckel (1962).

8.3 Cyclic process—comments and criticisms

The classical principles of freeze-drying are so well established that there is a certain scepticism at anything new. Since its debut in 1965, the cyclic process has aroused a number of comments and criticisms drawing into focus some of the basic tenets of freeze-drying, which are summarized as follows:

1. A firm belief in a vapour-pressure driving force based on the largest possible difference in temperature between evaporator and condenser.
2. A mistaken belief that a vacuum is essential to the maintenance of sublimation temperatures and that these should remain fixed at all times.
3. The necessity to work with a lower fixed pressure in secondary drying, in order to remove the last traces of residual moisture from the product.
4. The adage that "Energy for extra production is dearly bought", and that it is therefore wasted in the pressure cycles of the cyclic process.

8.3.1 TEMPERATURE AND PRESSURE CRITERIA

King (1970) in a review of freeze-drying of foodstuffs, discusses the cyclic process, pointing out that both heat and vapour transfer coefficients are highly pressure dependent and that the pressure change increases one whilst decreasing the other, and vice versa. This leads him to believe, from his curves of drying rate against absolute pressure (Fig. 8.4), that neither the thermal conductivity nor the

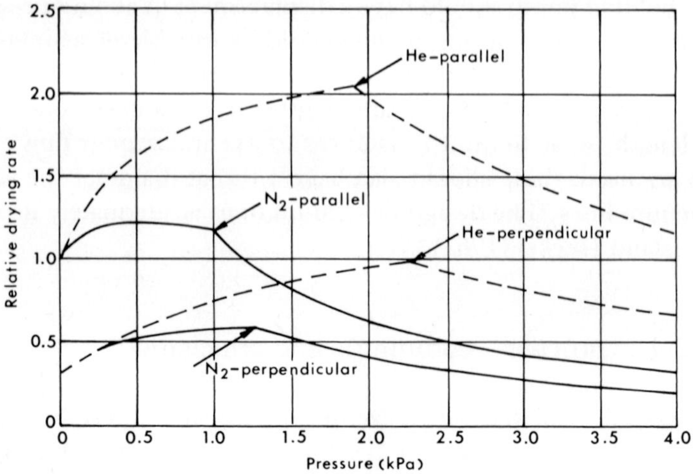

FIG. 8.4. Relationship between freeze-drying rates and absolute pressure for freeze-drying turkey (parallel and perpendicular meat fibres). Maximum surface temperature = 60°C; maximum sublimation temperature = −5°C. (After King, 1970.)

effective diffusivity will be at a maximum, and consequently improvements in drying rate will always depend on a compromise between the two coefficients. It is not appreciated that another and more important vapour transport mechanism is taking place simultaneously with the feeble diffusive flow during the pumping cycle. In Section 2.2.3 this has been called hydrodynamic flow, with a vapour flash commencing when the sublimation temperature is highest and occurs under an absolute pressure gradient. Thus the mechanism is outside the scope of point (1). Moreover, the effective vapour transfer coefficient for this would be commensurate with the sublimation pressure at the sublimation temperature.

King notes also that the sensible heat of the frozen layer increases during the high-pressure period when the vapour transfer is minimal, and this results in an increase in its temperature. This constitutes a warning of the possible danger of overheating and is a break with point (2). The answer is to operate the cyclic process under optimal conditions (Section 6.2). In another review, by Holdsworth (1968), fears have also been expressed about raising the pressure in a cycle too high, because of the possibility of melting the ice layer and causing pseudo-sublimation.

On a more optimistic note, regarding point (3) Meffert (1969) acknowledges that an intermittent pumping cycle in the final stages of drying would be advantageous as it would produce an evaporative cooling effect, facilitating vapour flow and thereby reducing the moisture content in the product. Such is in fact the case, as the drying rate curves (Fig. 6.4, Section 6.2) show the cyclic process for helium to be almost twice as fast as the normal process at this stage.

It should be mentioned here that in several early patents on evacuation of vessels containing adsorbed moisture and moist materials a method is described of breaking the vacuum with dry gas to atmospheric pressure once or twice before a final pumping-down stage achieves the desired vacuum (Kniepen, 1929; Takesaki, 1921).

8.3.2 PUMPING ECONOMY

In a detailed comment, Rowe (1970) examined the vacuum pumping requirements for pressure cycling. He has presumably in mind point (4) when he states that the extremes of the pressure in the cycle have been chosen for alteration in the drying chamber so that the lower will give optimal conditions of vapour flow through the dry layer (consistent with pumping economy) and the higher will give optimal conditions of heat transfer through the layer (also consistent with economy in pumping) to restore the lower pressure.

After discussing briefly optimal cycle times and continuous evacuation of the chamber against air admission by two alternately operated bleed-valves, Rowe considers, for example, a 1 m^3 chamber which can be pumped down from 2·7 kPa to 2·7 Pa in 1 min by a rotary vacuum pump (displacement 5 $m^3 h^{-1}$). The low-pressure bleed causes the pump-down time to be longer, and so either a slightly faster pump is chosen or this bleed is dispensed with, and it would not matter if the ultimate pressure were lower than 2·7 Pa. On the high-pressure side the displacement at 2·7 Pa would be unchanged and the high-pressure bleed would have to be about 21·5 l min^{-1} (STP). The chamber pressure would rise from 2·7 Pa to 2·7 kPa in slightly more than 7·5 s with the pump running.

No vacuum pumping difficulties would seem therefore to be involved in the cyclic process using air. Using helium is another matter, and Rowe prudently examines some of the problems which will arise if a process is contemplated in which the helium is recycled and becomes contaminated. These are: (1) recycling via an oil-sealed rotary vacuum pump, filter, absorber trap to remove moisture and reservoir; (2) limited recycling will be possible using an oil-free mechanical booster (Roots type) pump, but here the compression ratio of the pump is low (10 : 1). Rowe therefore envisages a booster pump without a backing pump to increase the compression ratio, connected to a reservoir which is several times larger than the drying chamber, to hold the helium temporarily until the commencement of the high-pressure part of the cycle.

Such a system as Rowe describes would be expensive and unwieldy and it is surprising that he has not considered employing an oil-free backing pump, of the modified carbon vane or the water-ring type, connected to a small reservoir which would only be a fraction of the size of the drying chamber.

King (1970) is critical of gas usage in the cyclic process, presumably on the grounds that it is wasteful. However, Kan and de Winter (1968) considered recirculating the gas within the vacuum chamber, but this would require circulating large volumes of gas at low constant pressure and some kind of blower working in a vacuum. The feasibility of the idea has, however, never been demonstrated.

8.3.3 POTENTIAL OF THE PROCESS

From a lengthy and tedious argument concerning net heat transfer in cycled and steady-state operations, King (1970) concludes that cyclic-pressure freeze-drying lacks the potential to become an attractive process.

For his analysis he compares the two operations by writing:

$$Q_{cyc} = \frac{1}{\tau l_d} \int_0^{\tau} \lambda(t)(T_s - T_i)\,dt \tag{8.3}$$

and

$$Q_{ss} = \frac{\lambda}{l_d}(T_s - T_i), \tag{8.4}$$

where Q_{cyc}, Q_{ss} = net heat-transfer rates for cycled and steady-state operations, τ = cycle period, T_s = surface temperature of dry layer, T_i = interface temperature, l_d = thickness of dry layer, t = time, λ = thermal conductivity.

In the integrand, the temperatures T_s and T_i depend on time directly, and the thermal conductivity λ depends indirectly on time because of the pressure.

King's argument is that $Q_{cyc} \equiv Q_{ss}$ and that therefore nothing is gained by cycling. This means that the difference $(T_s - T_i)$ on which the heating rate depends, would remain fixed even though T_s and T_i varied linearly about a mean value—a fallacy in interpretation, even though $(T_s - T_i)$ might conceivably vary in this way.

Greenfield (1972) has observed that T_s and T_i are out-of-phase sinusoidal temperature waves, with the amplitude of the first about twice that of the second, and that varies with pressure, which varies exponentially.

Consider next the example given by Mellor (1967b) which King uses. For a cycle with a 100-fold pressure variation (27 Pa to 2·7 kPa), assuming τ is twice the log cycle time, then by a formula similar to equation (4.30), Section 4.4, the log mean pressure equals 0·6 kPa.

The steady-state pressure was 53·3 Pa. At these pressures (Fig. 6.2, Section 6.2) for freeze-dried beef, $\bar{\lambda}(\tau) = 0{\cdot}04$ and $\lambda = 0{\cdot}033\ \mathrm{W\ m^{-1}\ K^{-1}}$.

Now $(T_s - T_i)$ minimum is about 33°C and the amplitudes ΔT_i and ΔT_s are 2°C and 4°C respectively.

Evaluating equation (8.3) by integrating two sine waves roughly 33°C apart, it is easily shown that $Q_{cyc} = 1{\cdot}56/l_d$, and from equation (8.4) $Q_{ss} = 1{\cdot}09/l_d$, giving an increase in heating rate of about 40 per cent or about 30 per cent for temperature waves in phase. In contrast, King has deduced by his analysis no increase for nitrogen and a 10–15 per cent increase for helium. Therefore, he suggested that there would be a negligible increase in drying rate over all pressure ranges for nitrogen and helium. Had King adopted a different method of averaging the heat input over a complete cycle, bearing in mind the length of the high-pressure period, and compared this with the heat input during constant-pressure operation, he would have obtained a different result, favouring cyclic-pressure operation (Greenfield and Mellor, 1974).

It will be noted that the range 30–40 per cent is for a pressure cycle increasing instantaneously and then falling back by pumping in a short finite time. Q_{cyc} could therefore be expected to increase still further for a ratio of high- to low-pressure intervals of 3 : 1, as was the case for freeze-drying beef.

King also states that entrainment of water vapour by the gas is likely to occur during pumping as well as agitation, a point well appreciated in view of some experiences with cyclic-pressure freeze-drying (fluidization of small granular particles having been observed as the vapour flashes over).

PART 3
APPLICATIONS

9 Laboratory Apparatus and Techniques

9.1 Introduction

As already mentioned in Chapter 1, freeze-drying originated because of a need to preserve biological specimens in the laboratory. Once a scientific curiosity, it took some time before the freeze-drying process came into use in practice; but the simple apparatus for freezing and subliming *in vacuo* soon gave way to more sophisticated freeze-dryers to meet the needs of microbiologists, virologists, histologists and chemists. The basic freeze-drying techniques, that influenced its early development, will now be dealt with. They fall into three types—distal, medial and proximal (Kekwick, 1953–54). In the distal type of unit (Fig. 9.1(a)) the condenser can be isolated to enable another adjunct process to be carried out in the evaporator chamber, and the vacuum can be broken with a perfectly dry gas. The medial type (Fig. 9.1(b)) is often fitted with a "bell jar" cover to permit the handling of specimens on the evaporator. The proximal type (Fig. 9.1(c)) is the ideal choice from the physical standpoint, since the close proximity of the evaporator to the condenser helps to accelerate the sublimation process. Rieutord (1960) has detailed the basic techniques.

9.2 Simple units with traps

The simplest type of laboratory freeze-dryer receives heat for sublimation from the walls of the vacuum chamber, and so the evaporator is just a tray or support for the specimen. Water vapour is trapped by a chemical absorber—for example, phosphorus pentoxide—or by a cold trap cooled by a mixture of "dry ice" (solid carbon dioxide) and alcohol. The vacuum chamber is evacuated through the trap to remove residual air which may have leaked in or come from the specimen during sublimation.

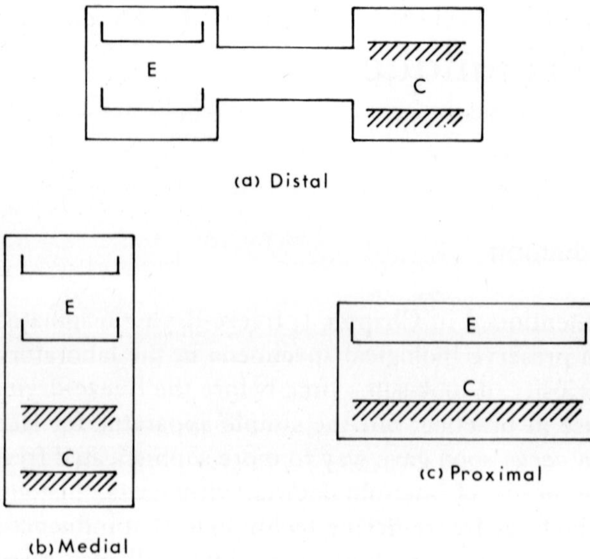

FIG. 9.1. Basic types of freeze-dryers. E, evaporator; C, condenser.

9.2.1 CHEMICAL TRAPPING TECHNIQUE

Barratt and Tatum (1950) described a simple laboratory technique for freeze-drying microorganisms which requires a minimum of equipment and materials. The sample is pipetted into a glass tube 18 mm diameter and 30 cm long closed at the bottom. A cotton plug is pushed down the tube to a point about 9 cm above the sample. A small amount of phosphorus pentoxide powder is placed on top and this is covered by a second cotton plug. The tube is partially immersed in a dry ice–alcohol mixture to freeze the sample, then evacuated with a vacuum pump and sealed off at the upper end. Next, the bottom end of the tube is placed in a cold bath (−5°C to −10°C). After about 4 hours the sample is dry and the tube is finally sealed at a point 5 cm above the sample. Many tubes can be processed simultaneously in this way. The technique is of the proximal type (Fig. 9.1(c)), except that it is characterized by a static vacuum.

9.2.2 COLD TRAPPING TECHNIQUE

Another simple freeze-drying technique, described by Holden and Scott (1951), employs a manifold holding 52 ampoules, the water vapour paths from all of which go to a common cold trap; this technique is essentially of the distal type (Fig. 9.1(a)) but with one condenser connected to many evaporators. The manifold and cold trap, made of 3 cm diameter glass tubing, are cooled with dry ice. The whole assembly is supported on a mobile table below which is a vacuum pump and the condenser can be lowered into the cooling mixture to trap the water vapour.

9.2.3 CENTRIFUGAL VACUUM-SPIN FREEZING TECHNIQUE

This technique was developed by Greaves (1944) to overcome a difficulty experienced with freezing by self-evaporation under vacuum of liquids such as protein solutions, plasma, broths and penicillin. Upon expelling dissolved gases, such liquids froth and bubble violently and then freeze with extreme suddenness. The difficulty can be avoided if the dissolved gases are removed by reducing the pressure carefully, but this is a tedious operation, and liquid explosions and erratic freezing may still occur.

In Greaves' method, samples in tubes are spun at 1450 rev. min^{-1} as in a centrifuge on a radius of 25 mm, the mouths of the tubes being inclined at 5° from the vertical towards the condenser (Fig. 9.2). The result is a frozen wedge of sample against the walls of the tubes and this gives an adequate surface for drying.

9.3 Freeze-dryers for microbiological research

9.3.1 AMPOULE FREEZE-DRYER

Many types of dryer for microbiological research have been described (Fry, 1954), but none meet fully the requirements for precise study of the effects of physical conditions during drying (Greiff and Pinkerton, 1954). However, Mellor and Ohye (1960) have described an ampoule freeze-dryer (Fig. 9.3) in which the conditions for freeze-drying

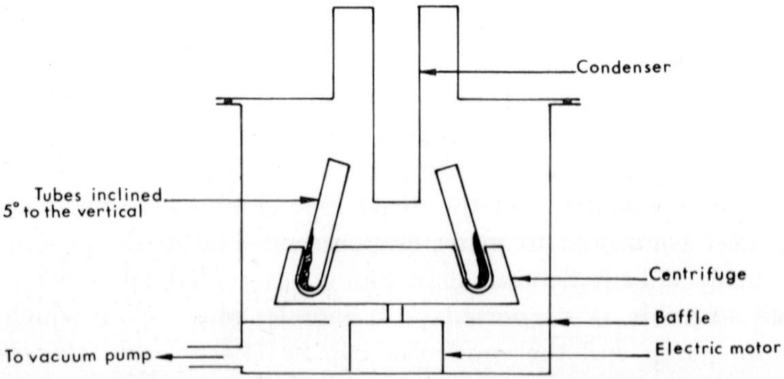

FIG. 9.2. Schematic diagram of the centrifugal vacuum-spin freeze-dryer.

suspensions of microorganisms can be varied, measured and controlled. In particular, provision is made for prefreezing the specimens *in situ* by placing them in thermal contact with the condenser with precise temperature control during drying. The rate of freezing may be varied by simple means and the specimen temperature may be controlled at any desired value in the range −50°C to 20°C. The dryer has a capacity of 72 ampoules and is characterized by simplicity of construction and operation.

9.3.2 METHOD OF COOLING

Specimens consisting of approximately 0·2 ml of suspension in each tube are frozen *in situ* before the chamber is evacuated in order to avoid frothing and the formation of surface films. Fairly rapid cooling and freezing are achieved by raising a tube holder containing the specimens into contact with the bottom of a cold condenser block so that cooling takes place by conduction along the metal. The tube holder is raised by means of a push-rod, shown in Fig. 9.4. The rate of cooling may be reduced when desired, by inserting a plastic disk of suitable thickness between the base of the condenser and the tube holder.

9.3.3 THEORY

Consider cooling of the volume element $2\pi rD\,\mathrm{d}r$ in the circular tube holder (Fig. 9.4) by conduction along a radius to the centre region in contact with the cold block. Using cylindrical co-ordinates and with the polar angle $\theta = 0$, the heat flux across the face at radius r is $-2\pi r\mathrm{D}\lambda_{\mathrm{me}}(\partial T/\partial r)$ and the heat flux across the face at radius $r+dr$ is

$$-2\pi rD\left[\frac{\lambda_{\mathrm{me}}\partial T}{\partial r}+\frac{\lambda_{\mathrm{me}}\partial}{r\partial r}\left(\frac{1}{r}\frac{\partial T}{\partial r}\right)\right].$$

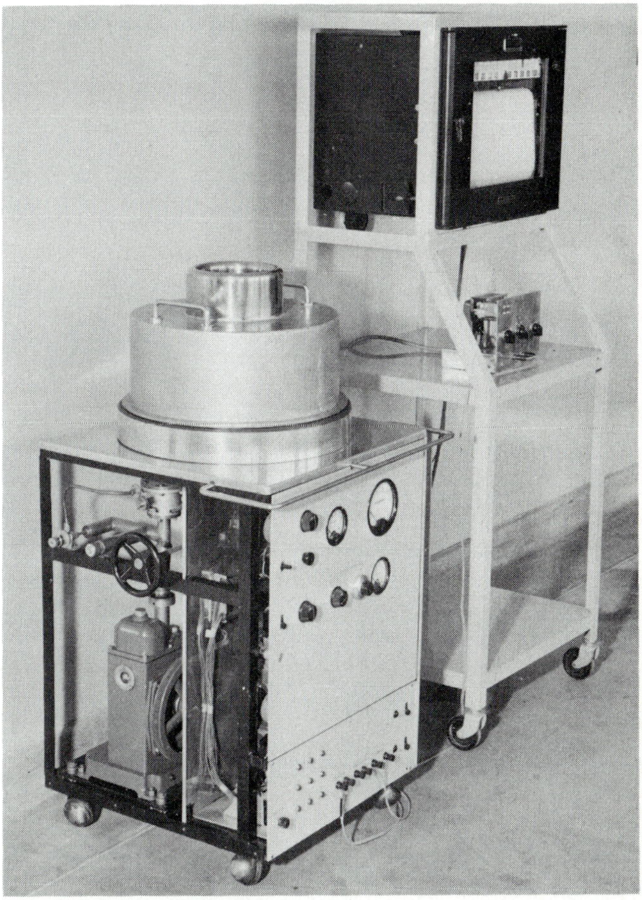

FIG. 9.3. An ampoule freeze-dryer for microbiological research. (Courtesy of Commonwealth Scientific and Industrial Research Organization, Division of Food Research, North Ryde, N.S.W., Australia.)

Thus the heat lost by the element becomes

$$H = 2\pi D \,\mathrm{d}r \lambda_{me} \frac{\partial}{\partial r}\left(\frac{1}{r} \frac{\partial T}{\partial r}\right). \tag{9.1}$$

The heat gained by the element at its top and bottom surfaces from the surroundings at zero temperature is given by

$$H = 4\pi r h T \,\mathrm{d}r, \tag{9.2}$$

where λ_{me} = thermal conductivity of the metal, and h = surface heat transfer coefficient.

Equating equations (9.1) and (9.2) gives

$$\frac{1}{r}\frac{\mathrm{d}}{\mathrm{d}r}\left(\frac{1}{r}\frac{\mathrm{d}T}{\mathrm{d}r}\right) - \frac{2hT}{\lambda_{me} D} = 0, \qquad a < r < b. \tag{9.3}$$

And with the boundary conditions:

$$\begin{aligned} T &= T_0, \qquad r = a, \\ \frac{\mathrm{d}T}{\mathrm{d}r} &= 0, \qquad r = b. \end{aligned} \tag{9.4}$$

According to Carslaw and Jaeger (1959), equations (9.3) and (9.4) have a solution for the temperature at $r = b$, and for the periphery of the tube holder given by

$$T = T_0 \frac{I_0(\mu b)K_1(\mu b) + K_0(\mu b)I_1(\mu b)}{I_0(\mu a)K_1(\mu b) + K_0(\mu a)I_1(\mu b)}. \tag{9.5}$$

The heat removed from the tubes in cooling, as given by equation (9.2), becomes

$$H = \frac{4\pi a h T_0}{\mu} \frac{I_1(\mu b)K_1(\mu a) - K_1(\mu b)I_1(\mu a)}{I_0(\mu a)K_1(\mu b) + K_0(\mu a)I_1(\mu b)}, \tag{9.6}$$

where I_0, I_1, K_0 and K_1 are modified Bessel functions and $\mu^2 = 2h/\lambda_{me}D$.

Equations (9.3) and (9.5) apply also to heating. In this case the tube holder is detached from the cold block and its base is heated electrically. For heating, the final temperature in the tube holder is really a balance between the self-cooling effect of the vacuum on the suspending fluid in the tubes and the heat input from the heater.

For the design of the tube holder, a copper disk was chosen 3·2 mm thick by 330 mm O.D., and 114 mm I.D. for the central portion in contact with the cold block on one side and the heater on the other. To position and provide thermal contact between tubes and holder, 72

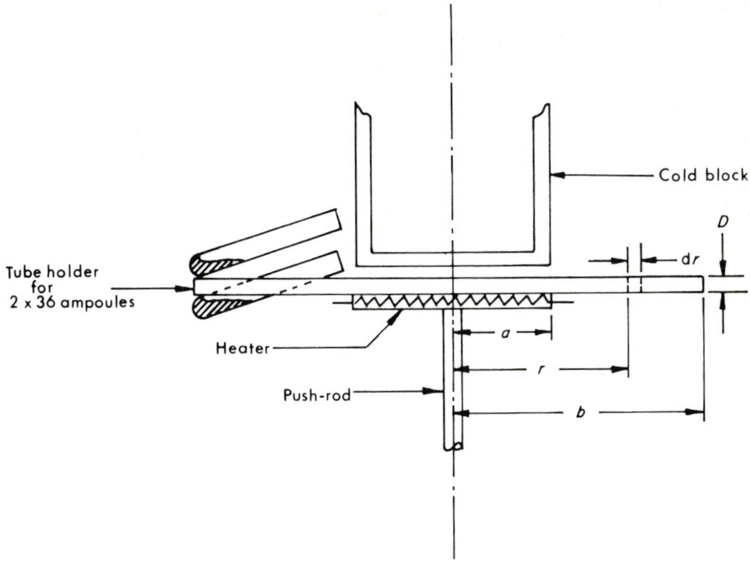

FIG. 9.4. Diagram of heater and cooler for ampoule freeze-dryer.

copper wells 10 mm diameter by 25 mm, in two layers, were radially spaced around the periphery and positioned at an angle of 5° to the horizontal, with the mouths of the tubes pointing toward the centre. The heater was a "Pyrotenax" cable. The total loaded weight of the assembly was 2·5 kg.

Assuming a cold block temperature of −70°C and the specimens cooling from ambient to −30°C, the cooling time from equation (9.5) after integrating for t was estimated at 10 min.

Typical temperature histories during cooling and freezing of 0·2 ml of suspension in 0·5 ml M-NaCl in tubes in the apparatus are shown in Fig. 9.5. Cooling times of 12 min for direct contact and 20–25 min with a plastic disk 1·6 mm thick were obtained. The curves show no plateaux at the freezing point of the solution, no doubt because the thermal capacity of the system is large compared with the heat of fusion of the liquid suspension. It is evident from these curves that fairly rapid freezing can be achieved and that substantial reductions in the rate of freezing can be obtained by inserting suitable disks of plastic material between the tube holder and the condenser.

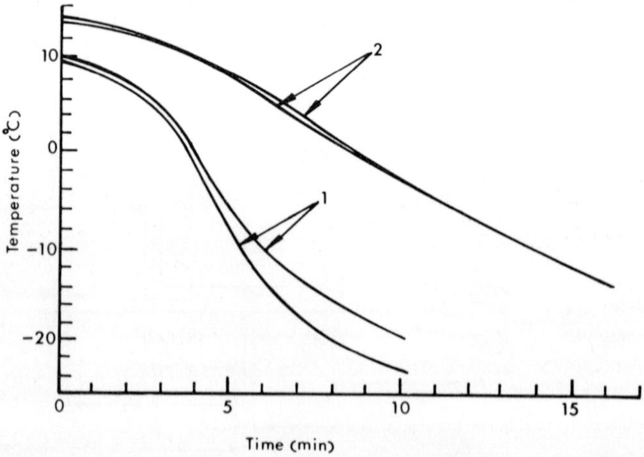

FIG. 9.5. Typical freezing curves for 0·2 ml of suspension in 0·5 M-NaCl with condenser temperature less than −70°C. (1) Tube holder in contact with condenser. (2) Tube holder in contact with condenser through plastic disk 1·6 mm thick.

9.3.4 MEASUREMENT AND CONTROL OF TEMPERATURES

The temperatures of the specimens in six tubes, of the tube holder and of the condenser are measured with thermocouples. These, and a short-circuited lead, are connected through a uniselector switch which may be operated either manually by a push-button or automatically by means of an automatic interval timer for use with a potentiometer or recorder. In order to minimize errors due to thermal conduction along the wires and to permit precise placing of the junctions at the deepest part of the layer of suspension in the tube, 40 SWG enamelled copper–constantan thermocouple wires are used in conjunction with heavier 26 SWG lead wires. The correct placing of junctions is important so as to determine the end-point of sublimation from the measurements.

The temperatures of the specimens are controlled indirectly by controlling the temperature of the tube holder to ±1°C. The output of a thermocouple with its hot junction in thermal contact with the tube holder is amplified through a rapid-action and sensitive electronic controller, Fig. 9.6, to operate an on-off relay in the heater circuit. During the sublimation phase the temperatures of the specimens are generally below that of the tube holder. This is due to the evaporative cooling effect, which is a function of the residual air pressure in the

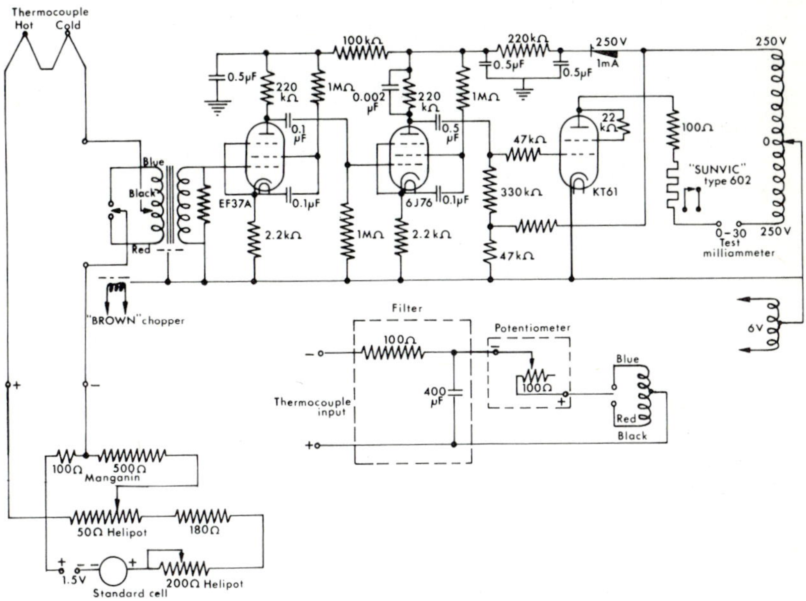

FIG. 9.6. Circuit diagram of electronic temperature controller (gain of amplifier = $2{\cdot}5 \times 10^5$).

drying chamber and of the properties of the suspension. Fairly close control of the amount of evaporative cooling was found feasible by manual adjustment of the bleed valve on the vacuum line. Control of the specimen temperature within the required operating limits could then be effected by heating the tube holder to maintain a constant temperature difference between it and the specimens.

Figure 9.7 shows typical temperature histories during drying of 0·2 ml suspension in 0·5 M-NaCl at two widely different temperatures, −6°C and −33°C, and indicates that in both experiments there was satisfactory control of the specimen temperature during the sublimation phase of drying. The rise in temperature at the end of this phase was quite sharp in the experiment, at −6°C. This type of curve was obtained when drying most types of suspension, whatever the temperature.

The curves for the −30°C to −33°C experiment in Fig. 9.7 fall into two groups, which are 3°C to 4°C apart in the sublimation phase. The higher-temperature curves were for tightly fitting tubes and the lower-temperature ones for loosely fitting tubes (the latter are generally used).

Figure 9.7 indicates that the temperatures of the specimens tend to be well below those of the tube holder during active sublimation. The

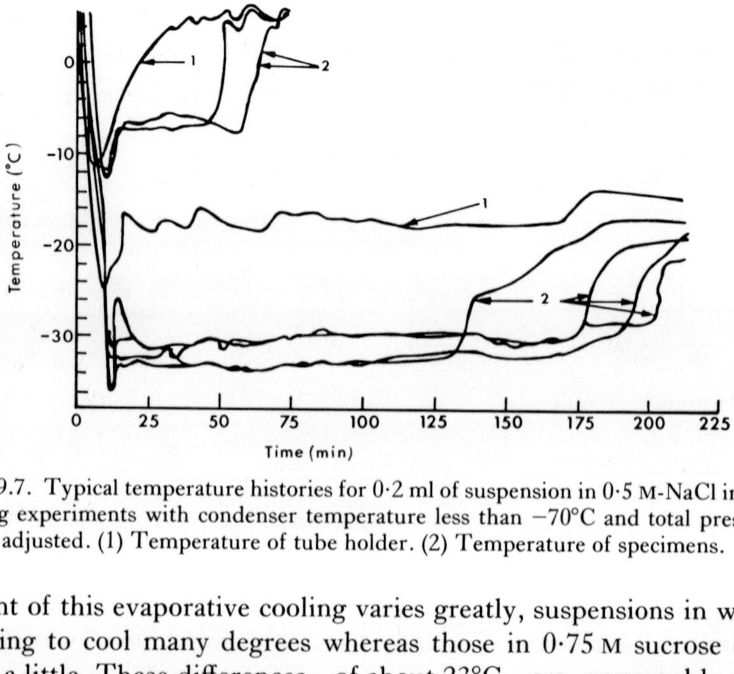

FIG. 9.7. Typical temperature histories for 0·2 ml of suspension in 0·5 M-NaCl in two drying experiments with condenser temperature less than −70°C and total pressure adjusted. (1) Temperature of tube holder. (2) Temperature of specimens.

extent of this evaporative cooling varies greatly, suspensions in water tending to cool many degrees whereas those in 0·75 M sucrose cool only a little. These differences—of about 23°C—are, presumably, due to differing resistances to the flow of water vapour within the specimens being dried and to the large thermal resistance, about 5·8 W m^2 K^{-1}, between the wells and the tubes at high vacuum.

It is desirable to be able to dry suspensions in different fluids at approximately the same temperature in one experiment, and two alternative methods of achieving this are available. Firstly, the thermal resistance of the apparatus may be reduced about thirty-fold by interposing a film of a suitable oil between the wells and the tubes. This method was found to reduce the difference in drying temperatures between suspensions in water and in 0·75 M sucrose when operating at a total pressure of 1·33 Pa to about 15°C. But this method alone will not reduce the temperature difference sufficiently. The second measure available is increasing the pressure of the residual air in the chamber, but this unfortunately reduces the rate of drying. The object must therefore be to find the most satisfactory compromise.

9.3.5 RADIATION HEATING EFFECTS

A common difficulty with small freeze-dryers is that radiation of heat from the warm walls of the drying chamber to the specimens being dried is relatively high, so that it is impossible to maintain the specimen at a constant low temperature when supplying electrical heat at a minimum rate. This difficulty is avoided in the present design by surrounding the tubes and tube holder by an envelope consisting of two concentric radiation shields.

When drying specimens with a very high internal resistance to the flow of water vapour it appears that radiation of heat from the walls of the dryer can, despite the radiation shields, be too great to allow the drying to proceed at −25°C or lower. This difficulty may be overcome by raising the tube holder so that it makes light contact, presumably through a layer of snow, with the condenser.

9.4 Developments in freeze-drying for histology

Freezing and drying of tissues for histological purposes is usually carried out on thin sections of plant or animal material at temperatures in the range −30° to −70°C. The process results in the development of a dry shell of honeycomb structure; consequently it takes much longer to dry a tissue than a liquid biological material of similar thickness.

9.4.1 MECHANISM

In a theoretical and experimental analysis of this effect and other factors controlling the rate of drying, Stephenson (1953, 1954) and Gersh and Stephenson (1954) considered first the formation of an ice crystal artifact during freezing of the tissue. They observed that after drying, embedding and staining, the tissue appeared to be a honeycomb formed by "walls" of protoplasm being compressed by fine ice crystals (up to a few microns in size) which separate out on freezing.

They attributed the slowness of drying to the obstruction which the honeycomb dry structure presents to escaping water molecules and they explained the mechanism of escape in terms of a diagram (Fig. 9.8) to illustrate possible pathways for the molecules. An escaping

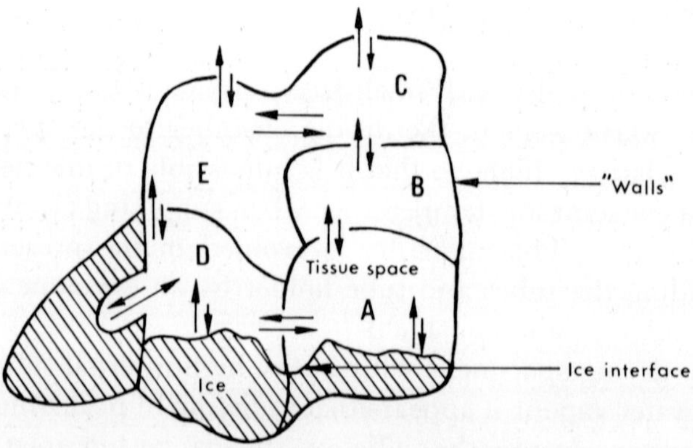

FIG. 9.8. Diagram of walls surrounding the subliming ice crystals. Length of arrows indicate the direction and strength of vapour flow. (After Gersh and Stephenson, 1954.)

molecule from an ice crystal enters a tissue space A partially occupied by that crystal and, after making numerous collisions with the walls of this space, is either recaptured by the ice crystal or passes through holes in the walls to neighbouring spaces. These may either be similar to A, such as D, or exterior spaces B, C, etc., but in every space the molecule makes further collisions with the walls until it escapes finally from the outside surface. For a molecule to escape there must be a pressure gradient between the saturated vapour pressure of ice p_i and the pressure at the surface of the tissue p_s. Pressures in A, B and C will be progressively less for net flow of water vapour but in A and D, E and C the average pressure will be the same, as there is no net flow of vapour between them.

9.4.2 DRYING RATE

Starting from classical sublimation theory, Stephenson derived an equation for the drying time of a piece of tissue in terms of the resistances in the dried shell and in the drying apparatus to the flow of water vapour. On the basis of Knudsen's relation for the sublimation rate $G = \alpha p_s(M/2\pi RT_s)^{1/2}$ the net flow through a dry shell 1000 layers of tissue cells thick could be expressed by

$$G_s < \frac{1}{1000}\alpha p_s\left(\frac{M}{2\pi RT_s}\right)^{1/2}, \tag{9.7}$$

where the subscript s denotes conditions at the subliming interface. In other words, for the resistance in the dried shell, if an ice crystal is assumed to have an average size of 1 μm, then the dried shell will have a thickness of 1 mm and the drying rate will be reduced by a factor of 1000 approximately. On the other hand, the resistance of the drying apparatus depends on whether the vapour flow through it is diffusive or free molecular. For simple tubular systems it can be calculated from well-known formulae.

Introducing the concept of escape probability f for the vapour molecules and assuming two components: a probability f_s that a water molecule reaches the surface of the dry shell without recapture by the ice interface, and a probability f_c that a water molecule reaches the vapour trap or condenser without re-entering the dried shell, it then follows that equation (9.7) can be applied more directly. The vapour flow rate in the presence of the dry shell is then given by $\alpha f_s p_s(M/2\pi RT_s)^{1/2}$ and is a maximum when there is no dry shell and $f_s = 1$. In the apparatus the flow rate is given by $af_c p_s(M/2\pi RT_s)^{1/2}$. The reciprocals of the three quantities for net flow conductance in series lead to

$$G = \alpha f_s p_s\left(\frac{M}{2\pi RT_s}\right)^{1/2}\frac{1}{1+f_s(1/f_c-1)}. \tag{9.8}$$

From equation (9.8) it is seen that the resistance of the apparatus is a minimum when $f_c = 1$ and $G = \alpha f_s p_s(M/2\pi RT_s)^{1/2}$ becomes the maximum drying rate in the presence of the dry shell. Stephenson calls f_s the "fractional drying factor" for a given dry shell thickness and interface temperature T_s. Determination of the resistances $1/f_s$, $1/f_c$, and the temperature T_s is therefore important.

9.4.3 RESISTANCES OF THE DRIED SHELL AND APPARATUS

The first resistance ($1/f_s$) is not easy to calculate but can be obtained by measurement; it is, for example, about 1000 for a dried shell of guinea pig liver 0·01 cm thick and about 3000 for 0·1 cm thick. In this range of shell thickness the fractional drying factor can be expressed by

$$f_s = \frac{a}{x^{\nu}}, \qquad (9.9)$$

where a = "drying coefficient" ($= 1{\cdot}2 \times 10^{-4}$), x = shell thickness, ν = "drying exponent" ($= 0{\cdot}55$).

Values for other examples might be expected to comply with this formula.

In some cases the second resistance ($1/f_c$) can be calculated. When the vapour flow in a tube is diffusive, $f_s \gg f_c$, the tube resistance is found from

$$\frac{1}{f_c} = \left[\frac{16\mu}{\pi p_s}\left(\frac{RT_s}{2\pi M}\right)^{1/2} \frac{A_i l}{f_s r^4}\right]^{1/2}, \qquad (9.10)$$

where μ = vapour viscosity, r = radius of tube, A_i = area of ice interface, l = length of tube.

When the flow is free molecular the resistance is found from

$$\frac{1}{f_c} = \frac{3}{8\pi} \frac{A_i l}{r^3}, \qquad (9.11)$$

9.4.4 INTERFACE TEMPERATURE

The sublimation rate in terms of the heat flux supplied to unit area of the interface is expressed by

$$G = \frac{Q}{\Delta H} = \frac{h}{\Delta H}(T_a - T_i), \qquad (9.12)$$

where h = overall heat transfer coefficient between the interface and surroundings, T_a = temperature of the surroundings (walls of drying chamber), T_i = temperature of the interface, ΔH = latent heat of sublimation.

The temperature difference between the surroundings and the interface, after substituting from equations (9.8) and (9.12), yields

$$T_a - T_i = \frac{\alpha f_s p_s}{h}\left(\frac{M}{2\pi R T_s}\right)^{1/2} \frac{\Delta H}{1 + f_s(1/f_c - 1)}. \qquad (9.13)$$

Stephenson (1953) found that with the wall of the drying chamber radiating at a temperature of −30°C to tissue with an appreciable dried shell thickness, the difference ($T_a - T_i$) was about 7·5°C; eventually the difference became less than 1°C with the wall at −50°C. From these data an approximate value of h can be calculated using equation (9.13).

Factors affecting the drying rate are: (1) wall temperature, (2) combined flow resistance of the dried shell and of the apparatus, and (3) heat transfer. These may be better understood by combining equations (9.8) and (9.13) and the Clapeyron–Clausius vapour pressure equation to give

$$G \simeq f\alpha p_a\left(\frac{M}{2\pi RT_a}\right)^{1/2} \exp\left[-\frac{\Delta HM}{RT_a^2}\frac{\Delta Hf}{h}\alpha p_a\left(\frac{M}{2\pi RT_a}\right)^{1/2}\right]. \tag{9.14}$$

Here, $\alpha p_a(M/2\pi RT_a)^{1/2}$ is the maximum sublimation rate at the temperature T_a of the walls of the drying chamber, $1/f = 1/f_s + 1/f_c - 1$ is the combined flow resistance, and $\Delta Hf/h$ is the ratio of the heat transfer and vapour flow resistances. However, Stephenson recommended calculating the value of G from equations (9.8) and (9.13), since ($T_a - T_i$) and h would be known by trial and error and not from the above equation.

9.4.5 DRYING TIME AND EFFICIENCY

The time required to dry a shell of thickness x is given by equation (9.15)

$$t = \rho \int_0^x \left(\frac{2\pi RT_s}{M}\right)^{1/2} \frac{1}{\alpha f p_s}\, dx, \tag{9.15}$$

where ρ is the specific gravity of the tissue and x the dry shell thickness. The minimum drying time is when $1/f_c = 1$.

Data attributed to Stephenson, on the drying time of slices of guinea pig liver as a function of temperature and thickness, are given in Fig. 9.9 and show the importance of these factors.

The drying efficiency for a given piece of tissue is defined as the ratio of the minimum to the actual drying time; it can be expressed in two ways, firstly, in terms of vapour pressures if temperature measurements are made or, secondly, in terms of the resistances in the dried shell and in the apparatus. The two expressions for efficiency η are:

$$\eta = 1 - \frac{p_c}{p_s} = \frac{1}{1 + f_s(1/f_c - 1)}. \tag{9.16}$$

From equation (9.16), since f_s is always small, η approaches a maximum value of 1 when the ratio of the resistances f_s/f_c becomes small,

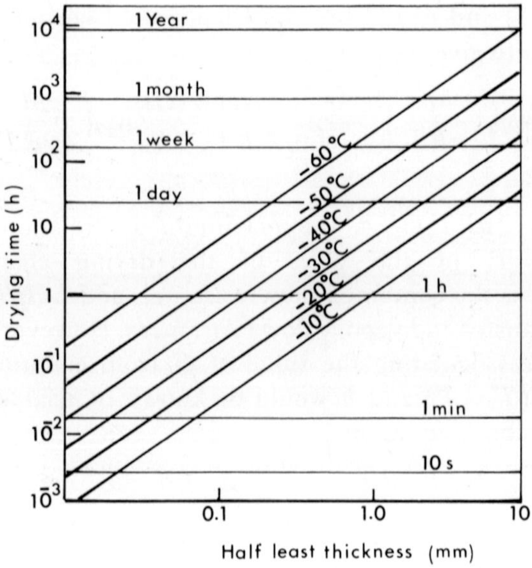

FIG. 9.9. Drying time of slices of guinea pig liver as a function of temperature and thickness (after Stephenson, 1953).

say, less than 0·1. Gersh and Stephenson comment that this would not be in agreement with many of the apparatus designs that have been described in the literature, largely because the high resistance of the dry shell in comparison with that of the apparatus has been ignored. They stress, however, that many authors have failed to specify the operating conditions of the apparatus and this adds to the confusion.

Stephenson (1954) computed a nomogram (Fig. 9.10) to show the relationship between efficiency, radius of the tube connecting the drying chamber and vapour trap, sublimation temperature and load factor. He gives, for example, $A_i = 100\ \text{cm}^2$, $1/f_s = 1000$, $l = 100$ cm

($f_sA_il = 10\ \text{cm}^3$), from which the radius required is found from the nomogram to be 2 cm if the flow is diffusive and 2.3 cm if it is free molecular, for an efficiency of 90 per cent and a sublimation temperature of −30°C. Now the mean free path λ can be expressed by

$$\lambda = 3 \times 10^{-3} \frac{f_c}{f_s p_s}, \tag{9.17}$$

which gives $\lambda = 0{\cdot}1$, and so in this case the flow is diffusive in a tube of radius 2 cm.

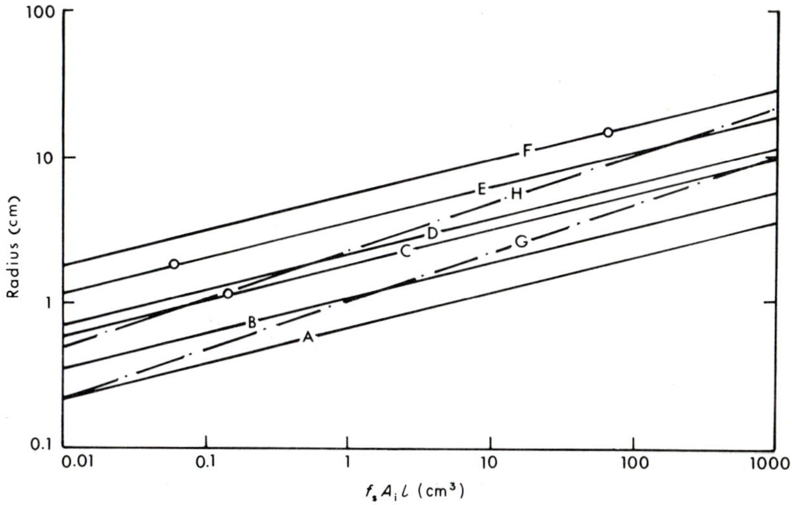

FIG. 9.10. Relationships between efficiency, radius of tube connecting drying chamber and vapour trap, sublimation temperature and load factor. —— Diffusive flow: $f_s/f_c = 0{\cdot}1$, (A) $T_s = -10$°C, (B) $T_s = -30$°C, (C) $T_s = -50$°C; $f_s/f_c = 0{\cdot}01$, (D) $T_s = -10$°C, (E) $T_s = -30$°C, (F) $T_s = -50$°C, —·— Free molecular flow: (G) $f_s/f_c = 0{\cdot}1$; (H) $f_s/f_c = 0{\cdot}01$. —o— Mean free path=radius. (After Stephenson, 1954).

To increase the efficiency from 10 per cent to 99 per cent for a given load and sublimation temperature, the radius must increase by a factor of 3·2 for diffusive flow and 2·2 for free molecular flow—hence this is a costly move. However, reducing the radius means a significant increase in drying time.

Loads of 1 cm³ or less are typical in practice and so tubing between 1 and 2 cm in radius will give good drying efficiencies, provided the length does not exceed 100 cm. As the flow is usually diffusive, bends in the tubing do not affect the efficiency. Glass equipment appears best for most purposes and stopcocks and valves should be adequately sized and prevented from freezing up.

9.4.6 TYPES OF APPARATUS

Several representative types of equipment that have been successfully used for histological studies are depicted in Figs 9.11–9.14; they

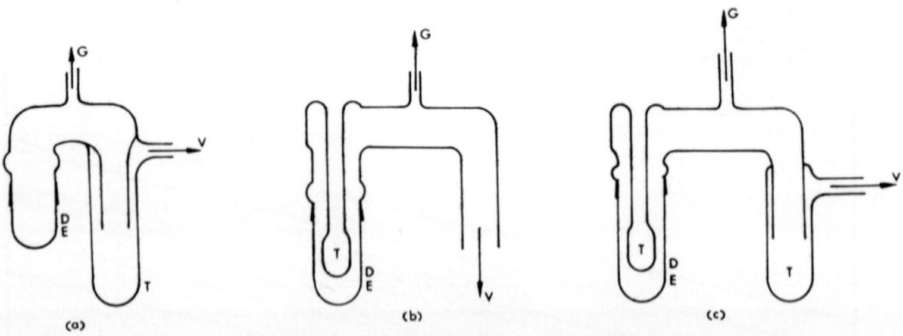

FIG. 9.11. Glick and Malmstrom's (1952) apparatus in three different forms.

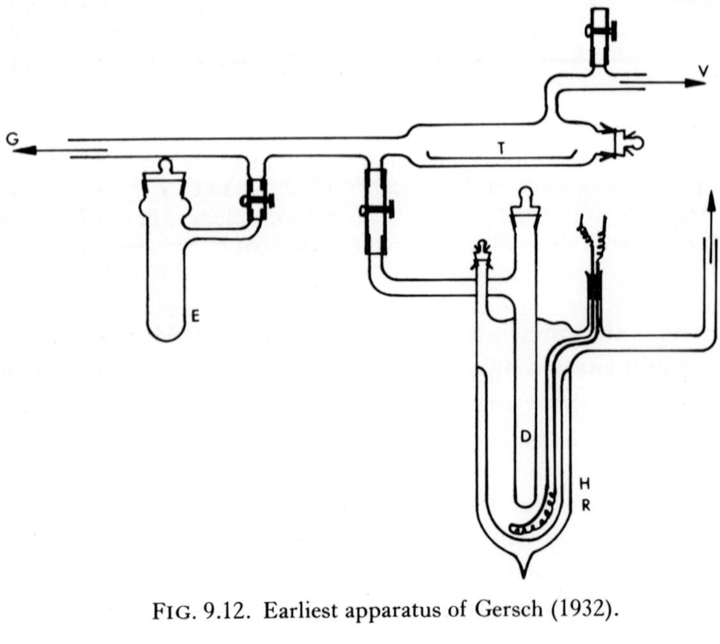

FIG. 9.12. Earliest apparatus of Gersch (1932).

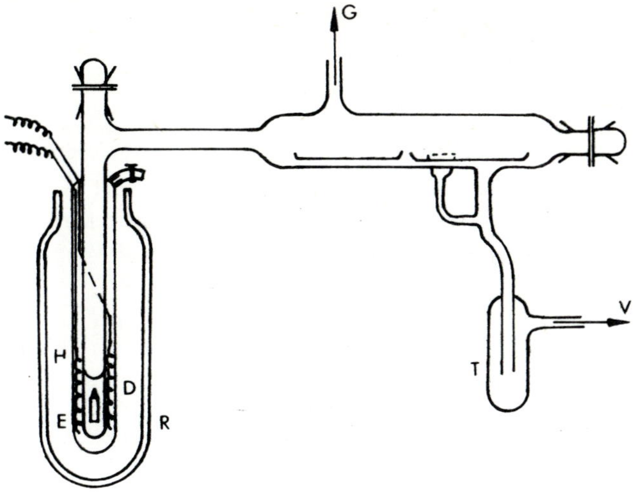

FIG. 9.13. Variable temperature control apparatus of Packer and Scott (1942).

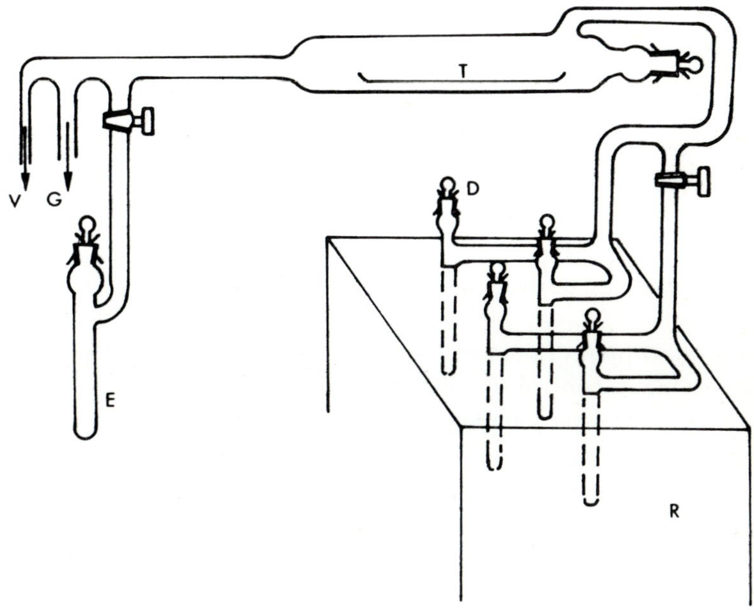

FIG. 9.14. Multiple drying chamber unit of Gersh (1948).

comply more or less with the design considerations already discussed. Additional features include variable temperature control (Fig. 9.13) and provision for freeze-drying multiple samples of tissue together (Fig. 9.14). In some of these freeze-dryers embedding is possible within the drying chamber. The component parts are labelled as follows: D, drying chamber; E, embedding chamber; G, to vacuum gauge; H, heater; R, refrigeration system; T, vapour trap; and V, to vacuum pumping system.

9.4.7 SUMMARY

Based on a purely classical approach, a theory has been developed for the design of apparatus for the freeze-drying of tissues for histological studies. It indicates that drying time, sublimation temperature and efficiency depend on two resistance factors, due respectively to the vapour flow in the dry shell and in the apparatus.

It was experimentally confirmed that at a given sublimation temperature the drying rate of guinea pig liver was lower by a factor of 1000 compared with the drying rate of a crystal of pure ice of similar size. This high dry shell resistance appears to be the dominant factor in the drying of tissues, overriding that of the resistance of the apparatus under design considerations unless the vapour flow between the drying chamber and vapour trap is restricted by the small bore of a stopcock. Generally, for maximum efficiency, the radius of the tubing should be between 1 and 2 cm.

9.5 End-point—control of final water content

For optimum preservation of some freeze-dried biological substances it is desirable to control the final water content by drying to a chosen and uniform level throughout, without over-drying at any stage. If the drying process, approaching equilibrium with that determined by the temperature of the condenser, is interrupted at a stage where the average water content is at some prescribed value higher than the final equilibrium value, the distribution of water in the sample will be non-uniform; that is, it will be lower near the surface than at the interior. It will become more uniform during storage of the sample, but the irreversible changes caused by drying will probably not be uniform throughout.

The relationship between the final water content at equilibrium and the relative vapour pressure of the dry sample, $p_{i,e}/p_{w,e}$($p_{i,e}$ = vapour pressure in equilibrium with ice at the temperature of condenser, $p_{w,e}$ = vapour pressure of liquid water at the temperature of dry sample when evaporation ceases) is expressed by the vapour-pressure isotherm or isostere at the final temperature of the sample. If the temperatures of the condenser and sample can be accurately controlled, then it will be possible to control the final water content without over-drying the sample (Sharp, 1953).

The practice of raising the condenser temperature when sublimation is complete does not achieve the desired results, because water vapour is absorbed by the over-dried portion of the sample. On the other hand, if the condenser is maintained throughout at the desired temperature level, equilibration of the sample may be too slow. The disadvantage of either alternative is that total pressure in the system would have to be raised to prevent rapid removal of water by the vacuum pump. As Mellor (1961) has shown, this disadvantage can be largely overcome if a sample of the substance is placed near, but not in contact with, a larger mass of ice which controls the evaporation by its contribution to the vapour pressure in the vicinity of the sample. This procedure is a rapid and sure way of obtaining a true end-point for research purposes, though at some sacrifice of drying efficiency.

One way to control evaporation experimentally is to evaporate a plug of ice and a small sample of the substance concurrently, with the water vapour from the plug passing through a frozen shell of the sample in a tube until evaporation from the latter has ceased. The apparatus is shown in Fig. 9.15. For comparison purposes, drying without the plug can be carried out either simultaneously or separately. The walls of the vacuum chamber act as the condenser surface when immersed in a dry ice–alcohol mixture. An electric heater is attached to the metal block and its temperature is controlled by a thermostat down to −45°C.

In a series of experiments using this method, solutions of ovalbumin were dried with 2 ml samples, shell-frozen in each of seven open-ended tubes. Plugs of ice with about 4 ml water were also frozen in seven smaller tubes and placed at the bottom of the holes in the block with open-ended tubes on top, as shown in Fig. 9.15. Also seven 2 ml samples were shell-frozen in closed tubes for one drying experiment, without the plug of ice, to control the evaporation. Temperatures were measured and are plotted in Fig. 9.16 for two typical experiments. In experiment (2), two tubes at a time for each process were removed at 90-minute intervals during the final stages of drying so that the approach to equilibrium could be estimated.

9.5.1 ANALYSIS OF EXPERIMENTAL DATA

These experiments showed that control of end-point depends on dynamic equilibrium between the vapour pressures of the plug of ice

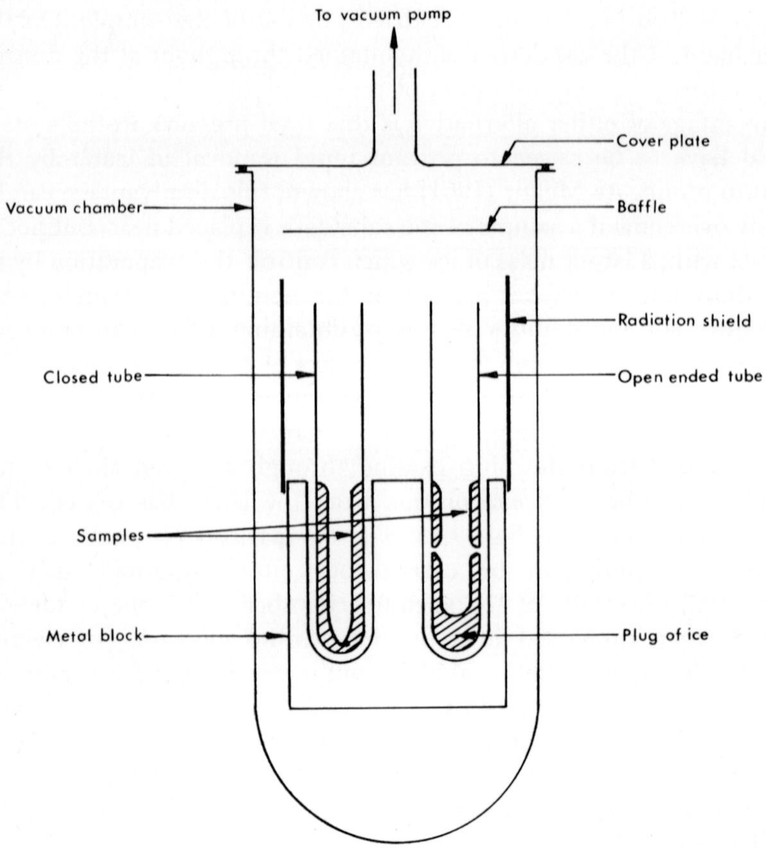

FIG. 9.15. Schematic diagram of essential parts of drying apparatus.

and of the sample after all the free ice in the sample has sublimed. Random motion of water molecules in the vapour zone between the plug of ice and the mouth of the tube is assumed. Departure from random motion was estimated by calculating the vapour-pressure gradient in the tube from the plug to its mouth. Using a total evaporation rate of $2{\cdot}9 \times 10^{-4}$ g s^{-1} and a mean free path of about 0·3 mm at 13·3 Pa absolute pressure the gradient was found to be about 1 per cent of the absolute (13 Pa approximately) for a typical drying curve at about −30°C (Fig. 9.16). The consistency with which water contents are attained after drying for 280 min is demonstrated in Table 9.1 for experiments (1), (3) and (4).

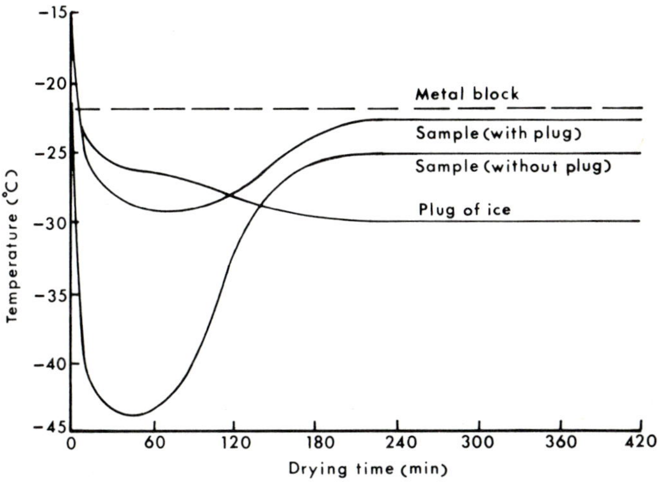

FIG. 9.16. Typical drying curves for ovalbumin: experiments (1) and (3).

Correlation between the final water content and vapour pressure was obtained from water adsorption isosteres for freeze-dried ovalbumin, using data obtained by Bull (1944) and Mellon *et al.* (1949).

Figure 9.16, which compares drying by this method with that without the plug of ice, shows that both give steady evaporation. Also, the results for experiments (1) and (2) in Table 9.1 indicate negligible further removal of water from the sample after its temperature has reached equilibrium at a little below that of the block.

TABLE 9.1

Estimated and experimental water contents for freeze-dried ovalbumin

Experiment	Drying time (min)	Vacuum pressure (Pa)	Mean temperature[a] at equilibrium (°C): Plug of ice	Sample	Vapour pressure at equilibrium (Pa): Ice at plug temperature $p_{i,e}$	Water at sample temperature $p_{w,e}$	Relative vapour pressure $p_{i,e}/p_{w,e}$	Water content (g water per 100 g dry protein): Experimental by weighing	Estimated from isostere
(Without ice plug) (1)	350 and 435	6·7 to 13·3	−78·0 condenser	−25·0	0·075	77·98	0·001	1·3 ± 0·3[b]	1·0 (approximate)
Controlled (with ice plug)									4·0 (approximate) not in equilibrium
(2a)	260	66·7	−44·5	−18·4	7·73	144·36	0·054	9.7 (means for 2 samples)	
(2b)	350	to	−44·8	−24·0	7·46	88·51	0·084	4·5 (means for 2 samples)	4·5
(2c)	435	13·3	−46·8	−26·8	5·87	68·52	0·086	4·7 (means for 2 samples)	4·6
Controlled (3)	280	13·3	−30·0	−22·8	38·12	98·51	0·387	10·0 ± 0·8[b]	10·2
Controlled (4)	285	53·3	−21·8	−14·0	86·78	207·95	0·417	9·7 ± 0·6[b]	10·3

[a] Temperature control to ±0·75°C.
[b] Standard deviation of the mean.

10 Drying Plant and Equipment for Freeze-Drying

10.1 Pilot freeze-dryers

10.1.1 PORTABLE UNITS

There is nowadays an increasing demand for freeze-drying units suitable for use in the laboratory and in the pharmaceutical and food industries to explore possibilities for the preservation of labile products, especially those of biological origin. The portable freeze-dryers now being manufactured commercially meet the physical requirements and are of a convenient size for developmental work on freeze-dried products in laboratories and factories throughout the world. Most designs incorporate self-contained facilities for refrigeration, heating and vacuum pumping. They can freeze dry batches consisting of from 5 to 20 kg of frozen product.

A typical portable unit (Fig. 10.1) consists of a cylindrical drying chamber 40 cm diameter by 60 cm mounted horizontally on top of a cabinet (1·07 m long, 0·8 m wide and 0·6 m high) which can be moved about on rubber-tyred swivel castors. One end of the chamber has a flanged opening, to which is fitted a swing cover similarly flanged but with a self-retaining O-ring vacuum seal and incorporating a viewing window.

Heating and cooling systems, essential to the freeze-drying process, are located inside the chamber. The heating system consists of a bank of five blackened, electrically heated radiant heaters (55 cm long,

25 cm wide) located in the central region of the chamber. Four metal trays to hold the product slide between the heaters, which are spaced about 6 cm apart. The cooling system consists of 9 m of 18 mm

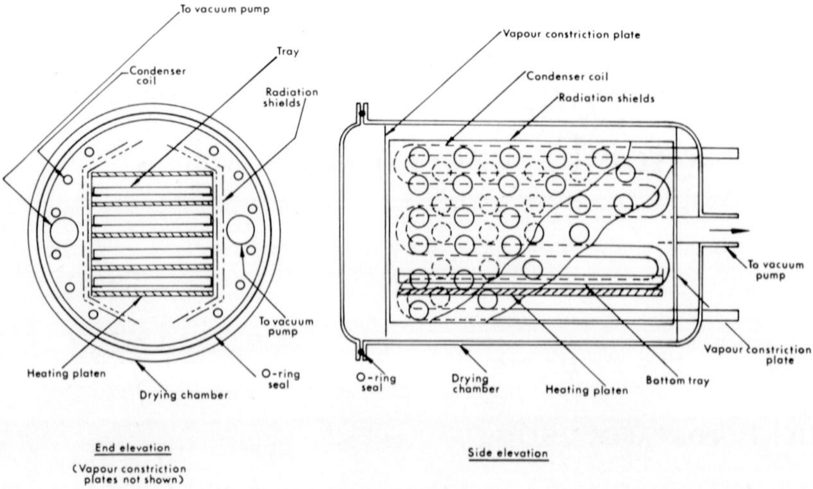

FIG. 10.1. Diagram of the essential features of a portable freeze-drying unit (after Mellor, 1966a).

diameter, cooling coil running the length of the chamber in zigzag fashion, with the nearest surface of the coil 12 mm from the inside wall. The space between the heating and cooling systems is shielded in such a way as to reduce heat exchange between heaters and coil yet allow water vapour to pass through.

In the cabinet beneath the chamber are the vacuum pump and mechanical refrigeration compressor, pipelines from which run to the cooling coil through special seals in the back of the chamber. The vacuum pump is a single-stage air-ballasted type having a displacement of 8·5 $m^3 h^{-1}$ and giving a pressure of 26·7 Pa. The refrigeration compressor is a 1·1 kW (R502 refrigerant) single-stage unit with sufficient capacity to reduce the coil temperature to −40°C under full load.

Heater temperature is controlled by means of a transistorized electronic device, and vacuum pressure is measured by a Pirani gauge. The instruments and controls are all panel-mounted, and the electrical leads to the back of the chamber are connected to demountable lead-in seals.

10.1.2 UNIT WITH PROTECTIVE GAS FACILITY

Much of the time spent in removing moisture towards the end of a freeze-drying process may be lost if the product adsorbs moisture when the vacuum is broken. Therefore, hygroscopic products such as freeze-dried fruit juices require unloading into a nearby glove-box or chamber containing a protective gas (Fig. 10.2). Products which are susceptible to lipid oxidation must be unloaded in the same way.

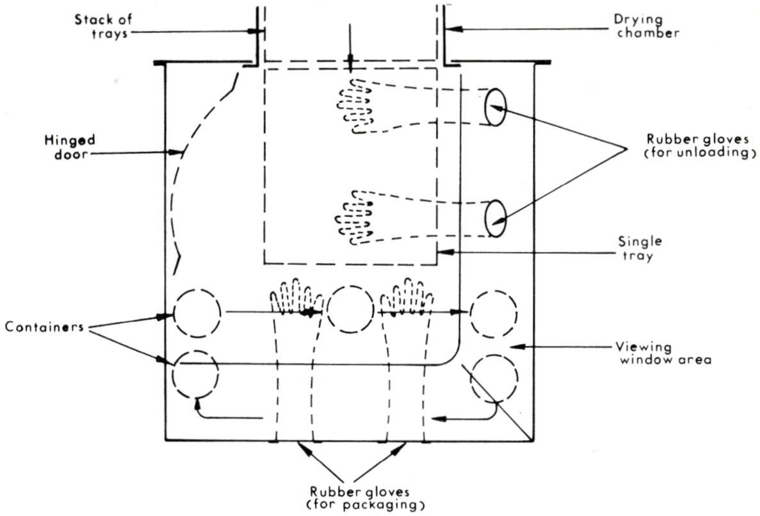

FIG. 10.2. Plan layout of a simple glove-box attachment for unloading and packaging freeze-dried products under sterile and protective gas conditions.

Dalgleish (1962) points out that the ratio of partial pressures of oxygen ordinarily encountered is 400 : 1, if the vacuum is broken with commercial N_2 (0·05 per cent O_2 by volume) and the product is then brought out into the atmosphere where the partial pressure of oxygen is 20·3 Pa. Oxygen uptake by the product would be quick and hence the need to provide an envelope consisting of a protective gas.

A pilot unit which can take a glove-box attachment on the front of the drying chamber for unloading trays of freeze-dried product is made in Germany. The unit has been designed with a view to obtaining data from trial runs simulating full-scale operations in tunnel freeze-dryers. Large tunnel freeze-dryers are constructed in Germany with fore-and-aft vacuum locks (see Section 10.4); the product is automatically discharged from the aft vacuum lock into a hermetically sealed room, either filled with a protective gas or provided with air-conditioning.

10.2 Tray freeze-dryers

Scaling-up of the type of freeze-dryer described in Section 10.1 is standard industrial practice in the United States, where equipment is provided for handling and loading the trays. There are two main types, depending on the type of condenser that is used. In one, the condenser plates are alongside the tray-heater assembly and in the same chamber; in the other, the condenser is in a separate chamber joined to the first by a wide-bore tube.

American plants of 1 tonne input capacity have shallow 120 cm by 90 cm aluminium trays which have been treated with a dark grey "hardcote" that combines chemically with the aluminium. The trays rest on 3 mm guides attached to the top surface of the heaters, which are 34 mm apart (Fig. 10.3). The tray-heater assembly is mobile, with couplings through which glycol is circulated to the heaters. The temperature of the heaters starts at 120°C and falls to 50°C during the

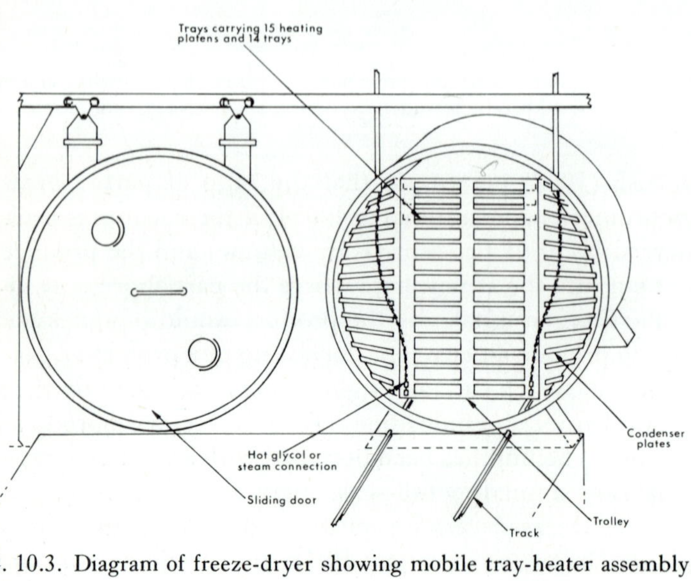

FIG. 10.3. Diagram of freeze-dryer showing mobile tray-heater assembly.

first 5 h of a 7·5 h run. The vacuum pressure is less than 27 Pa, and the condenser operates at −40°C. The internal refrigerated coils have been designed for batch runs with meat, fish and vegetables, principally for a company holding government contracts. A plant consisting of 14 batch units has been installed in Oregon, USA. Large numbers of similar chambers have been adapted for freeze-drying coffee; they are equipped with automatic facilities for freezing, filling the trays, loading and unloading, and an in-line tray-washing unit.

Another American plant has deep-ribbed anodized aluminium trays about 50 cm by 40 cm by 8 cm; the original design is covered by a form of patent protection. This applies also to the method of end-point moisture determination in the product, which is essentially a German barometric measurement of temperature. The external condensers are thus well suited to the adoption of this technique. With deep-ribbed trays, loadings can be more than twice the maximum for flat trays, e.g. 10 kg frozen material per square metre in the case of granulated coffee. Some American plants operate at lower heater and refrigeration temperatures.

10.3 Accelerated freeze-dryers

10.3.1 PLANT

In Britain, a single-cabinet plant for accelerated freeze-drying of meat and certain prepared fruits and vegetables has been built commercially to process 1·3 tonnes per day of fresh food to a moisture content of 2 per cent (dry basis) in three equal batches. The type of unit, and its associated equipment, is constructed to comply with government recommendations.

Each cabinet (Fig. 10.4) is 2·4 m diameter by 2 m long, with doors which seal the ends. It holds 30 trays having a total area of 32·5 m^2. The spacing between heating platens can be varied from 40 mm to 10 mm by hydraulic pressure (56 bar maximum).

The heating system consists of a vertical bank of 16 platens which can be heated or cooled with water. Hot water for heating is supplied at a maximum flow rate of 360 $l\ min^{-1}$ from a heat exchanger operated by dry-saturated steam at a pressure of 0·5 $kg\ m^{-2}$ gauge, and is sufficient to raise the temperature at 7°C min^{-1}. Another heat exchanger in the line serves to lower the temperature at about the same rate, with water at 18°C.

Instrumentation is provided in the form of vacuum gauges, and there is a thermocouple recorder for temperature measurement of the food trays and the platens.

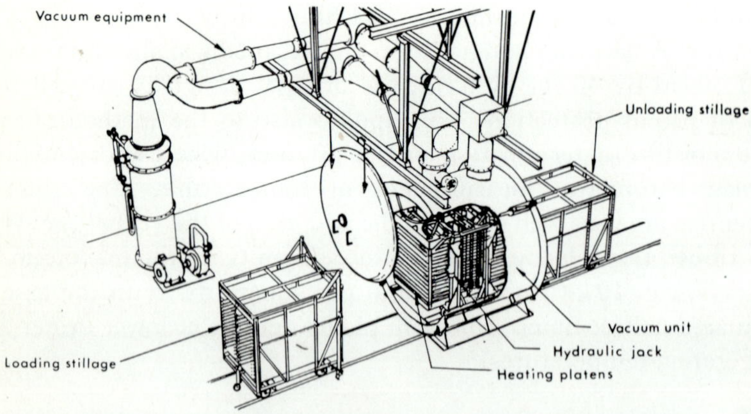

FIG. 10.4. Typical accelerated freeze-drying installation.

One firm has installed a number of these single-cabinet units as a multi-cabinet plant to dry an input of about 5·5 tonnes per day. Larger plants have been designed. In one model, intended for processing 17 tonnes of frozen food per day by semi-continuous operation, the frozen food will enter through a vacuum lock at one end of a tunnel and be taken out through a similar lock at the other end. A small secondary vacuum system will be required to operate these locks.

10.3.2 OPERATION

In this type of freeze-dryer, slices of food up to 18 mm thick are prepared and quick-frozen in the normal way, and loaded into the cabinet from an adjoining frozen store. They are spread on single sheets of expanded metal in shallow metal trays and other single sheets are placed on top, before finally being placed between the platens. Four-stage steam ejectors (Fig. 10.5), which can deliver up to

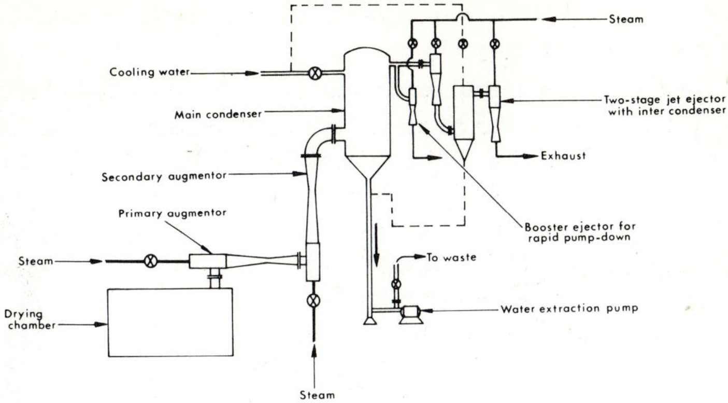

FIG. 10.5. Layout of a four-stage steam ejector.

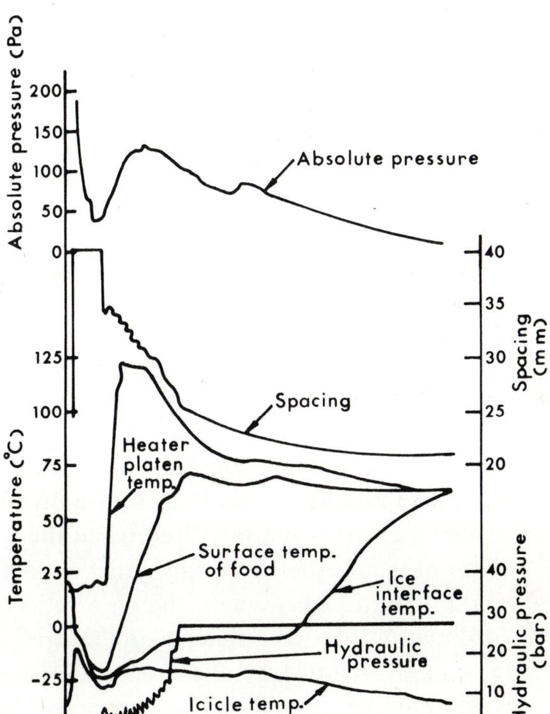

FIG. 10.6. Typical drying curves and other conditions in accelerated freeze-dryers. Food: cooked chicken.

160 kg h^{-1} of vapour, are then used to pump the sealed cabinet down to a pressure of 133·3 Pa. This takes about 6 min, after which the ejectors are used to pump water vapour directly. They work on dry-saturated steam at 0·5 kg m^{-2} gauge and consume on the average 500 kg h^{-1}. Water at 18°C and a flow rate of 8700 l min^{-1} is required for condensing. A total electrical load of 27 kW is involved.

Temperatures of platens and drying food for a typical process are shown in Fig. 10.6, and reflect the high surface temperatures employed after a dry layer of sufficient thickness has been formed. Drying times are of the order of 7 h in this type of plant and, allowing an hour for loading and unloading operations, it is possible to process three batches per day.

10.4 Tunnel freeze-dryers

With the application of freeze-drying to food processing, several new ideas concerning the operation and handling of large quantities of foodstuffs were introduced. Two important ones are: (1) a single measurement to represent the temperature of all the separate pieces of food being processed; (2) semi-continuous processing. These ideas have been successfully exploited in Germany.

10.4.1 "TEMP-PNEUMATIC" CONTROL

Systems for measurement and control of the ice interface temperature have to overcome the difficulty of the temperature probe embedded in the frozen layer being eventually uncovered when the ice sublimes. If the probe consists of a fine thermocouple junction positioned at the lowest point in the frozen layer, where the ice would be expected to remain until sublimation was complete, temperature control is satisfactory for, say, a single vial or tray of frozen material in a laboratory unit. But this method of control is at a serious disadvantage in industrial operations involving the processing of many separate pieces of frozen product together, and therefore for many such operations use is made of cam and programme controllers based on data obtained from previous trials.

A new principle of control and measurement of temperature by means of a variation in chamber pressure (Fig. 10.7), without measuring probes and the like, has been perfected by Neumann (1957). Called the "temp-pneumatic" system of regulation, it depends on

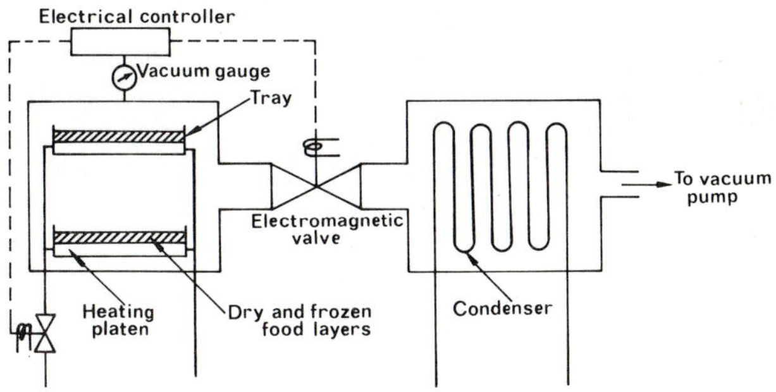

FIG. 10.7. Circuit diagram for the "temp-pneumatic" method of control.

raising the water vapour pressure in the drying chamber at regular intervals. When a large electromagnetic gate valve in a wide-bore pipe connecting the drying and condensing chambers is closed momentarily (5–10 s), a rapid rise in pressure in the drying chamber takes place as the lower vapour pressure in this space tends to adjust itself to that of the ice in the frozen food. The pressure is a function of the temperature of the ice, and hence can be monitored through an electrical controller to adjust either the heat supply to the drying food or the condenser temperature (in a mechanically refrigerated plant).

10.4.2 TUNNEL SYSTEM

German freeze-dryers incorporating "temp-pneumatic" control, can be built up from smaller tunnel sections, each capable of processing an input of 4 tonnes per day or 3 tonnes coffee extract per day.

Figure 10.8 shows schematically a plant consisting of a tunnel with vacuum locks at each end, one for loading deep-ribbed aluminium trays (510 mm by 165 mm by 75 mm deep) containing frozen lumps of food into the tunnel, and the other for discharging the freeze-dried product into an air-conditioned room where the dry product is automatically removed by machinery before being packed. In addition to lumps of food, slices or granules can be loaded into the trays to a depth of about 70 mm and, with four ribs per tray, such loose fitting pieces of food need only have a maximum dimension of 25 mm yet the normal shelf loading is extended by a factor of between two and three.

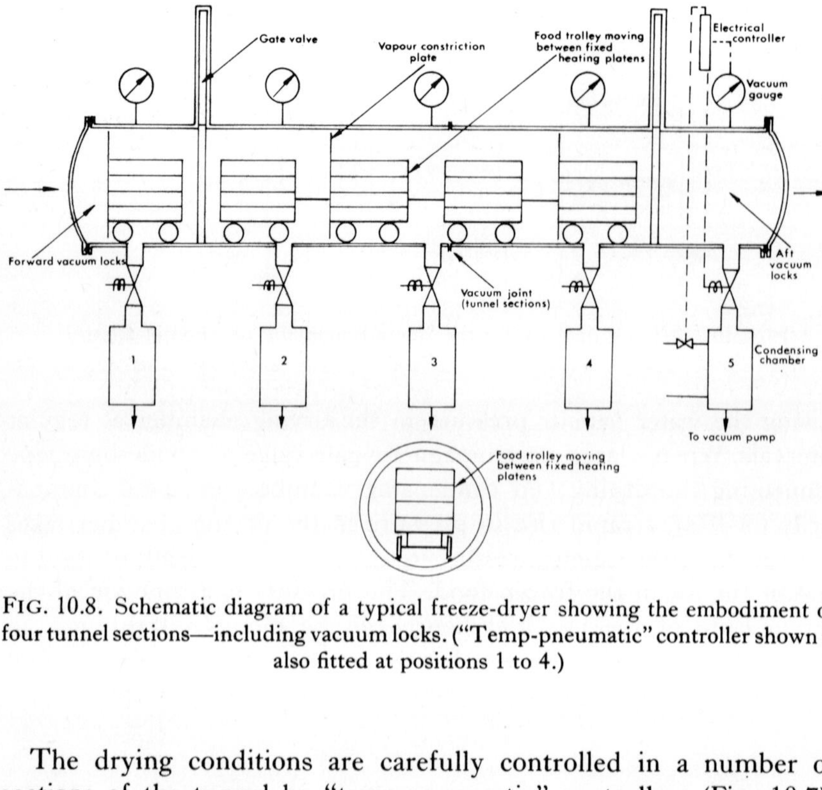

FIG. 10.8. Schematic diagram of a typical freeze-dryer showing the embodiment of four tunnel sections—including vacuum locks. ("Temp-pneumatic" controller shown is also fitted at positions 1 to 4.)

The drying conditions are carefully controlled in a number of sections of the tunnel by "temp-pneumatic" controllers (Fig. 10.7). Vapour constriction plates, fitting closely the inside walls of the tunnel yet allowing the trolleys to pass through, are at two locations in the main section of the tunnel, and gate valves shut off the locks from the main section. The tunnel is therefore virtually separated into five independent process zones. A monorail may be used as an alternative to trolleys.

During the period when the trolley is not moving, a tray lifting device causes all the 126 trays in each trolley to sit on top of the heaters below. The heaters have flat top surfaces and ribs underneath, through which vacuum steam circulates; they are cantilevered in pairs from both sides of the tunnel. Vacuum steam heating has several advantages, including a high latent heat of condensation and temperature control by means of pressure.

The refrigeration plant is unique in that a large aqua-ammonia absorption refrigerator is used instead of a compression plant, mainly because of the ease with which the refrigeration load can be varied by controlling the oil feed to the boiler which heats the absorber.

The total capacity of such a plant can be increased as the volume of business increases; large commercial plants for processing cottage cheese and coffee have been built up in this way.

10.5 Drying plant with "inner" heating

One convenient way of handling liquid foods for freeze-drying is to form a frozen shell of product on the surface of vertical tubes. Two types of plant utilizing this idea have been developed.

The first is a British process based on "inner" heating (Mason, 1963). Using this process a prototype plant (Fig. 10.9) has been built

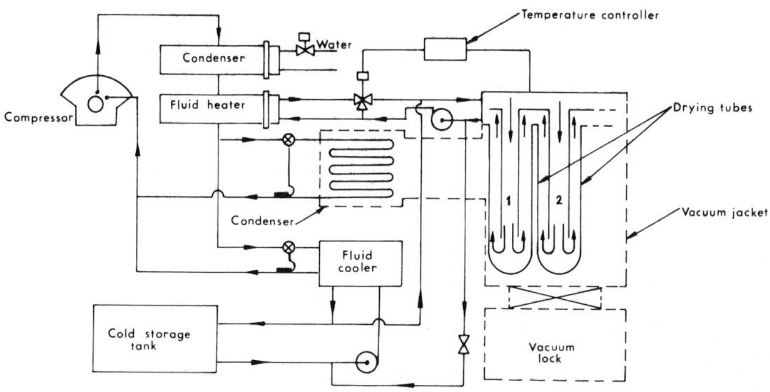

FIG. 10.9. Circuit diagram of freeze-dryer with "inner" heating (no's 1, 2, etc.—drying tubes located in vacuum jacket).

in which a shell of frozen product is formed on the outside of an array of 300 vertical tubes 1 m long through which cooling brine circulates. Frozen shells 6–10 mm thick are formed in about 10 min, and the remaining liquid is then drained. Next the frozen product is placed under vacuum and the tubes are gently warmed by circulating a warm brine through them. A refrigerated coil in a chamber connected to the jacket chamber is used to condense the water vapour given off by the product. The dried product is scraped off the tubes and collected in a hopper feeding a vacuum lock.

The circuit diagram (Fig. 10.9) shows a form of heat pump whereby the heat of compression in the refrigeration system is imparted to brine circulating in the drying tubes. Expansion of the liquid refrigerant cools the condenser of the freeze-dryer, also a brine storage tank via a fluid cooler. Cold brine can be utilized for freezing the shell and for adjusting the temperature in the drying tubes.

Table 10.1 gives the drying times and shell thicknesses in plants of 0·75 tonnes per day capacity for a range of products; for example,

TABLE 10.1
Data on trials in a prototype "inner" heating plant

Product	Solids content (%)	Shell thickness (mm)	Drying time (h)
Comminuted orange	11	6·3	7·25
Comminuted lemon	16	6·3	9
Milk	12·5	6·3	3
Egg	25	9·5	6
Coffee extract	24	6·3	7
Tea infusion	15	6·3	4·5

coffee extract can be freeze dried in 7 h or tea infusion in 4·5 h. The quality of the products should be good, judging from trials in the prototype plant.

The second type of process utilizing "inner" heating is a Dutch process, which is used commercially and in principle is the reverse of the first type. Here a layer 7–10 mm thick is shell frozen from the liquid inside the tubes, instead of outside, by condensing ammonia in the tube jacket. The frozen product is then freeze dried by condensing the ammonia still further, at a higher pressure.

Both methods aim to provide direct conduction heating through the ice layer by good thermal contact with the tube surface during drying. King (1970) pointed out that an impermeable skin which would hamper drying may be left behind after freezing. In the present system it could perhaps be removed by a rapid melt at the surface of the frozen shell with water flushing through the jacket. The difficulty could, however, be more than offset by the ease with which the vapour flows in radial channels in the product shell left by the freezing and drying process.

10.6 Vacuum-spray freeze-dryer

In the United States, a prototype vacuum-spray freeze-dryer (Fig. 10.10) has been developed for coffee extract, tea infusion or milk The product is sprayed from a single jet upwards or downwards in a cylindrical tower 3·7 m diameter by 5·5 m high (Thuse *et al.* 1968). The liquids solidify into small particles by evaporative freezing. In the

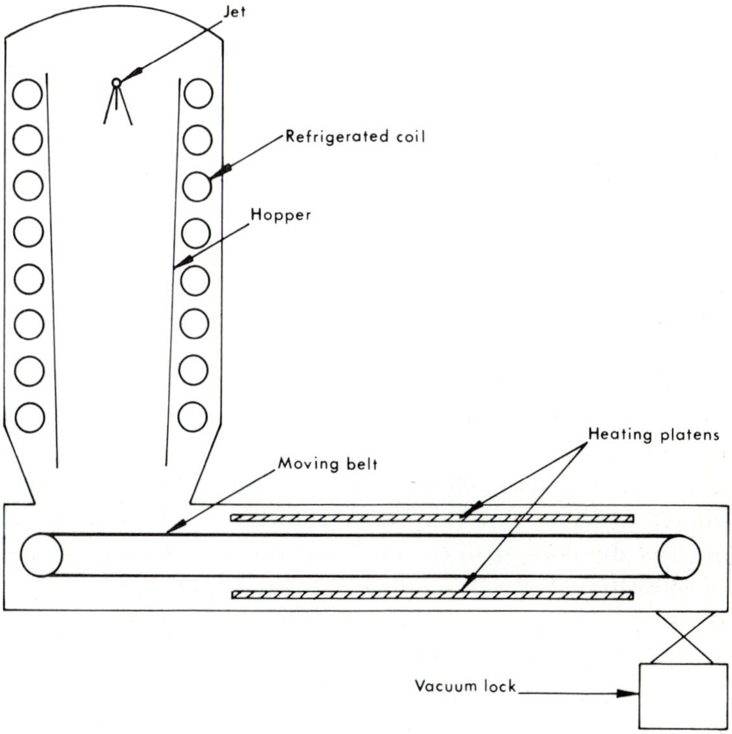

FIG. 10.10. Layout of a vacuum-spray freeze-dryer.

tower a refrigerated helical condenser is coiled between the inside wall and a central hopper, the latter collecting the partially dry powder as it falls freely to the bottom of the tower, which, in turn, is connected to a tunnel where the drying process is completed on a stainless steel belt travelling between radiant heaters. The product passes into a hopper that feeds a vacuum lock permitting intermittent removal of the product for packing. The whole plant operates under a vacuum of about 67 Pa.

In this plant, pre-concentration of the initial product up to 30 per cent solids can be used before freeze-drying. Frozen particles obtained by spraying into a vacuum are about 150 μm in diameter and lose about 15 per cent moisture in the initial evaporation. There is no sticking of these particles. Generally, sprayed freeze-dried coffee has less flavour than normal freeze-dried coffee and the product from this plant is no exception. However, it is hoped to improve flavour retention in the dried product by concentration before spraying into the tower.

10.7 Drying costs

Freeze-drying is costly because of the long drying times involved, and this factor has hindered the application of the technique to drying of materials in bulk. As a result, fixed costs tend to exceed running costs. Anquez (1966) has given some estimates of these costs for foodstuffs.

During the development of the accelerated freeze-drying process Forrest (1960) assessed the average drying costs from 40 proving runs on a range of foodstuffs, in order to find out whether the cost of processing them being quoted by equipment manufacturers was too optimistic. He used a plant with an imput capacity of about 1 tonne per day, operating for 6000 hours per annum. Estimates of fixed cost proved to be about one-third of running cost for a steam ejector system, and about one-half for a refrigeration condensing system. The total cost of drying was 3·5–3·8 p (Stg) per kg input, i.e. higher than that of quick-freezing (about 0·5 p (Stg) per kg in Britain at that time). Forrest found that the cost of preparing the food for freeze-drying was similar to that for freezing but that packaging costs were likely to be higher.

The situation was different a decade later; fixed costs then exceeded running costs by $1\frac{1}{2}$ to $2\frac{1}{2}$ times, as the following estimates for a cyclic-pressure freeze-dryer and for a vertical vibrational freeze-dryer show.

10.7.1 COST COMPARISON: CYCLIC AND VIBRATIONAL FREEZE-DRYERS

The comparison is based upon prices, operating capacities and labour requirements obtained from equipment manufacturers and their customers and upon two important assumptions: that (1) cyclic-pressure operation effectively gives a 30 per cent reduction in drying time, and (2) both systems have the same overall thermal efficiency.

Unit costs of utilities are assumed to be:

Power	2¢ per kWh
Steam	0·5¢ per kg
Cooling water	0·5¢ per 1000 l
Oil	$34 per tonne
Compressed air	$3·50 per m^3

10.7.2 COSTING

The annual capacity of each freeze-dryer is based on a 20 hour day and 250 working days per year, 4 hours per day being allowed for loading and unloading operations. Thus:
Annual capacity = 5000 × (rated capacity in kg ice per h)

a. *Fixed cost*

$$\text{Average fixed cost} = \frac{\text{Annual fixed cost}}{\text{Annual capacity}} \qquad (10.1)$$

assuming an annual fixed cost consisting of capital depreciation at 7·5 per cent per year, and loan charges at 8 per cent reducible by the formula

$$L = \frac{r_0(1+r_0)^n}{(1+r_0)^{n-1}} \tag{10.2}$$

where L equals the annual charge to repay \$1 loan and interest for n years and at rate r_0. Thus with $r_0 = 0{\cdot}08$ (8 per cent) for 10 years, $L = 0{\cdot}149$ and so the loan charge per \$ capital is \$0·049. Keey (1972) points out that formula (10.2) also gives the capital recovery factor (fraction of the original capital investment set aside each year over the working life of the plant) for comparing drying systems. An example shows that a plant would be preferred which is dearer to install but costs less to run.

Other fixed capital-dependent charges at 4 per cent of capital cost include maintenance, insurance and taxes.

b. *Running cost*

This consists of labour and utilities costs. Data on the thermophysical properties of foods and biological materials, required for estimating utilities cost, are not always to hand and so the calculation is based on only those properties pertaining to the frozen water content of the material.

Only one semi-skilled person is required to operate any of the dryers, at \$2 per hour. This should increase to \$4 per hour to cover supervision, operating supplies, payroll overhead, plant overhead and process control. Thus:

$$\text{Average labour costs per kg ice} = \frac{\$4 \times 24}{20 \times (\text{plant throughput per hour})}. \tag{10.3}$$

Preparation and packing costs are not included as these will depend largely on the nature of the product.

The utilities cost can be estimated from a heat and energy balance for 1 kg water undergoing freezing, sublimation, condensation and melting, and the breakdown is as follows:

1. *Refrigeration.* Heat extracted to freeze the water content of the material from 25°C to −30°C in 1 h: $25 \times 4{\cdot}19 + 335 + 30 \times 2{\cdot}10 \simeq 502$ kJ per kg ice.

Heat removed during condensation at −40°C: 2840 kJ per kg ice. The compressor power operating on ammonia is about 0·65 kW per kW (refrigeration), and thus the energy to be supplied to the compressor is $0{\cdot}65 \times 2840 = 1840$ kJ per kg ice.

2. *Heating.* Heat required to sublime ice at −20°C equals 2840 kJ per kg ice.

At the end of the drying cycle, the ice collected at the condenser has to be melted at −40°C and requires $335 + 40 \times 2{\cdot}10 \simeq 419$ kJ per kg ice.

3. *Vacuum pumping.* Electricity required for two-stage vacuum pumps equals 0·36 kWh kg^{-1} ice.

4. *Total energy*

	kJ per kg ice		kWh per kg ice
Refrigeration	502	≡	0·14
	1840	≡	0·51
Vacuum pumping		=	0·36
TOTAL		=	1·01
			kg steam per kg ice
Heating $\frac{2840+419}{2065}$		=	1·58

where the heat of vaporization of steam is 2065 kJ per kg at a pressure of 682 kPa.

Utilities costs			¢ per kg ice
Power	$1{\cdot}01 \times 2$	=	2·02
Heating	$1{\cdot}58 \times 0{\cdot}5$	=	0·79
TOTAL		=	2·81

Thus, allowing for a 20 per cent increase to cover thermal losses and other charges, the total utilities cost is 3·4 ¢ per kg ice.

Estimated cost of freeze-drying foodstuffs, based on rated capacities of some industrial freeze-drying systems being planned or in existence in 1970, is given in Table 10.2.

TABLE 10.2
Comparison of the cost of freeze-drying based on rated capacities and prices supplied by manufacturers of small industrial systems in 1970

Costing procedure	Cyclic freeze-dryer (3 multiple cabinets)		Vibrational freeze-dryer (vertical system)
Capital cost ($)	270 000		380 000
Rated capacity (kg h^{-1})	50·5		50
Water removal (2 runs per day) (kg per year)	252 000		250 000
Fixed cost (¢)			
Capital depreciation	2 025 000		2 850 000
Loan charges	1 323 000		1 862 000
Other charges	1 080 000		1 520 000
	2 runs per day	3 runs per day	Continuous
Average fixed cost (¢ per kg)	17·6	11·7	24·9
Running cost (¢ per kg)			
Labour	9·5	6·4	9·7
Utilities	3·3	3·3	5·5
Total cost (¢ per kg water)	30·4	21·4	40·1

11 Foodstuffs

11.1 Physical characteristics and preparation of foods for freeze-drying

11.1.1 SOLID FOODSTUFFS—MEAT AND FISH

Meat and fish could be described as semi-solid food, comprising muscular tissue characterized by labile proteins, interspersed with adipose tissue and about 1 per cent glycogen, and containing about 75 per cent water.

In meat, long bundles of cylindrical fibres and connective tissue make up the contractile muscle. The connective tissue, varying in shape and amount, separates individual fibres and surround the fibre bundles and the entire muscle. The individual fibres consist of extremely thin threads, the fibrils, and a complex matrix of liquid channels, the sarcolemma; they in turn are composed of shorter thin filaments, the actin and myosin proteins. There is also an ultra-structure consisting of some transverse channels and band-like structures. In fish the muscle composition is similar but in the form of blocks of muscle fibres—the myomeres—the blocks being separated by connective tissue. Connective tissue, the myocommata, is relatively impermeable to water vapour. Thus in both meat and fish the orientation of the muscle is important with respect to heat and vapour transport (Figs 11.1 and 11.2).

11.1.2 SOLID FOODSTUFFS—FRUIT AND VEGETABLES

A typical plant cell consists of a droplet of solution surrounded by an osmotic membrane and enveloped by a cellulose wall of a many-sided polyhedron, and with free water on the outside. Between adjacent cells is a pectic compound, the middle lamella, which cements the cells together. Immediately inside the cell wall is a fairly fluid protein, the

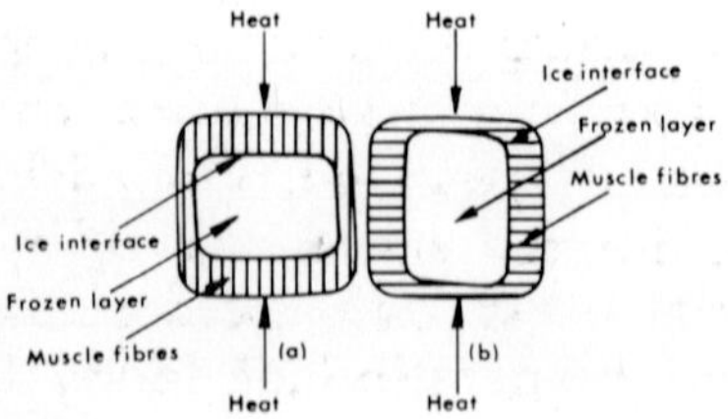

FIG. 11.1. Effect of muscle fibre direction on freeze-drying of lumps of meat.

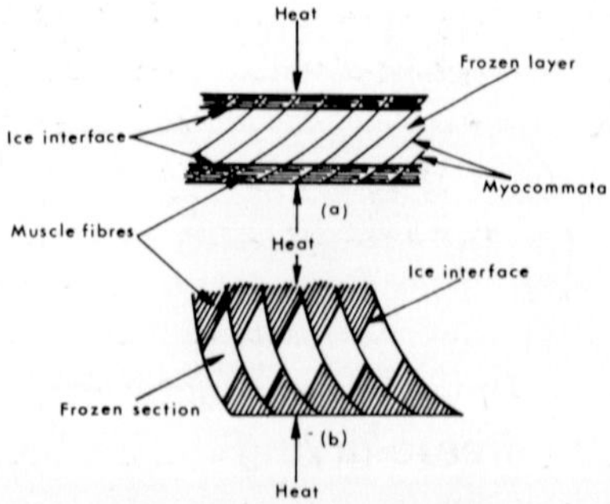

FIG. 11.2. Effect of muscle fibre direction on freeze-drying (a) fish fillet and (b) fish steak.

protoplasm and the protoplasmic inclusions; also there are the nucleus, mitochondria and other cell organs. And inside the semi-permeable osmotic membrane is the vacuole, containing an aqueous solution of many compounds.

The effect of structure in plant material, prepared ready for freeze-drying, on the shape of the ice interface is not so important. Plant material may be represented diagrammatically by Fig. 9.8 (Section 9.4) and, even if the ice interface were turned through 90°, its effect on heat and vapour transfer would be much the same.

Plant cells undergo respiration, with the oxidation of simple sugars to carbon dioxide and water to produce energy to maintain their organization, and have a turgor pressure balanced internally by the cell wall. It is the turgor pressure which is responsible for the firmness and crispness of fruit and vegetables. It is a desirable characteristic of a rehydrated freeze-dried product, ensuring that it does not become flaccid.

11.1.3 LIQUID FOODSTUFFS

Liquid foods have a non-specific form and contain much water, removal of which leaves rehydratable powders. Such foods may be divided into two main categories: (a) natural products and (b) extracts. Category (a) includes milk, egg white and yolk, while (b) includes fruit juices, and tea and coffee infusions. This is only a rough division, since an extract is obtained either by chemical processing (solvent extraction) or by physical separation (pressing, pulping or mixing), and milk may be separated from cream and whites from yolk in preparation for dehydration. However, as far as freeze-drying is concerned, extracts have come to mean solvent-extracted liquids varying in concentration.

The composition and concentration of liquid foods vary widely. Whereas fruit juices are low in solids content and high in reducing sugars, the reverse is true of coffee extract, while egg has a medium concentration. Most of the above examples primarily comprise carbohydrate, with the exception of tea extract which has a high tannin content, and egg which has a very low carbohydrate content.

11.2 General preparative treatments

11.2.1 MEAT

The heterogeneous nature of meat does not permit drying the bones, adipose tissue and muscle all together; bones dry very slowly and therefore necessitate a long drying period during which the adipose tissue could melt. For chops, it is desirable to arrange that moisture left in the bones at the end of the drying process can be removed by an in-package desiccant during storage, otherwise moisture transferred to muscle protein might reach too dangerous a level during equilibration. Because adipose tissue contains only 10–50 per cent water compared with 70–75 per cent for lean meat, some adipose tissue may melt on reaching a high surface temperature in the early stages of freeze-drying. This causes rendering and blocking of a portion of the porous dry meat by liquid fat, which inhibits vapour transfer and leads to thawing. Meat should therefore be trimmed of excess fat before being freeze dried.

Lean meat is prepared for freeze-drying by being moulded into a block, frozen, and sawn into thin slices 10–15 mm thick (see Section 11.6). The frozen block is sawn in such a way that, as far as possible, the cut surface is at right angles to the direction of the muscle fibres so that the ice interface moves with the grain during drying and remains parallel to the direction of heat flow, as shown in Fig. 11.1(a). Thus heat transfer and vapour transfer are facilitated. Drying is not so rapid where the heat flows across the grain, as in Fig. 11.1(b), because the vapour has difficulty in escaping. Rehydration also is slower in this case.

The same is true of fish steaks (Fig. 11.2(b)), where the receding ice interface initially develops a "saw edge" and the myocommata seal off the fibres near the centre of the steak.

Other raw-material variables such as the breed of animal, type of meat cut, fat content and ageing are discussed by Bengtsson and Bengtsson (1968) and Bengtsson (1975).

11.2.2 FRUIT

Varieties of fruit intended for freeze-drying should have high solids content, good colour and full flavour. Apples and pears should be peeled and cored and stone fruit split and the stone removed, then sliced or diced and dipped in 0·1 per cent sodium sulphite solution to prevent enzymic browning prior to freezing. Strawberries, raspberries and currants should be washed, plugged or strigged and, in some cases, the skins should be slit to improve the drying rate. Black currants, the skins of which have a low permeability to both water vapour and water, are best prepared by freezing followed by abrasion of the skins to reduce the thickness or to wear them away in patches.

11.2.3 VEGETABLES

Every kind of vegetable poses its own problems of preparation as well as of storage prior to drying. For example, peas are highly perishable and should be prepared for freezing as soon as possible after harvesting. But storage of Brussels sprouts is permitted for a few hours prior to drying and of potatoes for several months.

The main preparative operations for vegetables are: preliminary washing, peeling, trimming, cutting, final washing of cut material, and blanching.

Preliminary washing is necessary to remove adhering earth. Peeling is often overdone, causing serious loss of original material, for example, the calculated percentage volume lost for 76 mm diameter spherical potatoes ranges from 17·6 to 33 per cent for 1·6 and 3·2 mm thick peelings respectively. However, such losses may be reduced by continuous abrasive peeling or better by chemical peeling in hot caustic.

Trimming is usually a hand operation.

Root vegetables are cut into slices, strips or dice, and leafy vegetables are shredded. Exceptions are: Brussels sprouts, small carrots and pickling onions.

Leafy vegetables, spinach and peas are washed after cutting. With cut potatoes, the starch grains from ruptured cells left by cutting should be washed away to ensure good final quality.

Blanching for enzyme inactivation of freshly cut material exposed to air is done by scalding in hot water or steam. Scalding in steam gives less leaching loss of dry solids than scalding in hot water, about 3 per cent compared with about 13 per cent, and for freeze-dried products it is therefore preferred.

Onions and leeks are not blanched before drying since some enzymic action is important for the development of full flavour.

To reduce continuation of cooking as a result of scalding, hot vegetables should be cooled preferably in air. A further treatment in cold water would cause additional leaching.

11.3 Special preparative treatments

11.3.1 EXCIPIENTS

Freeze-drying and subsequent storage of some products causes loss of colour, flavour and texture, and these properties cannot be fully restored upon reconstitution. Could the solution to this problem be found in the field of medicine? Certainly it could if the medical term "excipient" were considered in the broadest sense in the context of freeze-drying (Mellor and Irving, 1969) as a substance which might provide protective action, thereby reducing irreversible changes, loss of consistency, colour and flavour during preparation, drying, storage and rehydration of freeze-dried products.

So far as freeze-drying of foods is concerned Bengtsson (1975) has stated unequivocally that the word "excipient" is more meaningful than the general term "food additive" that is in use at present, and he has published a list of excipients that have been used in preserving dried meat.

Some examples of special preparations employing excipients, reported by Mellor and Irving (1969), are summarized below.

11.3.2 SULPHITE TREATMENT FOR PEAS

Part of the normal preparation of vegetables for dehydration is a dip in a solution of sodium metabisulphite. The vegetables absorb a certain amount of sulphite as excipient, which protects vitamin C and plays a part in retarding the browning reaction, though in storage the treated vegetables gradually lose sulphite. Once freeze-drying of vegetables became customary, sulphite treatment of many vegetables was discontinued because lower moisture contents were now attainable (Gooding, 1962). However, in storage trials freeze-dried peas without sulphite treatment have deteriorated substantially in quality but with sulphite treatment good quality was obtained by ensuring that the excipient had actually been absorbed by the peas, by puncturing the skins through to the cotyledons.

The present author took two lots of peas which were blanched in water at 96°C for 80 s, cooled in water for 60 s, then drained, and frozen at −20°C. One lot was punctured with pins of 1·2 mm diameter before blanching and freezing and another left unpunctured. When sulphite treatments were included, blanched peas were dipped in a 0·3 per cent solution of sodium metabisulphite for 30–60 s. All the frozen peas were freeze dried by the cyclic-pressure process. Finally the peas were packed in cans, sealed under vacuum, and stored at 20°C and 37°C for 7 months.

The results for tasting tests carried out on all of the samples for colour, texture and flavour, are summarized as follows:

1. Sulphiting is important for storage at high temperature providing the peas are punctured and the moisture content is sufficiently low (below ~1 per cent).
2. The fact that the scores for punctured and sulphite treated peas stored at 20°C and 37°C were not significantly different points to possibilities for high storage temperatures.
3. Grading seems to influence only colour scores but puncturing is just as important, since during freeze-drying the cotyledons do not shrink if the skin of the pea is punctured to allow the water vapour to escape. Reconstitution is better also, and tasters tend to upgrade quality purely on appearance rather than on colour under such circumstances.

A comparison with grades of commercial frozen peas and freeze-dried punctured, sulphited and graded peas stored at 20°C shows that these freeze-dried peas are about equal to second-grade frozen peas.

11.3.3 CONTROL OF "BROWNING" IN FRUIT

Enzymic browning of pears and other fruits is rapid during preparation for freeze-drying. It has been partially controlled by dipping the freshly cut fruit in 0·5 per cent solutions of ascorbic acid. A more powerful excipient for this purpose was found to be 2-mercaptobenzolthiazole sodium, which could be used in smaller concentrations. Moreover, this substance, while inhibiting non-enzymic browning in storage did not, as did ascorbic acid, impart a pinkish colour to lighter coloured dried fruits, especially bananas.

To test the effect of the excipient, the present author took freshly picked pears which were peeled, sliced into 10 mm disks and cored; part of the test sample was dipped in 0·5 per cent ascorbic acid, part in 0·025 per cent mercaptobenzolthiazole solutions and part left untreated. After freeze-drying the products, which had a moisture content of about 4·5 per cent, were stored for 22 months at ambient temperature in aluminium foil–plastic laminate pouches sealed under nitrogen gas.

The effect of the mercaptobenzolthiazole treatment compared with the untreated product may be clearly seen from Fig. 11.3. The two top pictures show treated dried pears and the two bottom ones show untreated dried pears. In the treated disk there appear (top left) two small patches of darker colour; these are pieces of the skin of the pear still retaining their original green colour against the very light yellowish-green colour of the flesh. By contrast, the patches at bottom left

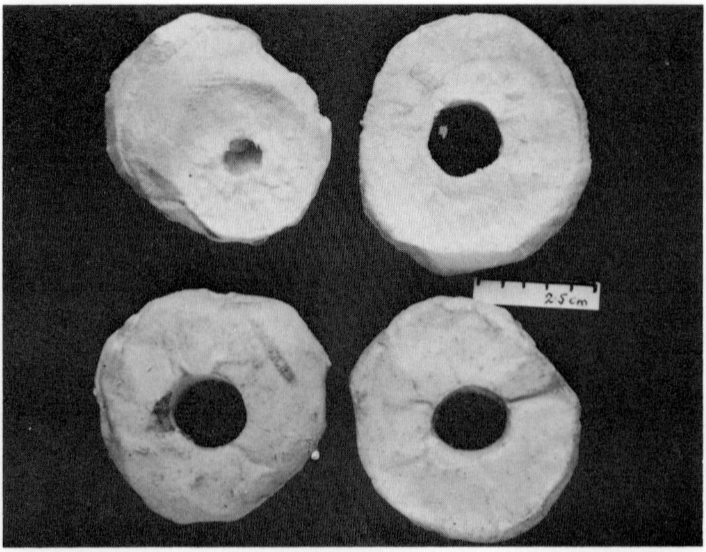

FIG. 11.3. Freeze-dried pear: treated with mercaptobenzolthiazole (top), untreated (bottom).

(untreated product) were dark brown against the greyish and speckled-coloured flesh. These colour effects were retained in the freshly rehydrated products, with the untreated products deteriorating rapidly compared with the treated. The results for the products treated with ascorbic acid were not much better than for the untreated products.

Mercaptobenzolthiazole might not be permitted as an excipient under health regulations.

11.3.4 PRE-SLAUGHTER ADRENALIN INJECTION FOR MEAT

Reconstituted freeze-dried meat is less tender and juicy than fresh or frozen meat and the effect is more pronounced after storage—a characteristic "woolly" texture on tasting. This is due to the serious loss of water-holding capacity of the muscle protein. Penny *et al.* (1964) found that the normal pH value (meat, pH = 5·8) had to be increased to an ultimate value of 6·8 to improve the water-holding capacity, and this was more effective than adjusting the pH of the water used for reconstitution. They achieved the pH adjustment by subcutaneous injection of adrenalin into rabbits, pigs and steers prior to slaughter. The excipient was injected 24 hours before slaughter and again 4 hours before, and it was found that the rehydration capacity or reconstitution ratio (amount or water regained on rehydration to initial water content) for the freeze-dried meat increased from about 2 : 1 for the control to 3·2 : 1 for the treated meat.

Other pre-slaughter excipients (insulin and glucoseoxidase) have been studied by Harper and Tappel (1957).

11.4 Pre-concentration processes

11.4.1 INTRODUCTION

If an aqueous liquid can be concentrated to reduce bulk prior to freezing and drying, then the overall cost of the de-watering processes should be reduced. The advantage of this procedure have been discussed by Monzini and Maltini (1973) for fruit juices and by Döge and Vêlebil (1968) for coffee as well. The procedure is being developed in six major areas for selective de-watering, viz. evaporation, pervaporation, crystallization and clathration (which involve phase changes) and osmosis and ultrafiltration (which do not).

The two main systems are multi-effect evaporation for nearly all fruit juices except orange juice, and freeze-concentration for heat-sensitive and delicately flavoured products.

Multi-effect evaporation is in general use for producing frozen concentrates, but there are some products where the heat sensitivity is such that temperatures in the range 70–100°C for only a few minutes can only be used, and depectinized juice becomes necessary for some concentrates. Severe volatile flavour losses and fouling of evaporator heater surfaces as a result of protein denaturation are also experienced.

Freeze-concentration is the only safe way of removing water and maintaining the conditions that permit the water, originally present, to separate in the form of ice crystals; it requires less energy than evaporation and retains all the volatile constituents, but the separation of the ice crystals from the concentrate is difficult unless the mother liquor occluded by the crystals can be removed by washing.

A third alternative—direct osmosis or reverse osmosis (ultrafiltration)—involves many difficulties with the cellulose acetate membranes that could be used with liquid foods, since they do not have the right sort of permeation and solute (aroma) retention properties. For further information on this new technique the reader is advised to consult Thijssen (1970), Thijssen and Middelberg (1966), and Michaels (1974).

11.4.2 MULTI-EFFECT EVAPORATION

Low cost multi-effect evaporation is easily the best developed process for de-watering liquid foods. However, as already explained, it has several serious drawbacks. One of these is fouling caused by high evaporator temperatures. In order to remedy this, lower temperatures and thin layers which wet completely the heater walls of the evaporator are used. The requirements are usually met by using either an expanded-tube design of a single-pass upflow evaporator (Morgan, 1967b) or a Wurling evaporator (Morgan, 1967a; and Carlson *et al.* 1967).

Heat damage to the product, another major trouble, is of course mitigated by the lower temperatures and by using shorter residence times, given by $\bar{t} \propto V/Ah$ for a specified amount of water removal and constant temperature difference, where V is the liquid volume in the evaporator, A is the heat exchange surface and h the overall heat transfer coefficient. Thus a large value of h, as with the Wurling evaporator ($h = 2{\cdot}9\ \text{kW m}^{-2}\ \text{K}^{-1}$), shortens the residence time.

The main trouble, overshadowing everything else, is that volatile flavour constituents can be lost during evaporation. Stripping, however, can be used advantageously to remove and collect the aromas from either the liquid or the vapour. They can then be concentrated and added back to the already stripped concentrate. Thijssen (1970) has described briefly steam stripping, working out that it takes about

$1/\alpha_e$ tonne of steam per tonne of feed to strip 90 per cent of the aromas in a sieve plate column or a packed column (α_e is the effective relative volatility of the aromas); he has also produced a formula for the total energy consumption on the basis of 90 per cent stripping and concentrating the aromas to about a 500-fold essence.

11.4.3 ANALYSIS OF VOLATILE AROMA LOSS

During evaporation volatile aromas and water vapour are driven off together. The degree to which the aromas are lost is determined by the volatility with respect to that of water. Thijssen (1974) has shown that the relative volatility of an aroma constituent at a given temperature is proportional to the product of the activity coefficient and the vapour pressure of that constituent, and is expressed by

$$\alpha_{a,w} = \frac{\gamma_a p_a^\circ}{p_w^\circ}. \tag{11.1}$$

Typical values of $\alpha_{a,w}$ for n-alcohols are tabulated in Table 11.1, which shows the same order of magnitude as most quality-determining aroma constituents in fruit juices, which are in the range $5 < \alpha_{a,w} < 50$.

TABLE 11.1
Boiling point, molecular weight and relative volatility of n-alcohols at zero concentration in water

Constituent	Boiling point (°C)	Molecular weight	Relative volatility $\alpha_{a,w}$
Methanol	64·7	32	8
Ethanol	78·4	46·1	11·5
Propanol	97·8	60·1	16·5
Butanol	117·0	74·1	27
Pentanol	138·0	88·2	30

The potential loss of aromas may be determined from Rayleigh's relation:

$$\frac{m_{a,0}}{m_{a,1}} = \left(\frac{m_{w,0}}{m_{w,1}}\right)^{\alpha_{a,w}}, \tag{11.2}$$

where $m_{a,0}$ and $m_{w,0}$ are the mass of aroma constituents a and of water w respectively in the feed and $m_{a,1}$ and $m_{w,1}$ the mass of aroma constituents a and of water w left after evaporation. Calculations from equation (11.2) show that a 50 per cent reduction in aromas is accompanied by only a 10 per cent water loss, assuming, say, $\alpha_{a,w} \simeq 7$.

11.4.4 FREEZE-CONCENTRATION

Freeze-concentration is the process whereby, to concentrate a solution, a slurry of ice crystals is produced by refrigerating the solution in a crystallizer, the ice crystals then being separated from the mother liquor by filter press, centrifuge or reflux wash column. Concentrates with solids contents of 30–40° Brix are usually produced.

Direct-contact (clathration) and indirect-contact crystallizers are being used. The direct method is still being developed and utilizes a compatible refrigerant such as carbon dioxide, nitrous oxide, Freon C318 or propane in direct contact with the solution; or it may use water vapour under vacuum, in which case vacuum pumps having a high capacity for water vapour are necessary. Using propane, with a solubility of about 200 ppm at 5°C in water, it would be necessary to strip the clathrate from the juice in a highly selective way—note that the activity γ(propane) > γ(volatile flavours).

Typical indirect-contact crystallizers use a vertical or horizontal cylinder equipped with a rotating scraper to remove the ice as it forms on the inside wall of the cylinder, the wall of the cylinder being cooled on the outside to below freezing point. In another form of crystallizer (Struthers' concentrator), not embodying an ice scraper, a slurry of crystals is stirred at temperatures that are carefully controlled so as to induce the growth of uniformly large ice crystals (Fig. 11.4). The ice that forms is separated in a continuous-screening centrifuge. This is a two-stage concentrator having de-watering facilities for washing the ice crystals.

There are two basic circuits for freeze-concentrators, namely single-pass and recirculation types.

The first, shown in Fig. 11.5, consists of a storage tank, pump, crystallizer and separator in which the flow rate of the juice is controlled. In this system the lowest practicable fresh juice concentration is about 12° Brix, below which it is difficult to scrape the ice from the walls of the crystallizer. Above 20° Brix ice crystal growth diminishes and only fine crystals are produced which are difficult to separate. The system is limited to final concentrations of about 30° Brix.

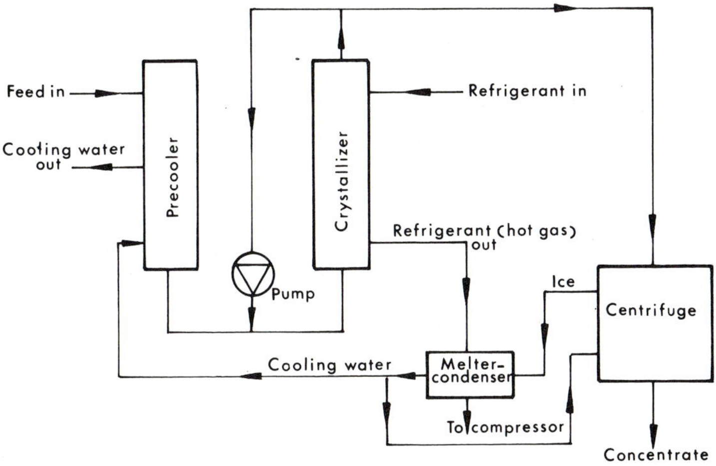

FIG. 11.4. Flow diagram of Struthers' concentrator.

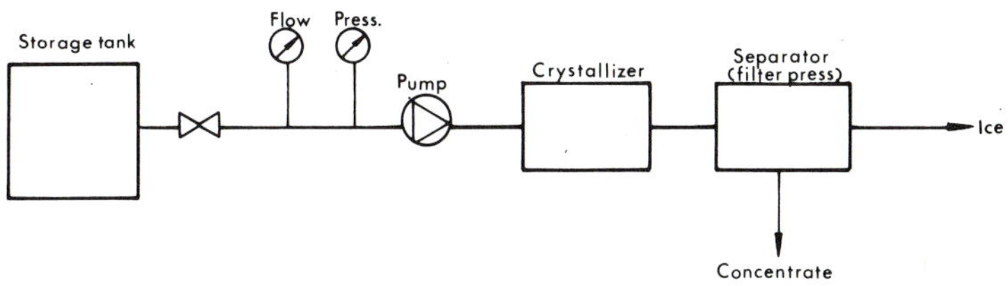

FIG. 11.5. Freeze-concentrator with single pass.

In the second basic circuit, shown in Fig. 11.6, the difficulties associated with low fresh juice concentrations, say, 6° Brix, and a limiting final juice concentration of about 30° Brix, are overcome by mixing the fresh juice in a circulation tank with some of the concentrate produced to make a semi-concentrate, which is again pumped to the crystallizer.

Losses occur during separation, and hence by adding another separating stage (centrifuge) and returning the small amount of juice recovered from the ice to the circulation tank gives better separation and therefore continuous freeze-concentration. Thus, final product concentrations of 30° Brix or greater can be produced from as low as a 6° Brix fresh juice concentration.

Figure 11.7 shows the degree of separation for a filter press in terms of the solids content of juice still held by the separated ice for different juice concentrations (° Brix). For example, it is about 3·7 per cent weight solids at 30° Brix even if some of the ice melts during centrifuging and washes the remaining ice.

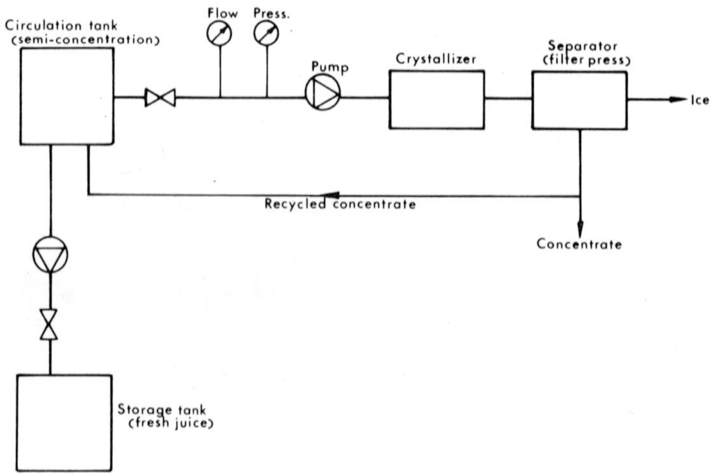

FIG. 11.6. Freeze-concentrator with recirculation.

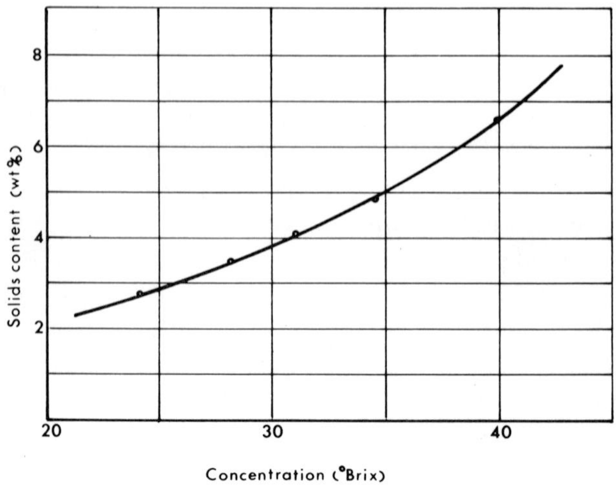

FIG. 11.7. Solids content of ice separated by filter-press.

In the Struthers' two-stage concentrator (Fig. 11.4) de-watering facilities for washing the ice crystals are obtained by just melting ice discharged from the centrifuge and feeding some of it back for washing the ice crystals, the remainder being fed into the tank to precool the juice. Since larger regular crystals are formed in the crystallizer under carefully controlled temperatures and by stirring with paddles, separation only requires one stage but with washing to prevent solids loss.

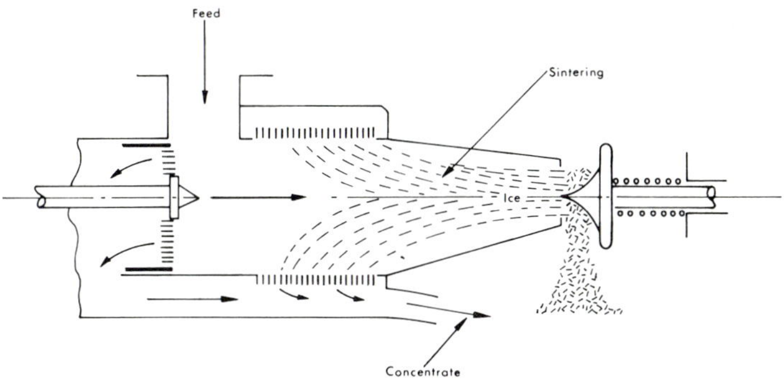

FIG. 11.8. Working diagram of the filter-press.

Eijkelboom (1966) used a model 6° Brix juice consisting of 6·3 per cent sugars (70 per cent sucrose and 30 per cent glucose) and 0·2 per cent citric acid, which represented 6° Brix fresh depectinized strawberry and raspberry juices fairly well, and found that in a continuous unit, having a capacity of about 40 kg h^{-1} final product, the losses amounted to about 13 per cent for about 220 kg h^{-1} fresh juice processed. The unit consumed a total power of 19·0 kW and at the same time 273 m^3 h^{-1} cooling water. The type of filter press that Eijkelboom used is illustrated in Fig. 11.8.

11.4.5 ANALYSIS OF CRYSTALLIZATION AND SEPARATION

The aqueous solution is frozen so as to form a thick viscous slurry consisting of relatively pure ice crystals and concentrate from the high proportion of the water present, and then the crystals and concentrate are separated from one another. If nucleation and growth of large ice crystals of very high purity occurs then freeze-concentration may be considered to be the most selective of all de-watering processes. However, this is only possible provided suspension freezing (paddle crystallizer) is substituted for layer freezing (scraped crystallizer), otherwise the separation will not permit a high enough proportion of concentrate with a low allowable solids loss. In suspension freezing the mixing of large and small crystals by the paddle provides a "ripening" effect, i.e. the large crystals grow at the expense of the smaller and the existence of small embryonic crystals depends on the amount of supercooling and the nucleating rate.

Thijssen (1974) points out that the nucleating rate increases with increasing solute concentration and is proportional to the square of the bulk supercooling ΔT^*, which is expressed by the Gibbs–Thompson equation

$$\Delta T^* = \frac{4\sigma_E T^*}{d\rho_i \, \Delta H}, \qquad (11.3)$$

where ΔT^* is defined as the lowering of equilibrium temperature of a spherical crystal in solution with diameter d with respect to equilibrium temperature T^* of the same solution containing a very large crystal, σ_E = surface free energy, ρ_i = density of the crystal and ΔH = heat of crystallization. In Fig. 11.9 this equation is plotted for critical

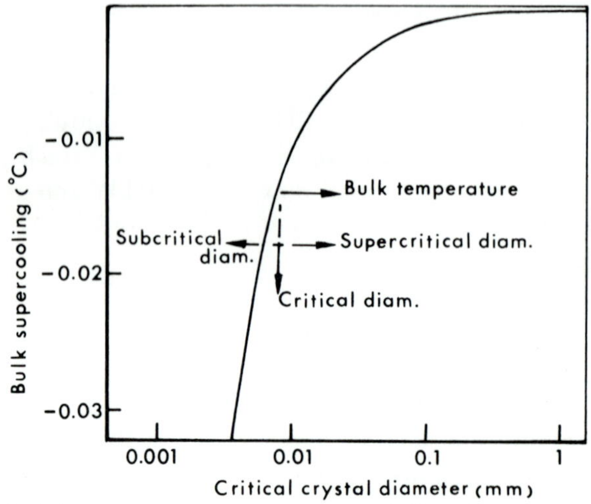

FIG. 11.9. Relation between bulk supercooling of a liquid and the critical diameter of the ice crystals (after Huige *et al.*, 1972).

diameters associated with certain bulk supercoolings; it shows that there are size limitations to crystal growth.

Now the final concentration which can be expected, starting with a given concentration and freezing point T_F, may be obtained from equation (2.37) (see Section 2.4), expressed in the incremental form

$$\Delta T = \frac{RT_F(T_F - \Delta T)}{\Delta H} \ln a_w, \tag{11.4}$$

where a_w is the water activity which may be considered to equal approximately the molar water concentration up to 50 per cent wt solids. Typical curves of freezing point versus concentration corresponding roughly to equation (11.4) are presented in Fig. 11.10.

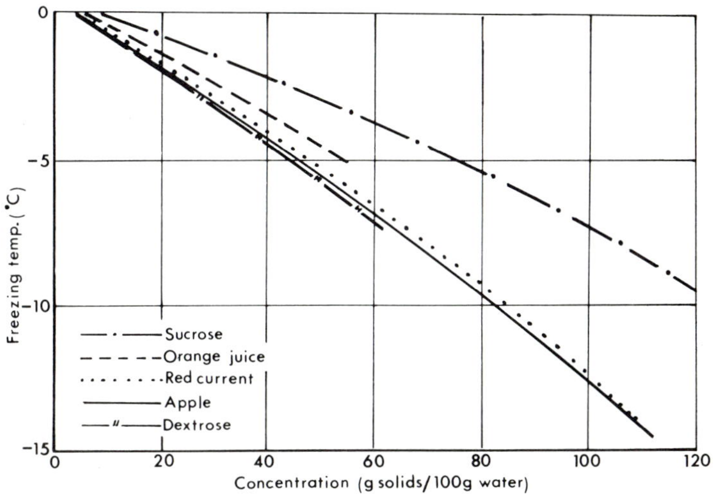

FIG. 11.10. Experimental freezing point curves of various sugar solutions and fruit juices (after Heiss and Schachinger, 1951).

They show, for example, that the cooling of apple juice with 9 per cent solids to −12·5°C results in an ice-phase separation of 90 per cent of the water originally in the fresh juice.

Ice-phase separation in practice always involves occlusions and adhesion of a proportion of the mother liquor to the ice crystals; the efficiency η_{sep} of the process has been determined by Hendrickson (1958) in the case of water desalination by freeze-concentration. His formula, which should apply also to fruit juices, is:

$$\eta_{sep} = w_{mix} \frac{w_{ml} - w_{is}}{w_{ml} - w_{ju}}, \tag{11.5}$$

where w_{is} = weight fraction of solids in ice slurry after separation, w_{ml} = weight fraction of solids in mother liquor after freezing, w_{ju} = weight fraction of juice before freezing, w_{mix} = weight fraction of ice in frozen mixture before separation.

From equation (11.5) it is found that the efficiency η_{sep} is 50 per cent for centrifuging, 71 per cent for vacuum filtration and 89–95 per cent for filter-pressing. To do better than this it is necessary to undertake counter-current washing combined with ice-crystal melting and refreezing.

That the process of separation in a reflux wash column can easily reduce the solids losses to less than 0·5 per cent, compared with 9 per cent for a filter-press and compared with that for centrifuges as well, seems to justify detailing the basic theory of this form of separation. Nevertheless, the theory outlined by Thijssen (1969) for the filter-press and for the basket centrifuge is worth mentioning, as it shows the limitations of these pieces of equipment.

The closed wash column (Fig. 11.11) is best suited to fruit juices. Here the slurry (ice crystals and concentrate) is force-fed at the top

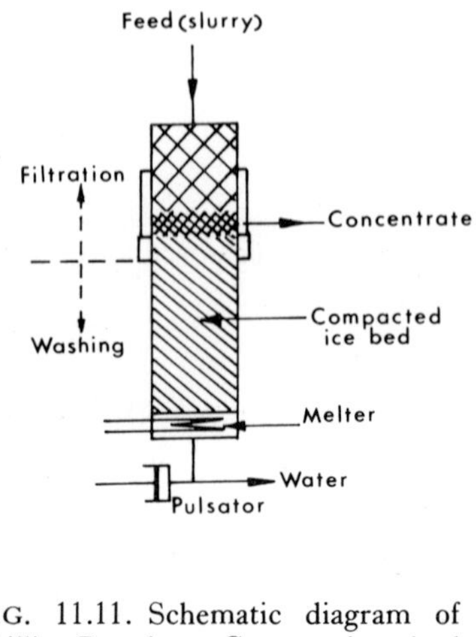

FIG. 11.11. Schematic diagram of Phillips Petroleum Company's pulsed wash column.

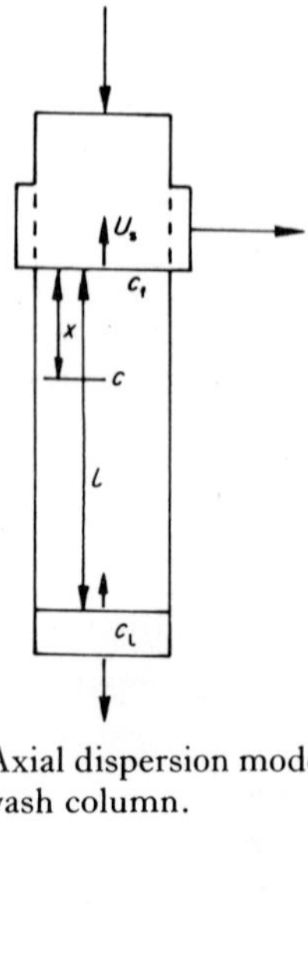

FIG. 11.12. Axial dispersion model of wash column.

and as the crystals reach the bottom they are melted through contact with a heater. Most of the melted ice runs out at the bottom of the column, but some is forced upwards by a pulsator through the descending crystal bed, for washing to remove occluded solids. In equilibrium, a concentration gradient develops down the bed, and at some convenient point in the wall of the column the concentrate is run out for collection.

Referring to the axial dispersion model of Thijssen (1969), shown in Fig. 11.12, the pressure gradient Δp is given by

$$\Delta p = \frac{U_s \mu l}{K}, \tag{11.6}$$

where l is the column length, μ is the viscosity, K is the bed permeability and U_s the superficial liquid velocity with respect to the movement of the compacted ice.

The differential equation for the model is

$$D_{con}\frac{d^2c}{dx^2} + U_s\frac{dc}{dx} = 0, \tag{11.7}$$

where c is the concentration at level x below the filtration section, D_{con} is the diffusivity and U_s the difference in velocity between upward wash and downward liquid carried by the ice.

With boundary conditions $c_l = 0$ and $c_0 = c_f$ (concentration of the slurry) the solution is

$$\frac{c_f - c}{c_f} = \frac{1 - \exp(-U_s x/D_{con})}{1 - \exp(-U_s l/D_{con})}. \tag{11.8}$$

Substitution from equation (11.8) with $l = 30$ cm, $U_s l/D_{con} = 570$, $D_{con} = 2\text{–}3{\cdot}5 \times 10^{-4}$ cm^2 s^{-1} (sucrose–water–ice), and $c_f = 0{\cdot}4$ g cm^{-3} gives a reduced concentration of $0{\cdot}023$ g cm^{-3} at $x = 0{\cdot}15$ cm. This distance is noticeably small and points to the fact that the wash column is a superior separator.

On the question of stability, Kuivenhoven (1966) has applied to wash columns for concentrating fruit juices Dumoré's (1964) criteria:

$$U_s \leqslant \frac{\rho_{sol} - \rho_w}{\mu_{sol}(\ln \mu_{sol} - \ln \mu_w)} gK_v, \tag{11.9}$$

where ρ_{sol}, ρ_w = densities of the solution and water respectively,

μ_{sol}, μ_w = viscosities of the solution and water respectively, K_v = permeability defined by d'Arcy's law, g = acceleration due to gravity. For example, $U_s \leqslant 0{\cdot}8$ cm h^{-1} for 30 per cent sucrose solution, which is too low to be practicable. Therefore, pulsation is needed. The best amplitude for pulsation is 0·5 cm, which gives 60 ideal mixings in cascade in a wash column of length 30 cm and diameter 29 cm.

11.5 Vacuum cooling and freezing

11.5.1 PRINCIPLE OF "PUMPING TO FREEZE"

If biological material is considered to have very moist surfaces, then reducing the pressure from atmospheric to below the high vapour pressure existing at the surface of the material causes water from the moist surfaces to evaporate rapidly to form vapour which expands greatly. The water cools first by adiabatic evaporation and then freezes, provided the reduced pressure is well below the saturated vapour pressure of ice (610·5 Pa). Evaporative or vacuum cooling and freezing ("pumping to freeze") involves two basic events. First, when the absolute pressure is sufficiently reduced a flash of vapour takes place. This is known as the flash point. Secondly, the evaporation of moisture requires a large quantity of latent heat, and this is extracted from the material, which is thereby cooled. In a rapidly pumped system the flash point occurs a few minutes after starting the pump; the material then cools rapidly with the loss of a small quantity of surface moisture. When the freezing-point of the material is reached a longer pumping period begins during which the latent heat is being extracted and the freezing process takes place. Therefore, most of the water evaporated is pumped off during this stage.

The overall fraction of water δ evaporated from 1 kg mass to cause cooling and freezing in the range 25°C to −20°C, can be obtained from:

Heat lost by the water = Heat gained by the vapour.

This gives

$$H_1 + \Delta H_w(1-\delta) = \delta\,\Delta H_i - H_2, \qquad (11.10)$$

where ΔH_w = latent heat of crystallization of water, ΔH_i = latent heat of vaporization of ice, H_1, H_2 = sensible heats before and during freezing respectively.

The values of ΔH_w and ΔH_i are 0·334 and 2·51 MJ kg^{-1} respectively, and those of H_1 and H_2 are 0·105 and 0·042 MJ kg^{-1} respectively, so, from equation (11.10), $\delta = 0{\cdot}17$. However, δ would have to be greater to freeze foods, because water is in contact with a large thermal mass. Rolfe (1963) gives, for example, a tray load of 12·75 kg m^{-2} cooked haddock which takes 1 min to cool from 25°C to −2°C in a freeze-drying cabinet and a further 2·5 min to freeze at −3°C and sub-cool to −20°C with the loss of 21 per cent water. As seen from the above discussion, this loss is 4 per cent higher than if water alone were evaporatively frozen. Mellor (1964) gives another example of this, relating to the vacuum cooling of mushrooms.

The principle of cooling by pumping can be better understood from Table 11.2, which gives the flash point, heat of vaporization and specific volume of the vapour for a range of pressures from atmospheric down to 100 Pa.

TABLE 11.2
Properties of water vapour at various pressures

Pressure (kPa)	Flash point (°C)	Heat of vaporization (MJ kg^{-1})	Specific volume of vapour (m^3 kg^{-1})
101·3 (atm)	100	2·27	1·75
2·5	21·1	2·45	54·3
1·3	1·1	2·47	96·9
0·7	0·1	2·49	189·1
0·29	−10·0	2·51	455·6
0·13	−19·8	2·51	1096·8

11.5.2 OTHER METHODS OF EVAPORATIVE FREEZING

Several sophisticated liquid injection methods of evaporative freezing have been described by Folsom (1946). They include "explosive freezing" *in vacuo*, freezing droplets on a flexing belt, "spray freezing" from a revolving disk, and "film freezing" on scraped rollers. Spray freezing of orange juice has been used also by Levinson and Oppenheimer (1944); and Greaves (1962) has described nozzle and high-speed spinning disk and nozzle and low-speed drum methods in vacuum chambers. This is also the well-known method of "centrifugal freezing" (Greaves, 1944) under reduced pressure in order to suppress frothing of the liquid.

In a review of these methods, Rowe (1970) explains that the partial pressure of air over water in a cryostat can be above 610·5 Pa and still evaporative freezing will occur, provided the rate of evaporation is not balanced by the heat gain from the surroundings. Evaporative freezing has been demonstrated in the laboratory where the absolute pressure was 2·2 kPa and the partial air pressure 1·6 kPa, and in some regions of the world on clear nights when the dew point of the air was very low.

11.5.3 PUMPING SYSTEMS FOR VACUUM FREEZING

In practice the pumping system for a vacuum freezer might consist of a combination of a Roots pump, a single-stage rotary vacuum pump and a water-cooled condenser (Fig. 11.13). The high-capacity Roots

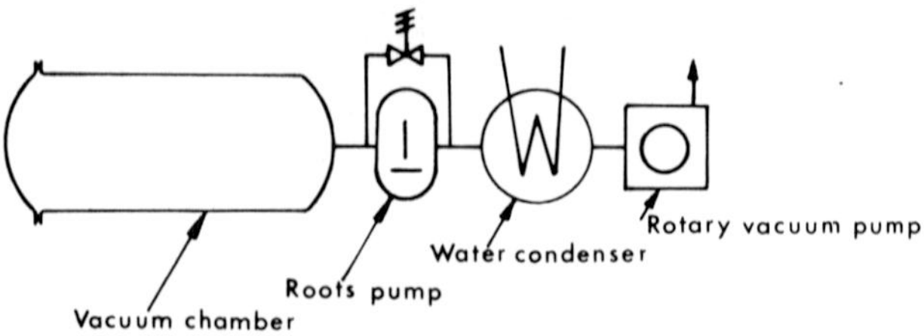

FIG. 11.13. Layout of a direct pumping vacuum system.

pump transfers the considerable volume of water vapour to the condenser, and the remainder (a mere trace) which escapes is extracted by the gas-ballasted rotary pump. The Roots pump would need to be fitted with a by-pass and pressure-relief valve, as shown in Fig. 11.13, to function just before the flash point.

An alternative arrangement is a refrigerated coil system (Fig. 11.14) consisting of a coil operating at a temperature below −20°C and of smaller capacity than a freeze-dryer. Occasional defrosting of such a system, after a run by water sprays, could be necessary unless identical condensers were worked alternately in parallel. The chamber could be pumped by a single-stage rotary vacuum pump.

11.5.4 CONTINUOUS BELT FREEZER

For production freezing, a continuous belt freezer is very useful. The one shown diagrammatically in Fig. 11.15 has been successfully used

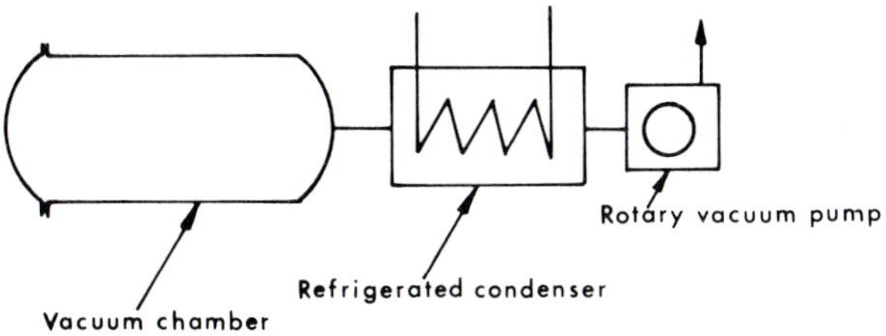

FIG. 11.14. Layout of a refrigerated coil and vacuum pumping system.

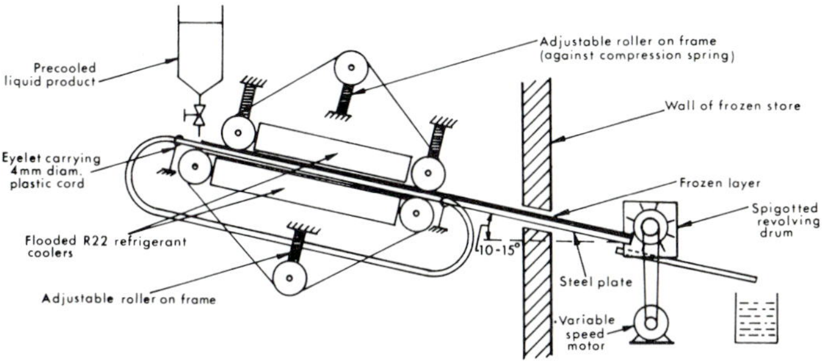

FIG. 11.15. Layout of a continuous belt freezer.

for freezing fruit purées and coffee extracts. It operates in a room at ambient temperature adjoining a frozen store and consists simply of two wide teflon plastic belts which can be tensioned to slide over flooded R22 refrigerant coolers and adjusted to provide a gap between them. In the gap two continuous 4 mm diameter plastic cords move with the belt to prevent overflow of a pre-cooled liquid product from the edges of the belt. The pre-cooled liquid product is continuously poured from a tank onto a steel plate passing through the wall of the frozen store and, as the edge of the frozen layer comes up against a grinder (a spigotted drum), it is broken up into granules. Another steel plate after the grinder is vibrated to assist the flow of frozen granules entering the bucket.

The two coolers operate in the range −20°C to −40°C. The moving belts and grinder are driven by variable-speed electric motors.

The design of equipment has sufficient flexibility in respect of temperature range and speed of freezing to permit freezing and granulation of many different liquid products without affecting the colour of the final product and without producing too much frozen powder (fines).

Using a continuous belt freezer, granulated frozen strawberry has been obtained by pre-cooling to 15°C, freezing to −14°C and grinding at −25°C. The product has a fairly well-defined granular shape and uniform colour. Coffee granules are understood to have light feathery edges and a colour varying from straw to brown but foamed coffee concentrate is believed to be a better product. With a unit having a throughput of 200 kg h^{-1} it takes only 2 min to freeze a 4–5 mm thickness of coffee concentrate.

11.6 Rate of freezing

In the freeze-drying process the rate at which the preliminary freezing is carried out is particularly important, since it determines the size and location of the ice crystals, and so affects the structure and hence the water and vapour permeabilities of the freeze-dried product. Another consequence of freezing is the occurrence of a pellicle in some products, which affects the permeability. Thompson *et al.* (1962) have observed that muscle fibres in meat may be folded or blocked at the surface by a gelatinous barrier if the meat is sliced before being frozen. This makes drying and rehydration slower. Smithies (1966) recommends that the best loin muscle should be trimmed and laid out flat, moulded carefully to a roughly circular block, and frozen. After freezing, steaks should be cut 10–15 mm thick from the block with a bandsaw. Quast and Karel (1968) have observed an impermeable surface layer when liquid foods are frozen.

Small biological specimens take several minutes to freeze, and slabs of food take about an hour. Longer times are needed to freeze blocks of food that are to be eventually sliced by sawing, or broken up into lumps so as to obviate the problem with the pellicle.

It is of both practical and theoretical interest to be able to calculate the rate of freezing of blocks of food having slab and cylindrical shapes, in order to secure the best compromise on the effects of the rate of freezing for freeze-drying. The simplest and most widely used approximate method of calculation is due to Plank (1913). Modified calculations have included sensible heat effects (Cowell, 1969; Slatter and Jones, 1972). The last two authors consider that many methods of calculating the rate of freezing are unreliable, and others making use of approximations are better, though more difficult to use.

In order to obtain Plank's (1913) formula for the freezing time in cylinders and slabs of material, it is assumed that:

1. Biological substances contain a high percentage of free water (exceeding 74 per cent) which may be regarded as entirely that percentage of water for purposes of calculation—the smaller percentage of solids having a negligible effect in most working ranges of freezing temperature.
2. The freezing point depression is only small (1°C to 2°C) and the material to be frozen is initially at the freezing point.
3. The sensible-heat effects of the solid and liquid phases are small and can be neglected; usually the specific heat of the frozen material can be ignored.

Solutions to the problem of calculating the rate of freezing will now be given which are much simpler than the original analysis due to Plank.

11.6.1 FREEZING IN A CYLINDER

The static freezing process in a cylinder, shown in Fig. 11.16(a), begins with heat extraction from the surface by the cooling medium with a heat-transfer coefficient h_2. Meanwhile, latent heat of crystallization of ice is being removed at the moving interface separating solid from liquid phases, and it may be assumed that the transfer coefficient h_1 here is still finite. The sensible heat and latent heat are therefore conducted through the solidifying medium to the outside surface of the cylinder with a thermal conductivity λ.

Consider axial symmetry in the cylinder with its temperature T a function only of the radius r; the differential equation of heat conduction is

$$\frac{\mathrm{d}}{\mathrm{d}r}\left(r\frac{\mathrm{d}T}{\mathrm{d}r}\right)=0. \tag{11.11}$$

At the interface the heat flux is

$$\lambda\frac{\mathrm{d}T}{\mathrm{d}r}=h_1T, \quad r=r. \tag{11.12}$$

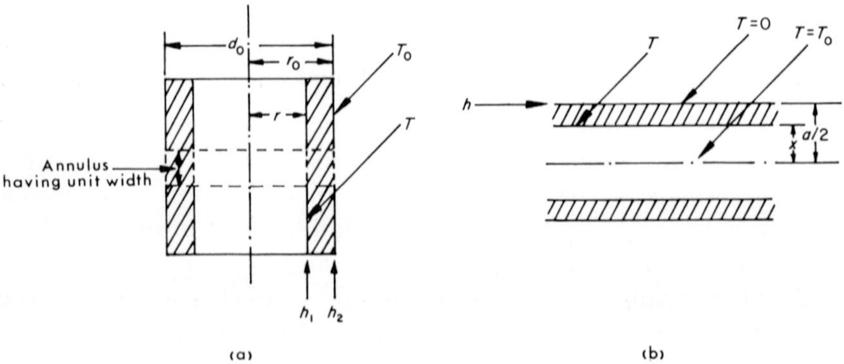

FIG. 11.16. Diagrams showing the moving interface for freezing in a cylinder (a) and in a slab (b).

And for a uniform and constant surface temperature T_0, the surface heat flux is

$$\lambda\frac{\mathrm{d}T}{\mathrm{d}r}=h_2(T_0-T), \quad r=r_0. \tag{11.13}$$

Equation (11.11) has a solution $T=A+B\ln r$ where A and B are constants. Substitution from equations (11.12) and (11.13), after eliminating A, yields

$$B=\frac{T_0}{\lambda/h_1r+\lambda/h_2r_0+\ln(r_0/r)}. \tag{11.14}$$

Assuming an annulus of frozen solid with unit width, its outside surface area at $r=r_0$ is $2\pi r_0$ and its inside moving interface area at $r=r$ is $2\pi r$. Thus the outside heat flux is given by

$$Q=2\pi r_0\lambda\left(\frac{\partial T}{\partial r}\right)_{r=r_0}=\frac{2\pi\lambda T_0}{\lambda/h_1r+\lambda/h_2r_0+\ln(r_0/r)} \tag{11.15}$$

after taking the derivative with respect to r of the solution $T = A + B \ln r$ and substituting B from equation (11.14).

If ΔH_v = latent heat per unit volume of the substance the rate of advance of the interface, $\mathrm{d}r/\mathrm{d}t = Q/2\pi r \Delta H_v$, whence

$$\frac{\mathrm{d}r}{\mathrm{d}t} = \frac{2\pi\lambda T_0}{2\pi r \Delta H_v [\lambda/h_1 r + \lambda/h_2 r_0 + \ln (r_0/r)]}. \tag{11.16}$$

Integrating equation (11.16) in the range $0 \leqslant r \leqslant r_0$ gives the freezing time

$$t = \frac{\Delta H_v}{T_0}\left(\frac{r_0}{h_1} + \frac{r_0}{2h_2}\right) + \frac{\Delta H_v}{T_0}\int_0^{r_0} \frac{r}{\lambda} \ln \frac{r_0}{r}\, \mathrm{d}r. \tag{11.17}$$

Now

$$\int_0^{r_0} \frac{r}{\lambda} \ln \frac{r_0}{r}\, \mathrm{d}r = \frac{r_0^2}{4\lambda}$$

and writing $1/2h = 1/h_1 + 1/2h_2$; $r_0 = \mathrm{d}_0/2$, the freezing time for a cylindrical body is given by

$$t = \frac{\Delta H_v}{T_0}\left(\frac{1}{4}\frac{d_0}{h} + \frac{1}{16}\frac{d_0^2}{\lambda}\right). \tag{11.18}$$

In practice, since the cooling curves for an average-size body which has either a rectangular parallelepiped shape or a cylindrical shape are almost equivalent (Carslaw and Jaeger, 1959), equation (11.18) can be applied to the freezing of blocks without loss of generality. Otherwise the freezing time may be expressed by

$$t = \frac{\Delta H_v}{T_0}\left(P\frac{d_0}{h} + Q\frac{d_0^2}{\lambda}\right), \tag{11.19}$$

where P and Q are two fractions for which the thickness ratios β_1, β_2, according to the size of the block, are given by the nomogram (Fig. 11.17).

In the circumstances where freezing takes place under almost static conditions apart from thermal convection currents in the liquid being frozen, equation (11.18) applies and a value for h has to be estimated. Otherwise for agitated freezing conditions, especially in the liquid, h_1 becomes infinite and so $h = h_2$.

11.6.2 FREEZING IN A SLAB

The rate of freezing of a slab depends on the temperature and velocity of the cooling fluid, the thickness of the slab, and the thermal properties of the substance undergoing freezing.

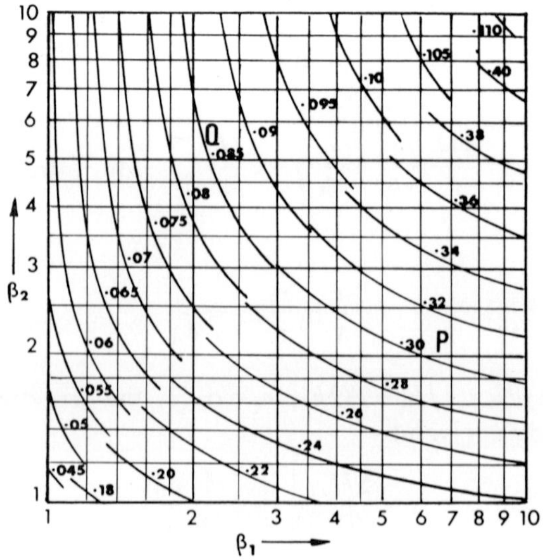

FIG. 11.17. Chart giving constants P and Q in Plank's equation (β_1 and β_2 are interchangeable) (after Ede, 1949b).

Consider the steady-state temperature T in a slab of thickness a and with temperature 0°C outside and freezing point T inside (see Fig. 11.16(b)).

The equation for heat conduction in the slab is

$$\frac{d^2T}{dx^2} = 0, \tag{11.20}$$

which has a solution $T = Ax + B$, where A and B are constants.

Assume for the boundary conditions:

$$T = T_0, \qquad x = 0, \tag{11.21}$$

$$\lambda \frac{dT}{dx} = -hT, \qquad x = x, \tag{11.22}$$

where h = surface heat-transfer coefficient.

Substituting equations (11.21) and (11.22) in the solution gives

$$A = -\frac{hT_0}{\lambda(1 + hx/\lambda)},$$

$$B = T_0. \tag{11.23}$$

The heat flux $-\lambda(\mathrm{d}T/\mathrm{d}x) = \Delta H_v(\mathrm{d}x/\mathrm{d}t)$, whence $hT_0(1 + hx/\lambda)^{-1} = \Delta H_v(\mathrm{d}x/\mathrm{d}t)$ and on solving for t over half the thickness $a/2$ of the slab, the freezing time is given by

$$t = \frac{\Delta H_v}{T_0}\left(\frac{1}{2}\frac{a}{h} + \frac{1}{8}\frac{a^2}{\lambda}\right). \tag{11.24}$$

11.6.3 SENSIBLE-HEAT EFFECTS

Consider sensible heat and latent heat together before and after the advance $\mathrm{d}x$ of the interface, and a new set of conditions, namely T_F = freezing point, T_0 = initial liquid temperature, and T_c = cooling medium temperature. The total heat per unit area can be easily determined from Fig. 11.18.

Before the advance of the interface it is

$$H = -\left(\frac{a}{2} - x\right)\rho c_f \frac{T_F + T_c}{2} + x\rho c_L \frac{T_F + T_0}{2} + x(\Delta H_v + \rho c_f T_F - \rho c_L T_F). \tag{11.25}$$

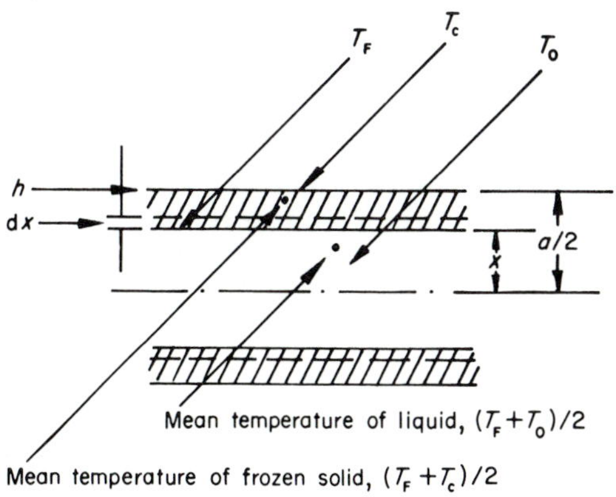

FIG. 11.18. Diagram for determining difference in enthalpy for freezing in a slab.

After the advance it is

$$H+\mathrm{d}H=-\left(\frac{a}{2}-x-\mathrm{d}x\right)\rho c_{\mathrm{f}}\frac{T_{\mathrm{F}}+T_{\mathrm{c}}}{2}+(x+\mathrm{d}x)\rho c_{\mathrm{L}}\frac{T_{\mathrm{F}}+T_{0}}{2}$$
$$+(x+\mathrm{d}x)(\Delta H_{\mathrm{v}}+\rho c_{\mathrm{f}}T_{\mathrm{F}}-\rho c_{\mathrm{L}}T_{\mathrm{F}}). \tag{11.26}$$

Thus the difference in total heat becomes

$$\mathrm{d}H=-\mathrm{d}x\left(\Delta H_{\mathrm{v}}+\rho c_{\mathrm{f}}\frac{T_{\mathrm{F}}-T_{\mathrm{c}}}{2}+\rho c_{\mathrm{L}}\frac{T_{0}-T_{\mathrm{F}}}{2}\right), \tag{11.27}$$

where c_{f} = specific heat of frozen solid, c_{L} = specific heat of liquid, $\bar{\rho}$ = mean density of solid–liquid phases.

This means that the enthalpy above the level at the frozen state is greater than before, when sensible heat is included, and from equation (11.27) the heat energy for freezing is now

$$\Delta H_{\mathrm{v}}+\rho c_{\mathrm{f}}\frac{T_{\mathrm{F}}-T_{\mathrm{c}}}{2}+\rho c_{\mathrm{L}}\frac{T_{0}-T_{\mathrm{F}}}{2}$$

per unit volume.

Substituting this term for ΔH_{v} in equation (11.24) gives the corrected formula for the freezing time as

$$t=\frac{\Delta H_{\mathrm{v}}+\rho c_{\mathrm{f}}\dfrac{T_{\mathrm{F}}-T_{\mathrm{c}}}{2}+\rho c_{\mathrm{L}}\dfrac{T_{0}-T_{\mathrm{F}}}{2}}{T_{\mathrm{F}}-T_{\mathrm{c}}}\left(\frac{1}{2}\frac{a}{h}+\frac{1}{8}\frac{a^{2}}{\lambda}\right) \tag{11.28}$$

This simplified analysis neglects to take into account the actual surface temperature of the frozen solid and its variation during the advance dx of the interface, and consequently overestimates the freezing time by a few per cent. However, it makes the calculation easier and gives a result which, according to Slatter and Jones (1972), may even be more accurate.

11.6.4 NOMOGRAPH

For practical purposes a nomograph is a useful tool (Anon, 1968). Figure 11.19 is a typical nomograph showing the relationship between thickness of a slab, air velocity, air temperature and freezing time; it is based on equation (11.28). Proceeding horizontally along the dashed line from a point where the half thickness is given as 20 mm to the curve of air velocity at 3 m s^{-1}, then vertically to the air temperature lines, it is seen that the freezing times for slabs, 40 mm thick, are 4·4

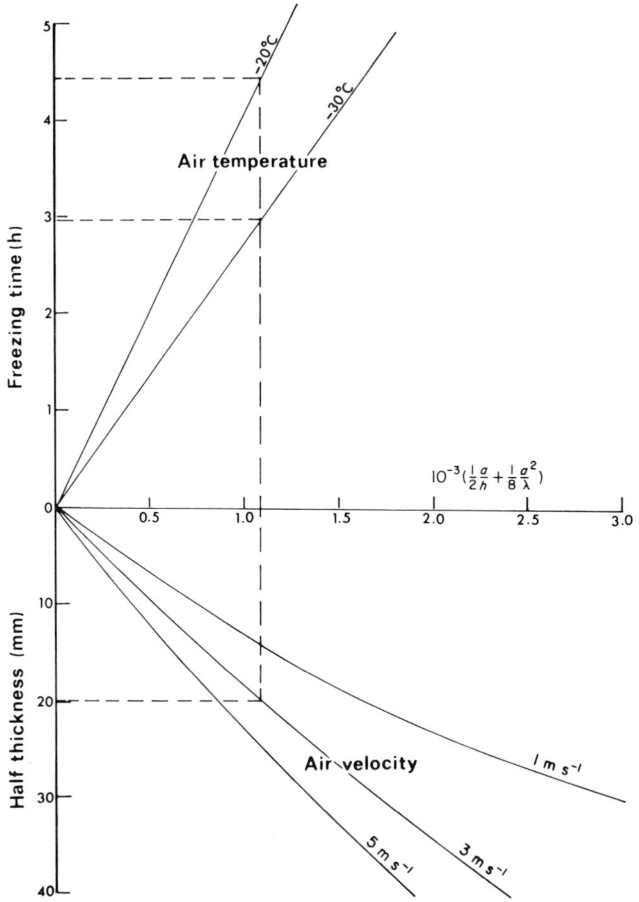

FIG. 11.19. A typical nomograph for slabs freezing in air.

and 2·9 hours at −20°C and −30°C respectively. The nomograph has been constructed so as to apply to the freezing of slabs of meat with the following properties:

$\rho = 1156 \text{ kg m}^{-3}$ (water content: 74 per cent for lean meat)

$\lambda = 1{\cdot}6 \text{ W m}^{-1} \text{ K}^{-1}$

$\Delta H_v = 209 \text{ MJ m}^{-3}$

$c_f = 1{\cdot}55 \text{ kJ kg}^{-1} \text{ K}^{-1}$

$c_L = 3{\cdot}44 \text{ kJ kg}^{-1} \text{ K}^{-1}$,

meat with a freezing point $T_F = -2°C$ and initial temperature $T_0 =$ 20°C being assumed. Earle and Fleming (1967) have given surface heat-transfer coefficients ($\text{W m}^{-2} \text{ K}^{-1}$) of 13·4, 21·3 and 28·7 for airspeeds (m s^{-1}) of 1, 3 and 5 respectively, these being the usual data for blast-freezing operations.

The central axis of the nomograph shown in Fig. 11.19 may be easily adjusted to account for variations between calculated and experimental freezing times, if this should be necessary. Adjustments to the axis of a similar nomograph (Anon, 1968) were made for the experimental results, involving two types of packaging materials used in the blast-freezing of cartoned meat, and were found to be quite satisfactory.

Increasing the fat content of meat reduces the water content and hence the latent heat. At the same time the thermal conductivity is reduced, but the overall effect is for freezing times to fall as the amount of fat rises; for example, 15 per cent fat gives 17 per cent reduction in freezing time.

12 Effects of Freeze-Drying

12.1 Shrinkage in drying

12.1.1 MECHANISM

Drying a body may result in undesirable shrinkage of the material. Raising the drying temperature increases shrinkage and raising the drying rate reduces it, but a change in one is accompanied by a change in the other and so the effect on shrinkage almost cancels out. Lowering the temperature, as in freeze-drying, should reduce shrinkage; and the data obtained by Luikov (1950) for freeze-dried apples between −5°C and −25°C confirms that this is the case. Shrinkage in freeze-drying is usually much less than in ordinary drying, occurring in highly concentrated solutions or retarded by the sublimation losses not compensating for the heat gains. Shrinkage depends on the mobility of the interstitial concentrate, which is very viscous but insufficiently rigid in the frozen state. At low temperatures the mobility is low, so that no changes occur to the porous structure left by the subliming ice; however, if the temperature reaches a critical level—the "collapse" temperature—the mobility of the concentrate is sufficient to cause flows and sags, eventually closing off or disrupting the pores. Sometimes intense evaporation can be induced under vacuum drying conditions in a non-rigid concentrate, with the result that expansion —not shrinkage—occurs in the form of puffing and frothing.

In structured foods shrinkage is less than in liquids since the membranes and cell walls help to prevent it, and so collapse temperatures are higher. For instance, the intact tissue of apple slices can be satisfactorily freeze-dried below −25°C whereas apple juice needs to be dried below −42°C.

12.1.2 BEHAVIOUR IN FOODS

Data on the collapse temperature (Table 12.1) set an upper limit to the sublimation temperature for freeze-drying (MacKenzie, 1966; Ito,

TABLE 12.1
Collapse temperatures for different solutes

Substance	Collapse temperature (°C)		
Sucrose	−24[a]	−32[b]	−22[c]
Glucose	−40	−40	−40·5
Lactose	−19	−31	−21
Dextran	−2	−10	—
Orange juice	—	—	−20
Sodium chloride	−22	—	—
Coffee extract	—	—	−20

[a] After Ito (1971).
[b] After MacKenzie (1966).
[c] After Bellows and King (1972).

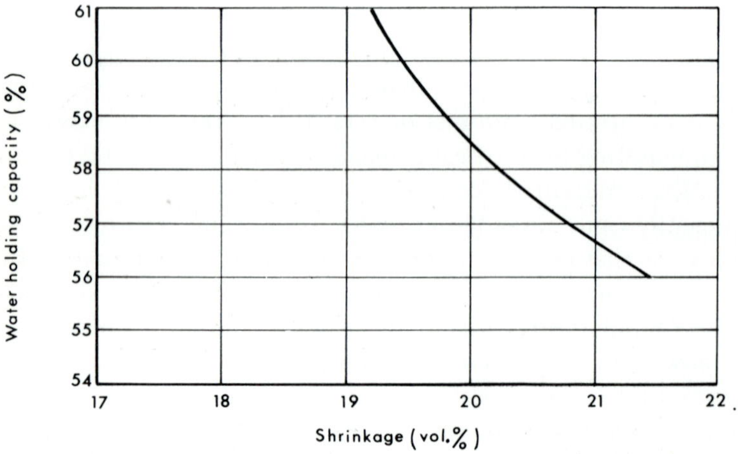

FIG. 12.1. Relationship between shrinkage and water-holding capacity of meat.

1971; Bellows and King, 1972). Rey's (1964a) data for the maximum temperature of incipient melting may also be interpreted as collapse temperatures (Section 13.1). Karel (1972–73) pointed out that shrinkage also depends on the moisture content and can occur in areas away from the ice interface and also may vary from place to place.

Beke (1973) observed that shrinkage depends significantly on heating temperature and on the duration of the heating process, and he determined a correlation between shrinkage and water-holding capacity in meat (Fig. 12.1). He also observed a correlation between water-holding capacity and heating temperature (Fig. 12.2). Table 12.2 shows that for freeze-dried pork shrinkage is 19 vol. per cent at an optimal heating temperature of 55°C.

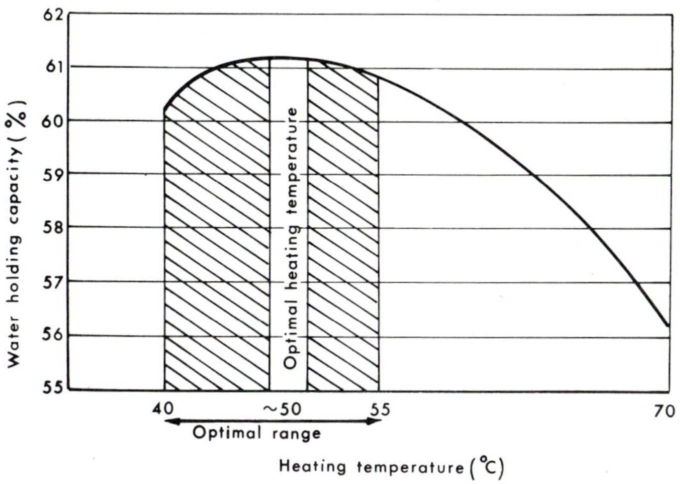

FIG. 12.2. Correlation between water-holding capacity and heating temperature.

TABLE 12.2
Shrinkage and water-holding capacity of freeze-dried pork (after Beke, 1973)

Property	Temperature (°C)		
	40	55	70
Average shrinkage (vol. %)	19·6	19·1	21·4
Water holding capacity (%)	58·8	60·9	56·1
Correlation coefficient	−0·76	−0·88	−0·67

Further consequences of higher collapse temperatures have been attributed to (1) radiant heating instead of heating by conduction through the ice; (2) removal of the surface skin formed upon freezing; and (3) low initial solids content.

12.1.3 ANALYSIS

King (1973) compared predicted and observed collapse temperatures for fruit juices and coffee extract (Table 12.3). The predicted values in

TABLE 12.3
Comparison of predicted and observed collapse temperatures for fruit juices and coffee extract (after King, 1973)

Product	Initial solids content (%)	Average solute mol. wt.	Collapse temperature Predicted (°C)	Collapse temperature Observed (°C)
Orange juice	23	277	−30	−24
Sweetened Concord grape juice	16	261	−33	−33·5
Grapefruit juice	16	228	−34	−30·5
Lemon juice	9	194	−36	−36·5
Prune juice	20	180	−42	−35
Apple juice	22	194	−40	−41·5
Pineapple juice	10	180	−40	−41·5
Concord grape juice	16	180	−41	−46
Coffee extract	25	—	—	−20

Table 12.3 are for temperature in terms of viscosity measurements of the residual concentrate for which the temperature values may be assumed from the ice-equilibrium freezing curves for solutions of various sugars. The viscosity can be determined on the basis of the energy dissipation equation for contraction of a cylindrical pore in an infinite medium, giving the time t_c required for surface-tension forces to overcome the gravitational forces causing flow of the interstitial concentrate, namely

$$t_c = \frac{\mu_{con} r}{2\sigma_{con}}, \tag{12.1}$$

where μ_{con} = viscosity of the concentrate, σ_{con} = surface tension of the concentrate, r = pore radius.

If t_d is the time to complete drying to the point where the flow of the interstitial concentrate stops after the passage of the ice interface then, if $t_c < t_d$, collapse occurs and, if $t_c > t_d$, it does not. Thus, calculations using equation (12.1) with $1 < t_d < 1000\,s$ in practice show that collapse occurs when $\mu_{con} < 10 - 10^4\,kN\,s\,m^{-2}$ (critical range).

King has also determined the collapse temperatures for 25 per cent fructose solutions containing added gelatin or pectin (Table 12.4).

TABLE 12.4
Experimentally observed collapse temperatures for 25 per cent fructose solutions with additives

Amount of additive (%)	Collapse temperature (°C)
None	−44
1 (gelatin)	−41
2 (gelatin)	−28
1 (pectin)	−34

The main conclusions that can be drawn from the data are: (1) equation (12.1) is sufficiently accurate for use in conjunction with other data for predicting collapse temperatures, (2) lower molecular weight sugar solutions have substantially lower collapse temperatures and (3) the addition of small amounts of gelatin or pectin to fructose solution results in an increase in collapse temperature without altering the normal physical characteristics of the solution—perhaps it explains why citrus juices with a naturally occurring soluble pectin content have higher than predicted collapse temperatures.

Haas *et al.* (1972) have investigated how the solubility of a gas in the water of biological tissues under pressure during freezing (typically 7 MPa) helps to reduce shrinkage and to improve the rate of drying. King (1974) ascertained that pressurized freezing also has certain advantages for freeze-drying, and determined the maximum effectiveness of certain gases such as air, methane, carbon monoxide and ethane in reducing shrinkage. These gases should have a Bunsen coefficient of solubility $\beta' \geq 1{\cdot}5\,m^3\,m^{-3}$ and a permeability $D_{g,w}\beta'_{gas} < D_{g,w}\beta'_{air}$, where $D_{g,w}$ is the diffusivity of the gas in water.

12.2 Flavour retention

12.2.1 PHYSICAL MECHANISM

Flavour retention is remarkably better in freeze-dried foods than in other dehydrated products, in spite of the high volatility of the aroma constituents and the fact that processing is done under vacuum. Flavour retention has been especially studied in model systems such as pectin gels, malto-dextrin, starch and cellulose, polyvinylpyrrolidone and (sometimes) apple slices. The addition of test volatiles such as acetone, n-alcohols and esters to these model systems has been monitored during the freeze-drying process by gas chromatography or radioactive labelling.

As would be expected, in the course of freeze-drying, water and volatiles continue to evaporate together until a certain water content is reached, then loss of volatiles ceases but loss of water continues. At the end of the drying process the observed values for the retention of volatile compounds differ widely, for they depend on the relative volatility. Finally, re-humidification above the cricital moisture content causes evaporation of volatiles again. For example, Chirife and Karel (1973) found that re-humidification from 11 per cent to 75 per cent relative humidity produced almost total loss of retained 2-propanol from freeze-dried starch.

Good volatile retention, therefore, is possible above a certain critical moisture content and it has sometimes been known to exceed 90 per cent. Also it has been shown by Saravacos and Moyer (1968) and Sauvageot *et al.* (1969) to depend on the nature and amount of dissolved solids, and to increase with decreasing freezing rate (Sauvageot *et al.*, 1969; Flink, 1969; Flink and Karel, 1970a, b).

The extent to which aroma constituents are lost in freeze-drying has been investigated by gas chromatography technique; by Spanyār *et al.* (1966) for coffee extract; by Soboleva and Popovsky (1971) for blackberry and apricot; and by Popovsky *et al.* (1965) for strawberries.

A satisfactory physical explanation for the fact that volatile flavour compounds are strongly held by very thin slabs of freeze-dried material, such as coffee extract and fruit products, was proposed by Menting and Hoogstad (1967a,b) who discovered two important contributory factors; firstly that the molecular diameters of the volatile compounds are linearly dependent on their diffusion coefficients, as shown in Fig. 12.3; secondly that volatile and water diffusion coefficients are related to moisture content, as shown in Fig. 12.4.

This discovery led Thijssen and Rulkens (1968, 1969) to develop a selective diffusion theory of volatile retention for predicting optimal

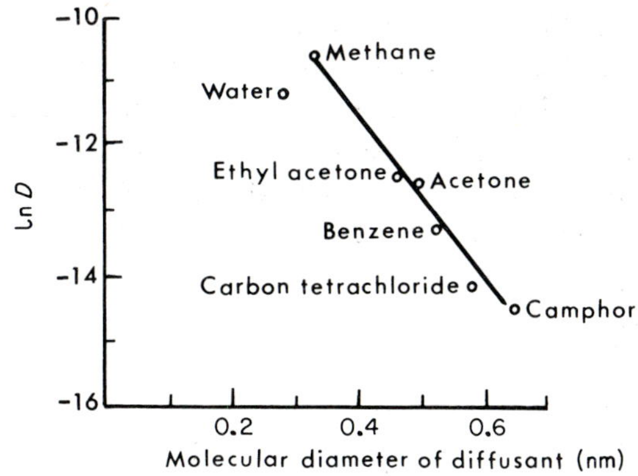

FIG. 12.3. Variation of diffusion coefficients of water and organic volatiles in malto-dextrin (moisture content: 20·1 per cent) against molecular diameter of the diffusant according to Menting and Hoogstad (1967a).

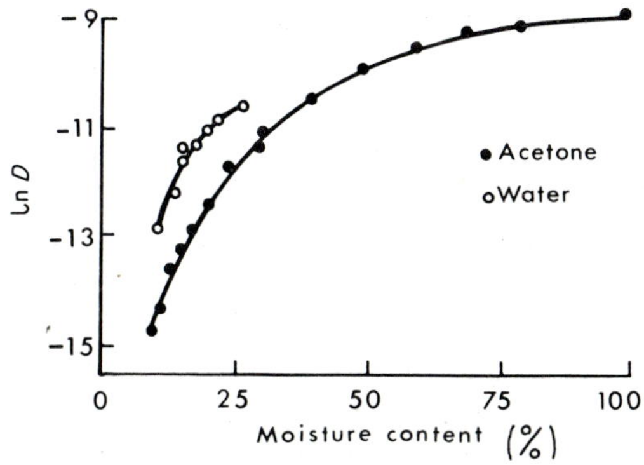

FIG. 12.4. Dependence of the diffusion coefficients of water and acetone on moisture content of malto-dextrin according to Menting and Hoogstad (1967a).

processes, and King and Chandrasekaran (1973) found it necessary to extend the analysis to a ternary diffusion process because the volatiles, water and dissolved solids all have quite different rates of diffusion. Another theory, not necessarily in conflict with theirs, has been proposed (Karel and Flink, 1969; Flink and Karel, 1970a); it is that the mobility of the volatiles involves their entrapment in the microregions of the dry product. Earlier, Rey and Bastien (1962), using added test volatiles (e.g. acetone) to Earl's balanced salt and glucose solution, had suggested that the volatiles migrating from the substrate were being adsorbed by the active surface of the dry layer. Even the formation of a surface skin impermeable to volatiles has been suggested. However, these theories have since been disproved by the observation that volatile retention is uniform throughout the dry material, with no further loss occurring when the dried mass is disintegrated on being subjected to mild heat and evacuation. Also adsorption isotherms of organic volatiles on dry carbohydrates do not suggest that they are more strongly adsorbed than water. Nevertheless, the adsorption theory still persists, since Bartholami *et al.* (1974) working on mushroom flavour compounds, and Chirife and Karel (1973) working with cellulose model systems, have demonstrated readsorption of volatiles in the dry layer and inclusions of volatiles between polymer chains. Also Chandrasekaran and King (1971b) tested volatile retention in slush-ice evaporation and obtained good results.

12.2.2 ANALYSIS OF SELECTIVE DIFFUSION

In this analysis the rates of diffusion of water and volatiles are different, with the diffusion coefficients of the volatiles decreasing much more steeply than those of water, as the water content decreases during the drying process, until they are virtually non-existent at low values of residual moisture.

When drying commences from the frozen state, the aroma constituents remain in the unfrozen liquid and eutectic phases until the ice crystals, which are essentially pure, disappear. Then the constituents are set free and diffuse along with residual moisture into the gas pores of the dry layer under temperature and concentration gradients. The temperature and concentration profiles have been discussed by Bomben *et al.* (1973). Thijssen and Rulkens (1969) have shown that, for the water–vapour flux,

$$N_w = k_w(c_{w,i} - c_w) = -\frac{D_w}{1 - c_{w,i}/\rho_w}\frac{\partial c_w}{\partial x}. \quad (12.2)$$

And for the flux of an aroma component j it holds that

$$N_j = k_j(c_{j,i} - c_j)$$
$$= -D_j\left(\frac{\partial c_j}{\partial x}\right)_{x=0} \qquad (12.3)$$

The relations between $(c_w)_{gas}$ and c_w and $(c_j)_{gas}$ and c_j are given by

$$K_w = \frac{(c_w)_{gas}}{c_w},$$
$$K_j = \frac{(c_j)_{gas}}{c_j}. \qquad (12.4)$$

The ratio of k_j to k_w is the relative volatility $\alpha_{j,w}$, given by

$$\alpha_{j,w} = \frac{k_j}{k_w}$$
$$= \frac{\gamma_j p_j^o}{\gamma_w p_w^o}. \qquad (12.5)$$

Now it can be assumed that $c_j = 0$, since the relative volatilities of most aroma components have high values (see Table 12.5) and thus the

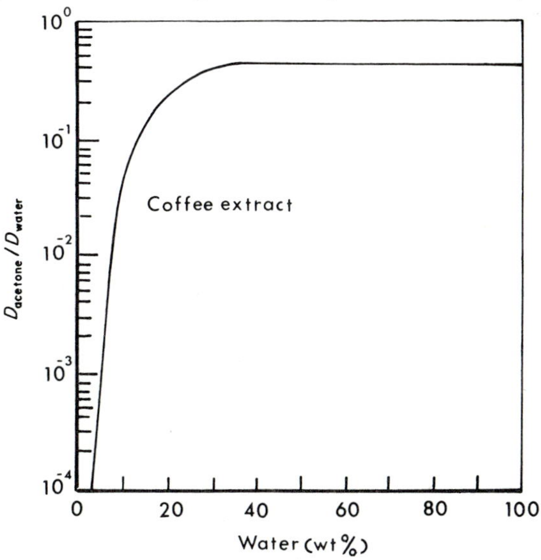

FIG. 12.5. Variation of the ratio of diffusion coefficients with dissolved solids concentration for coffee extract at room temperature.

TABLE 12.5

Vapour pressure, activity coefficients and relative volatilities of some aroma compounds at infinite dilution and in 70 wt per cent sucrose (after Chandrasekaran and King, 1971)

Aroma compound	Temperature (°C)	Vapour pressure p_j^0 (kPa)	Activity coefficient γ_j		Relative volatility $\alpha_{j,w}$	
			0% sucrose	70% sucrose	0% sucrose	70% sucrose
Methanol	25	12·67	1·53	—	7·84	—
Ethanol	25	7·83	3·37	6·7	7·79	15·5
Ethyl acetate	20	9.84	65	300	269	1200
Isobutanol	25	2.33	42·9	—	22·5	—
Acetal	25	3·67	150	—	173	—
n-Hexanol	25	—	1010	5700	390	2200
n-Butyl acetate	20	1·13	1270	7500	689	4070
Ethyl 2-methyl butyrate	20	—	5060	40 000	1160	9170
2-Hexanol	25	—	1160	—	146	—
n-Hexanol	25	0.13	676	—	31·20	—
n-Hexyl acetate	20	—	23 100	350 000	1580	23 900

concentrations in the liquids and gas–liquid interface can be neglected, i.e. the aroma constituents escape simply by molecular diffusion.

Diffusion coefficients in aqueous solutions are strongly dependent on water concentration, especially at low water content. For example, the ratio $D_{acetone}/D_{water}$ as a function of concentration in coffee extract is shown in Fig. 12.5. Below about 10–15 wt per cent the ratio becomes so small that the extract in fact can be considered as being selectively permeable for water only.

Consider the model of a drying slab as represented in Fig. 12.6 comprising three separate parts: ice phase, liquid phase and cylindri-

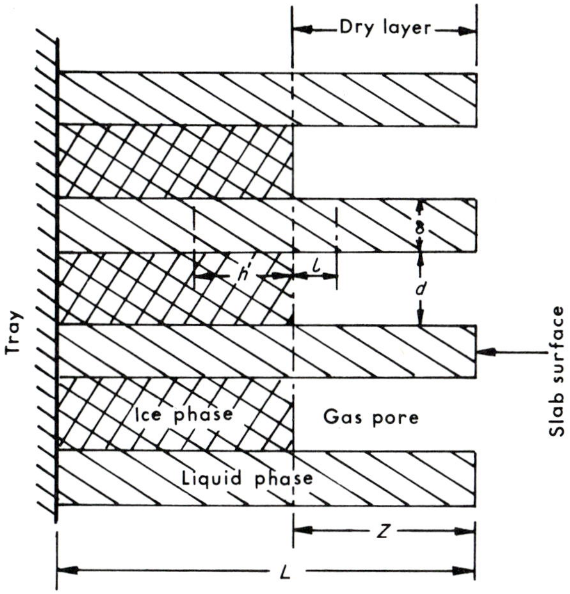

FIG. 12.6. Schematic representation of a drying slab (according to Thijssen and Rulkens, 1969).

cal gas pores. Above the sublimation front the aromas can escape from the liquid phase only because the ice phase is impermeable. Furthermore, they can only escape within a distance l in the dry layer above the sublimation front where the residual moisture has decreased below the critical level of 10–15 per cent; above this level they are immobilized. Now, to a good approximation, the distance l is inversely proportional to the velocity v_s of the sublimation front and directly proportional to the diffusion coefficient of water in the liquid phase, thus

$$l \propto \frac{D_w}{v_s}. \tag{12.6}$$

From the analysis of diffusion for the flux of aromas above about 50 per cent the length l is directly proportional to the square root of the dimensionless Fourier number $D_j t_s/\delta^2$ and inversely proportional to the drying surface.

At the interface the aroma loss of component j per unit drying surface is

$$N_j \propto (1-\varepsilon) c_j l \left(\frac{D_j t_s}{\delta^2}\right)^{1/2}. \tag{12.7}$$

Now the diffusion time t_s (time required to sublime the ice to a thickness l) is directly proportional to the length l and inversely proportional to v_s, i.e.

$$t_s \propto \frac{l}{v_s}. \tag{12.8}$$

Substitution from equations (12.6), (12.7) and (12.8) leads to

$$N_j \propto (1-\varepsilon) c_j \frac{D_w}{\delta v_s^2} (D_j D_w)^{1/2} \tag{12.9}$$

which in unit time becomes

$$N_j \frac{v_s^2}{D_w} \propto (1-\varepsilon) \frac{c_j}{\delta} (D_j D_w)^{1/2}. \tag{12.10}$$

Below the sublimation front, especially when the solution is not entirely frozen, as with coffee extract, aromas diffuse to the point where evaporation occurs. Schematically the aroma concentration profile is represented in Fig. 12.7 and may be approximated by a step function. Taking the concentration in the increment z to $(z+h')$ equal to $c_{j,z}$ and beyond, it is equal to $c_{j,0}$ originally. The penetration depth h' is chosen so that the two shaded areas are equal (Fig. 12.7); h' is proportional to D_j and inversely proportional to v_s, i.e.

$$h' \propto \frac{D_j}{v_s}. \tag{12.11}$$

Taking a mass balance in the frozen part and applying equations (12.7) and (12.9) it can be shown that the concentration of aroma component c_j in the dried layer at $y = z - l \simeq z$ is given by

$$\frac{c_{j,z}}{c_{j,0}} = \frac{ah'}{1+ah'} \exp\left[-\frac{(1+ah')z}{h'}\right] + \frac{1}{1+ah'}, \quad z \leq L - h', \qquad (12.12)$$

where a, for aroma loss above 50 per cent, is given by

$$a \propto \frac{1}{\delta}\left(\frac{D_w}{D_j}\right)^{1/2}. \qquad (12.13)$$

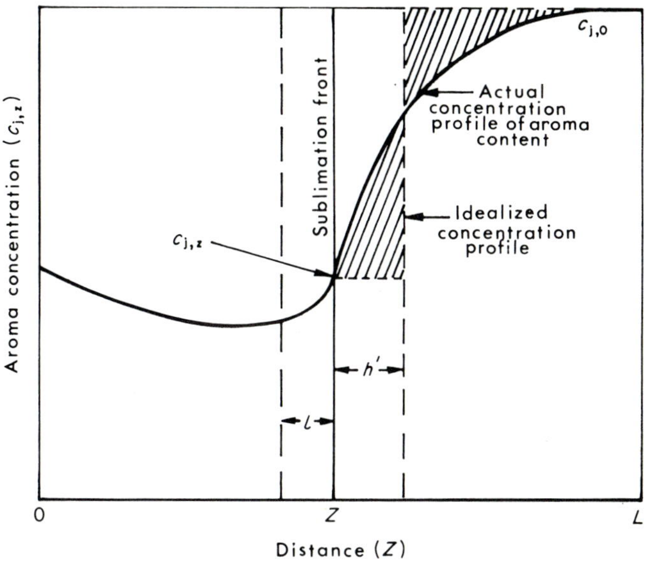

FIG. 12.7. Schematic representation of the aroma concentration profile in a drying slab.

Denoting the concentration $c_{j,0}$ by $c_{j,z}$ at $z = L - h'$ the concentration at $L - h' \leq z \leq L$ becomes

$$\frac{c_{j,z}}{(c_{j,z})_{z=L-h'}} = \left[\frac{L-z}{h'}\right]^{ah'}. \qquad (12.14)$$

Calculated aroma retentions from equations (12.12) and (12.14) for four different values of a and h', together with the experimental values are plotted in Fig. 12.8. The theory therefore gives good predictions of observed behaviour.

12.2.3 PREDICTION OF VOLATILE RETENTION

The main conclusion that can be drawn from the selective diffusion theory is that aroma retention strongly increases with drying rate.

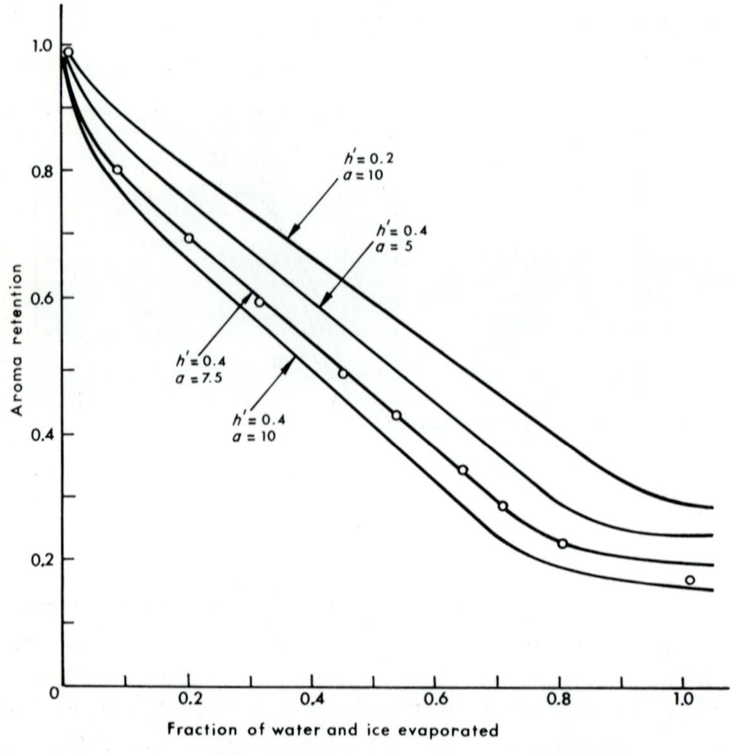

FIG. 12.8. Calculated aroma retentions with progress of drying for different values of a and h'. Experimental points are obtained for a 10 wt per cent pectin L.M. gel at 30°C surface temperature.

Consequently, it can be expected that retention in a fast drying bed of granulated particles would be higher than in a slab of the same thickness, or for any accelerated freeze-drying process. The dependence of volatile retention on a number of physical parameters for slabs and granules has been studied by Rulkens and Thijssen (1972).

This problem has been examined also by King and Chandrasekaran (1973) on the basis of a three-component diffusion model where volatiles, water vapour and dissolved solids all have different diffusion coefficients.

In addition to the water-vapour flux (equation 12.2) there is the volatile flux which Chandrasekaran (1971) gives as

$$N_j = -D_j\left(\frac{\partial c_j}{\partial x}\right)_{x=0} - \left[D_{j,w} - D_w\left(\frac{c_j/\rho_w}{1-c_w/\rho_w}\right)\right]\frac{\partial c_w}{\partial x}. \qquad (12.15)$$

A third equation, due to Chandrasekaran and King (1971a) for the binary diffusion coefficient for the volatile–water vapour system, is given by

$$\frac{D_{j,w}}{c_j} = \left[\frac{1}{c_w}\frac{\partial \ln \gamma_j}{\partial \ln c_w} + \frac{1}{\rho_j(1+c_w/\rho_w)}\left(1+\frac{\partial \ln \gamma_w}{\partial \ln c_w}\right)\right](D_j - \mathrm{D}_w)$$
$$+\frac{1}{c_w(1-c_w/\rho_w)}\left(1+\frac{\partial \ln \gamma_w}{\partial \ln c_w}\right)D_{j,w}. \qquad (12.16)$$

Combining the three equations gives

$$N_j = -D_j\frac{\partial c_j}{\partial x} - \frac{c_j}{\mathrm{c}_w}\left\{\left[\frac{\partial \ln \gamma_j}{\partial \ln c_w}(1+c_w/\rho_w) + \frac{c_w}{\rho_j}\left(1+\frac{\partial \ln \gamma_w}{\partial \ln c_w}\right)\right]\left(\frac{D_j}{D_w}-1\right)\right.$$
$$\left.+\left(1+\frac{\partial \ln \gamma_w}{\partial \ln c_w}\right)\frac{D_{j,w}}{D_w} + \frac{c_w}{\rho_w}\right\}N_w. \qquad (12.17)$$

The change in sign of $D_{j,w}$ is striking, as the measurements by Chandrasekaran and King (1971a) (Fig. 12.9) show. At about 40 per cent sucrose, $D_{j,w}$ becomes negative, returning to positive values at about 80 per cent sucrose. The effect is for the coefficient term of N_w (term in braces) in equation (12.17) to change sign. When it becomes positive the volatile loss rises; and when negative, the volatile loss falls. Thus, a locally inward flow of volatiles from the drying surface takes place at high drying rates, building up to an internal maximum volatile retention after a certain time. This is borne out by the results obtained by King and Chandrasekaran for water and ethyl acetate distributed in a gelled sucrose solution (Fig. 12.10).

There is, then, the possibility of predicting higher volatile retentions by combining a negative value of N_w at some intermediate water content and a high water vapour pressure difference for drying.

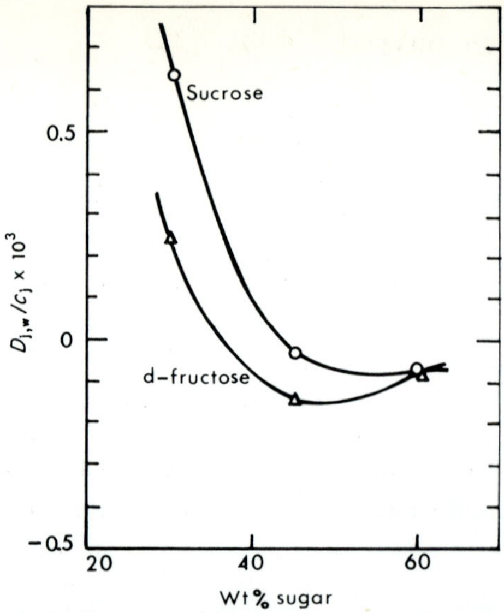

FIG. 12.9. Diffusion in a volatile–water vapour system (after Chandrasekaran and King, 1971a).

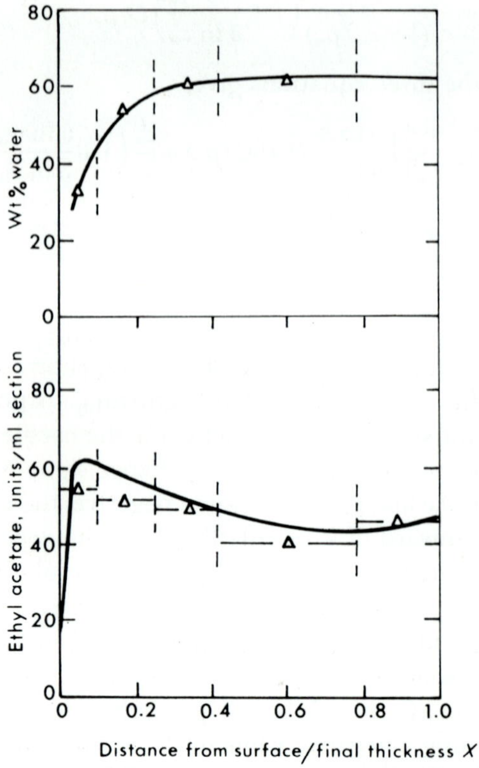

FIG. 12.10. Water and ethyl acetate distribution in a gelled sucrose solution after drying at 25°C and 133·3 Pa pressure. Initial concentration 25 wt per cent sucrose, 0·5 vol. per cent ethyl acetate and 1·9 wt per cent gelatin. (After King and Chandrasekaran, 1973.)

12.2.4 VOLATILE ENTRAPMENT

King (1970) considers that volatile retention by carbohydrates during freeze-drying is a reflection on the way in which the substance is prefrozen, and thus is reminiscent of Rey's devitrification treatment of a frozen substance in order to bring about a change from an amorphous to a crystalline state (see Section 13.1). This is in conflict with Flink and Karel's (1970a,b) idea, which stresses the importance of an amorphous structure for the retention of volatiles. They postulate that the volatiles are trapped in the microregions of an amorphous structure where the H-bonds between carbohydrate hydroxyls and water are replaced by H-bonds between carbohydrate molecules themselves. These bonds have to be broken to release the entrapped volatiles and this can occur through either crystallization or humidification to saturation values.

Upon humidification, volatiles retained by an amorphous carbohydrate are released because the adsorbed moisture disrupts the H-bonds holding the carbohydrate molecules together in the microregions. Thus, structural collapse due to moisture is important; however, in another experiment at −18°C, Flink and Karel (1969) demonstrated that structural collapse by itself was sufficient to release volatiles. This result throws doubt upon the selective diffusion model of volatile retention because it suggests that diffusion coefficients would have to be independent of water content, which they are not.

Flink and Karel (1972) demonstrated the amorphous state rather than the crystalline in a lactose model system containing 2-propanol. When the lactose was humidified and allowed to crystallize it released, besides water, the propanol, and this release commenced as the moisture content reached twice the monolayer value; thereafter the volatile and water loss rates remained the same. Although the reorientation of the microstructure affects volatile retention, Thijssen (1974) believes that an explanation can equally be found in terms of the physical hindering of the mobility of the aroma constituents. It would appear therefore that the entrapment concept requires some further proof. Earlier the effect had been believed to resemble closely the "locking in" of flavours by saccharides, which has been observed for citrus oils (Schultz *et al.*, 1956).

Among other observations, Spiess (1964) found that volatile retention in various foods was independent of temperature until the point of incipient melting was reached; but this is only an indication of the effects of loss of volatiles in evaporation.

12.3. Substrate loss—theory of particle entrainment

The disappearance of particles from the substrate during the course of freeze-drying is not uncommon. Pease (1947) observed the loss of K^+ and Cl^- ions from a freeze-dried myosin sol in 0·5 M potassium chloride under the electron microscope. Stein and Rogers (1950) showed that microorganisms and viruses could be cultured from the condensate water at the end of freeze-drying runs, and Cowan (1951) observed cross-infection in the vacuum chamber where different organisms were being dried together. Cowan (1951) and Thorne (1953) considered that the carry-over occurred during evaporative freezing when vaporization was extremely fast and sudden. Luikov (1969) thought that ablation of the substrate takes place due to the sudden increase in volume on sublimation. Anticipating shock waves and the entrainment of ablated particles by the vapour stream, he suggested that actual drying rates also involve substrate losses. Entrainment of particles is not surprising in view of the large expansion in volume of the vapour—for instance, 1 g water at a pressure of 13·3 Pa produces 9·5 m^3 water vapour. Greaves (1954) has suggested that, if the drying rate cannot be reduced, some neutral binding agent such as a protein should be added to the suspending fluid in order to make the freeze-dried substrate more solid and less susceptible to fragmentation and consequent substrate loss.

In freeze-drying, sublimation occurs mainly in the transition flow regime in the pressure range 7–133 Pa. It may be therefore instructive to deal with this phenomenon using Dauvois' (1970, 1971) analysis, by writing the transition flow equation for the case of particles entrained by a gas stream in terms of two parameters: (1) a dimensionless number ϕ, which is a function of Knudsen number; and (2) an equation for the limiting pressure in transition flow in terms of the diameter of the particles.

Based on Dauvois' measurements, Fig. 12.11 is typical for particle velocities in terms of pressure; and for the first parameter it leads to:

$$\text{Molecular flow:} \quad v = 0{\cdot}545\phi\frac{d^2\rho}{\mu_g} \qquad (12.18)$$

$$\text{Transition flow:} \quad v = 0{\cdot}545 \frac{d^2 \rho}{\mu_g}. \tag{12.19}$$

When $\phi = 1$ Knudsen number $Kn = 0{\cdot}18$, as in equation (12.21). Elsewhere the flow rate can be represented empirically by:

$$\text{Viscous flow:} \quad v = 0{\cdot}78 \frac{d^{1\cdot14} \rho^{0\cdot714}}{\mu_g^{0\cdot728} \rho_a^{0\cdot286}}. \tag{12.20}$$

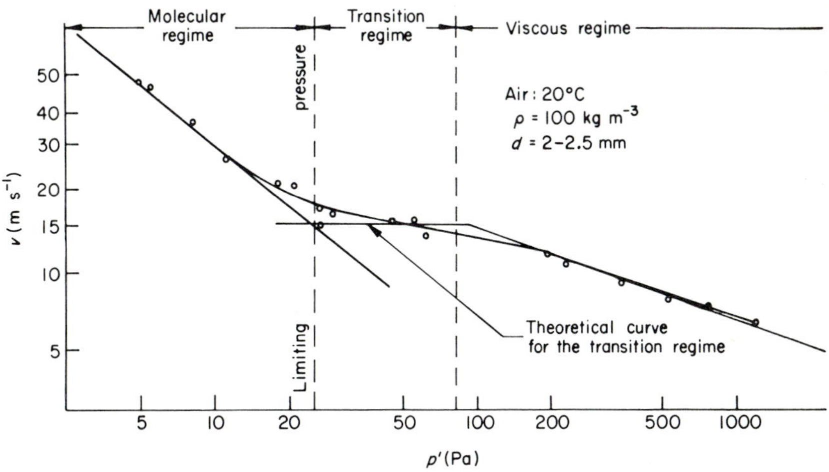

FIG. 12.11. Variation of particle velocity with air pressure for the molecular, transition and viscous flow regimes (after Dauvois, 1971).

In the pressure range 7–133 Pa and for particles ranging in diameter from 0·7 to 2·8 mm Dauvois verified that

$$\phi = 3{\cdot}3 Kn^{0\cdot7}. \tag{12.21}$$

Concerning the second parameter, he established that the limiting pressure depends on particle diameter but not on flow rate (as Fig. 12.12 shows, for particles with the same diameter and different densities) and he expressed it by

$$p' = \frac{\lambda}{0{\cdot}18d}, \tag{12.22}$$

where λ is the mean free path at 133 Pa and where $\lambda = 2{\cdot}3 \times 10^{-20} T / p\sigma^2$. He was able to verify equation (12.22) for air (Fig. 12.13) and also for carbon dioxide and argon but not for helium.

In general, equations (12.18–12.20) should apply to water vapour and therefore to freeze-drying conditions.

In order to examine the flight of particles in a fluidized bed at the bottom of a vertical tube fitted with a non-escape bulb, Dauvois

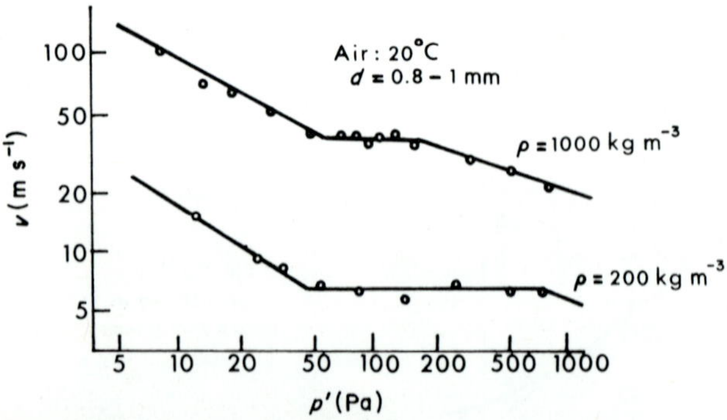

FIG. 12.12. Variation of particle velocity with air pressure for particles with the same diameter and different densities (after Dauvois, 1971).

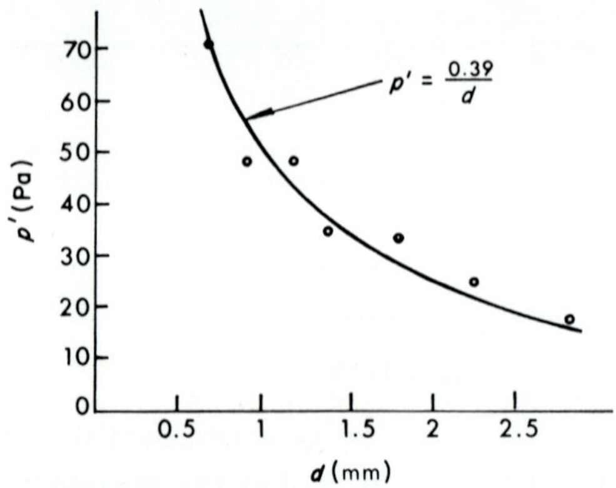

FIG. 12.13. Experimental data for verifying equation (12.22) for air (after Dauvois, 1971).

measured the gas velocity in a graduated glass burette in which oil from a connecting reservoir could be pressurized from a gas cylinder. The same gas could be made to fill the burette and flight tube by opening a connecting valve on the gas cylinder. He also simulated the flight of particles coming from a freeze-drying run. However, from the sketch of the apparatus (Fig. 12.14) it is not clear whether the entrained particles were collected along with water vapour in the trap or simply allowed to fall back into the sublimator. Also the sublimator does not appear to have a non-escape bulb.

It is noted that the ratio of equation (12.18) to equation (12.19) gives ϕ explicitly, hence the values are found by calculating molecular flow

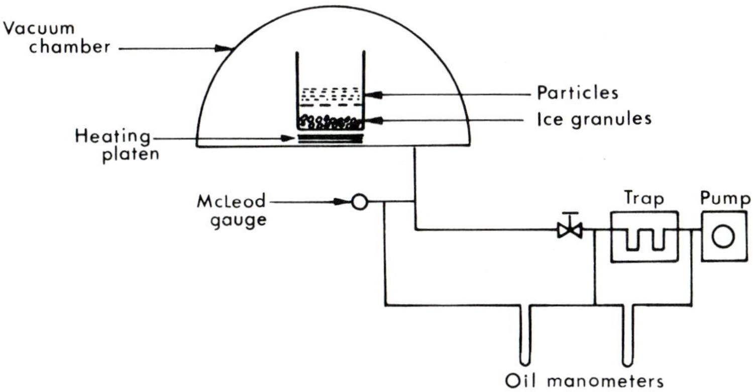

FIG. 12.14. Layout of a sublimator for examining the flight of particles from a fluidized bed.

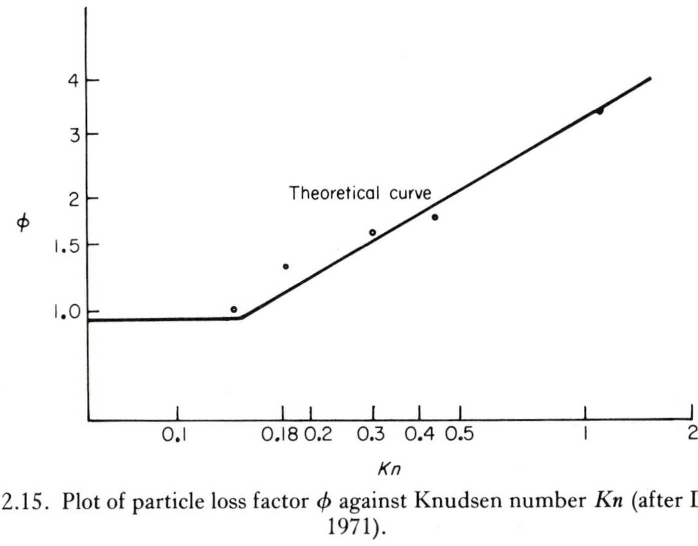

FIG. 12.15. Plot of particle loss factor ϕ against Knudsen number *Kn* (after Dauvois, 1971).

velocities on the one hand and observed transition flow velocities on the other. Parameter ϕ is important so far as freeze-drying is concerned since it shows that, for a given particle diameter and pressure, the specific sublimation rate without the flight of particles is 2–3 times higher than that estimated for the transition flow regime. This is obvious from the experimental curve in Fig. 12.15.

The theory has been found to apply to heterogeneous particles in freeze-drying. Measurements on a 35 per cent solids milk concentrate show that half the apparent volume of the bed disappears, indicating that the loss varies according to the specific volume of the particles. Over the specific density range 385–770 kg m^{-3} for the particles, the results show that the flight of particles occurs when their diameters range from 0·3 to 0·4 mm and the sublimation rates from 5 to 13 kg m^{-2} h^{-1}. The author has not been able to differentiate quantitatively substrate losses from sublimation weight losses for this range.

Substrate losses, involving ions and microorganisms, might be similarly treated, bearing in mind that the particles are lighter and the sublimation rates generally lower than those previously considered.

12.4 Rehydration

With some foods the freeze-drying process is carried out more easily than the rehydration process. This is because the drying process imposes irreversible changes and water is unable to penetrate fully a collapsed porous structure or to imbibe water as water of reconstitution. Besides, water molecules loosely bound to protein molecules may not return on rehydration to the precise positions they occupied prior to freezing. Several problems of rehydration capacity have been briefly dealt with elsewhere (Sections 11.3 and 11.6) and there are many more in the literature on the subject, notably the discovery of Wismer-Pedersen (1965) that a neutral solution of EDTA increased the rehydration capacity of meat without affecting the water removal during freeze-drying. In medicine and biology also the same is true: Greaves (1964) recommends the suspending fluid *Mist. Desiccans* for freeze-dried microorganisms and, amongst other things, addition of polyvinylpyrrolidone (PVP) or Dextran to aid reconstitution.

Several examples of freeze-dried foods where rehydration is particularly difficult will now be dealt with.

12.4.1 "PRONASE" RECONSTITUTION OF OYSTERS

Oysters are relatively easy to freeze dry but one of the most difficult products to store when dried and, in addition, whole oysters are extremely difficult to reconstitute in water (Mellor and Irving, 1969). Fresh oysters are composed of 87 per cent water and 13 per cent solids, of which 6·2 per cent is protein and 1·2 per cent fat. Oxidation of fat in freeze-dried oysters is very rapid, with off-flavours developing within a few hours. The protein, on the other hand, needs to be digested to a certain extent with a proteolytic enzyme to ensure successful reconstitution in water. One such proteolytic enzyme from *Streptomyces griseus*, known as "Pronase", has been studied by Nomoto and Narahashi (1959). A very small percentage of the enzyme added to the water is a better aid to reconstitution of a dry oyster than is puncturing with a pin, or prior evacuation of the air from the product to aid penetration of water.

The present author has adopted the following method. Chopped and whole oysters were freeze dried and carefully removed from the freeze-dryer under an atmosphere of dry inert N_2 gas; they were then packed in cans *in vacuo*. Reconstitution was carried out in shallow dishes containing water or a 0·1 per cent solution of Pronase P, so that rehydration took place from the bottom upwards through the pieces. Some whole oysters were pricked before drying, others after drying, and some whole ones were rehydrated under vacuum.

There was not much difficulty in reconstituting chopped oysters in cold water provided the pieces were small. Reconstitution took some minutes, depending on size, but in boiling water it was generally rapid. Whole oysters were extremely difficult to reconstitute in cold or warm water, sometimes taking up to 2 h during which time off-flavours developed. Pricking and rehydration under vacuum had little or no effect.

When a 0·1 per cent solution of Pronase P was used whole oysters were reconstituted completely in about 15 min. The slight digestion resulting from the hydrolysis of the peptide bonds in 0·5 g protein estimated for each oyster was sufficient to allow the water to penetrate through the bottom and up to the top surface of the oyster, where the skin was tough and glaze-like in appearance. This did not impair their structure or make them too soft for eating. Even if oysters were allowed a longer time for reconstitution excessive digestion was not apparent. The oysters reconstituted under these conditions were similar in appearance to fresh ones and, provided they were eaten within a short time, appeared to be of reasonable texture and flavour.

12.4.2 INHOMOGENEOUS STRUCTURE AND REHYDRATION

In the freeze-drying of fruit and other materials containing large quantities of sugar there are difficulties in obtaining a homogeneous dried product, and consequently the product cannot be readily rehydrated. Mellor (1954) reported non-uniformity in freeze-dried apricot and peach purées in the form of frothed material and the formation of gummy films, and described a method of protecting the surfaces of the purées in trays by forming thin ice layers above and below the material in the freezing stage. A homogeneous dried product was obtained which rehydrated easily, but the ice-layer method increased the amount of ice to be sublimed substantially—thinner ice films formed by spraying water were not tried.

Rolfe (1958) found that attempts to evaporative-freeze raw beef steaks led to case hardening, with the result that freeze-drying and rehydration proved impossible. However, spraying the meat with water led to improvements. Another method of reducing the formation of a pellicle during evaporative freezing is by interposing the foodstuff between perforated flat sheets, which are pressurized but allow the water vapour to escape.

Improved rehydration capacity of freeze-dried citrus juice powders is attributable to the addition of "flavedo" to lemon juice and gelatin to orange juice as a stabilizer. Flavedo is an extract from the rind of the fruit. Campbell *et al.* (1943–44) found that the addition of flavedo to lemon juice improved rehydration capacity and also improved the rate of freeze-drying and the flavour, and added a source of vitamin C and carotenoid colour. They added 1 per cent flavedo before freeze-drying.

In the case of orange juice, added gelatin proved superior to pectins and other natural substances, giving a much better rehydration capacity and flavour.

12.4.3 REHYDRATION CAPACITY

Rehydration capacity is defined as the ratio of the weights of water regain to loss during freeze-drying, and is determined by rehydrating a weighed sample of the freeze-dried product to the point where there is no excess surface moisture left after removal with a sponge and then reweighing. The test may be looked upon as a simple indication of a good freeze-drying process.

Goldblith *et al.* (1964) studied the rehydration capacity of freeze-dried shrimps. They found that high drying temperatures, longer pre-cooking times at high temperatures and high rehydrating water temperatures all contributed to a lowering of rehydration capacity. Beke (1969) has studied the rehydration capacity of freeze-dried pork; King *et al.* (1968) freeze-dried turkey; Karel (1963) freeze-dried salmon; and Goldblith (1964) freeze-dried haddock.

13 Miscellaneous Aspects of Freeze-Drying

13.1 Process control—crystallization and melting points

Trials with the objects of determining drying conditions, measuring process parameters and obtaining satisfactory products are an essential part of freeze-drying technology. However, in some instances trials can be dispensed with, especially for freeze-drying of liquid substances, by determining the start of incipient melting by differential thermal analysis (DTA) and the temperature of complete ice crystallization by resistance measurements.

13.1.1 DIFFERENTIAL THERMAL ANALYSIS (DTA)

The DTA method, as outlined by Rey (1960a), compares the thermodynamic behaviour of a frozen substance during the course of rewarming with that of a frozen reference liquid, usually distilled water under identical conditions. The procedure is based on Vuillard's (1957) method, used in mineral chemistry.

Deviations from zero temperature difference between the unknown liquid and the reference liquid (water) are recorded against temperature during the course of cooling to −196°C, and during the course of reheating at the rate of 1°C min^{-1}. (The existence of a supercooling effect in most liquids prevents precise measurements during cooling; thus measurements during rewarming are preferred.)

A typical melting curve (Fig. 13.1, for orange juice) is one which circumvents the supercooling phenomenon. Others are shown in Figs 13.2 and 13.3. Rey and Bastien (1962) attribute the exothermic peak at

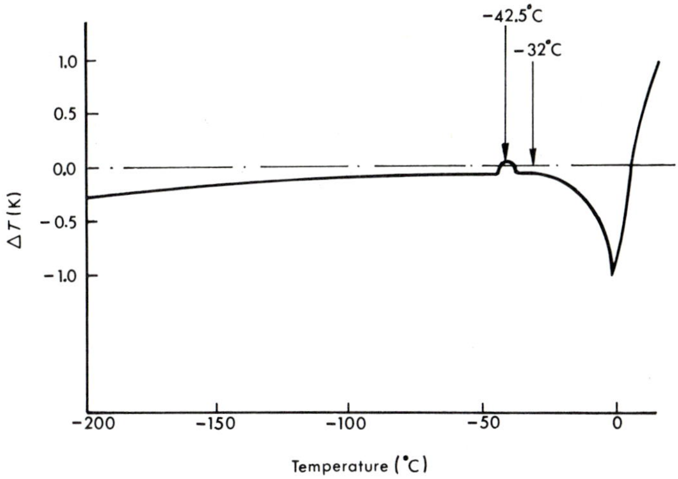

FIG 13.1. Differential thermal analysis of orange juice (reference solution, distilled water) (after Rey and Bastien, 1962).

−42·5°C in Fig. 13.1 to devitrification (recrystallization) because ice does not initially crystallize upon cooling down to the eutectic point; instead it ends up as a vitrified or glass body without releasing heat. Rey and Bastien reported a more pronounced peak for Earle's balanced salt solution with 50 per cent glycerol added. Recrystallization temperatures are usually 20°C to 30°C below the melting point, as indicated in Table 13.1, being more pronounced for carbohydrates and glycols and insensitive to initial concentration. (Luyet (1960) contends that this last statement refers to the concentrates formed by removal of water during ice crystallization and not to the initial concentration itself.)

13.1.2 RESISTANCE MEASUREMENTS

The electrical resistance method is based on a proposal by Greaves (1954) for measuring the electrical conductivity of a continuous network of channels containing interstitial fluid which tightens up

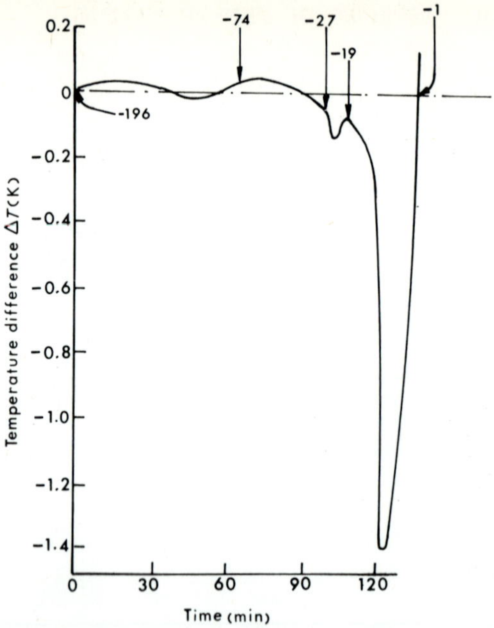

FIG. 13.2. Differential thermal analysis of Earle's balanced salt solution (after Rey, 1960a).

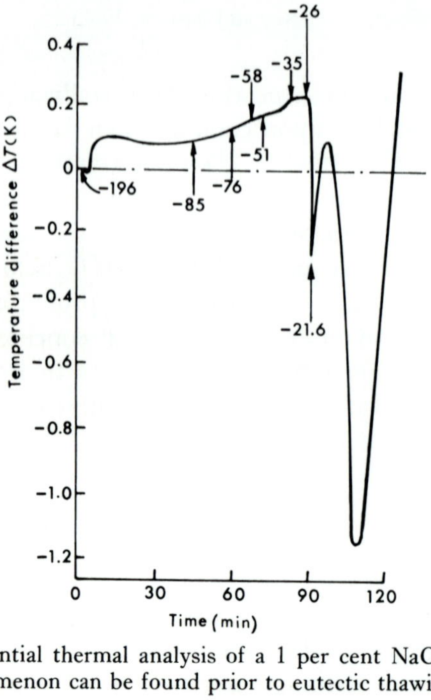

FIG. 13.3. Differential thermal analysis of a 1 per cent NaCl solution (case where exothermic phenomenon can be found prior to eutectic thawing) (after Rey, 1960a).

TABLE 13.1
Recrystallization temperatures (after MacKenzie, 1966, and Rey, 1964a)

Substance	Temperatures of incipient melting Min. (°C)	Temperatures of incipient melting Max. (°C)
Biologicals		
Earle's balanced salt solution	−27	−19
Horse serum	−34	−17
Suspending fluid for microorganisms (dextran 1%, sodium glutamate 5%)	−41	−19
Seric gonadotropin 1% in 10% solution of glycerine in water	−11	−8
Human vitreous	−23	−19
Ground pork adrenals	−37	−27
Ground beef lungs	−30	−20
Ground beef spleen	−29	−20
Sugars		
Dextran	−10	
Fructose	−48	
Glucose	−41	
Raffinose	−27	
Sucrose	−32	
Foods		
Strawberries	−32	−20
Orange juice	−43	−18
Apple juice	−43	−23
Grape juice (Muscat)	−45	−27
Lemon juice	−22	−19
Peach juice	−21	−18
Apricot juice	−29	−22
Precooked foods		
Homard à l'américaine		
meat	−38	−32
sauce	−47	−47
Coq au vin		
meat	−41	−24
sauce	−51	−32
Tournedos sauce madère		
meat	−37	−25
sauce	−48	−48
Miscellaneous		
Gelatin	−11	

around the ice crystals as they form and significantly increases the resistance.

During freezing, the concentration of a solution increases sufficiently to cause the electrical resistance to rise, because the narrowness of the veins of interstitial fluid decreases in proportion to the progress of crystallization. Microscopic examination shows that narrow veins persist in frozen solutions—for example, in a frozen sodium chloride solution down to −40°C. This resistance rises quite steeply and uniformly until it levels off at around −45°C when the frozen liquid has attained the temperature of complete solidification. On slow rewarming such a solution freezes entirely and the resistance is inflected sharply at the beginning and more slowly in the neighbourhood of −20°C, when the mass appears to thaw (Figs 13.4, 13.5 and 13.6). The path of the curve, following successive freezing and rewarming cycles, is always the same, and consequently a control point can always be prescribed and relied upon.

13.1.3 HEAT TREATMENT

As far as freeze-drying is concerned there is no definite ice crystal structure which corresponds to any given freezing temperature; rather

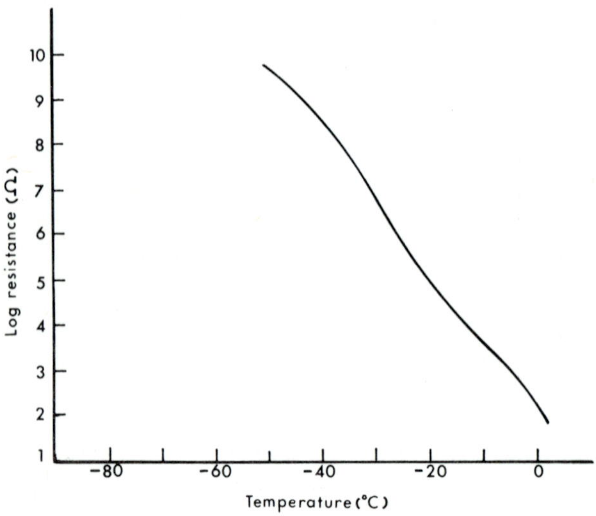

FIG. 13.4. Electrical resistance of orange juice at low temperatures (rewarming) (after Rey and Bastien, 1962).

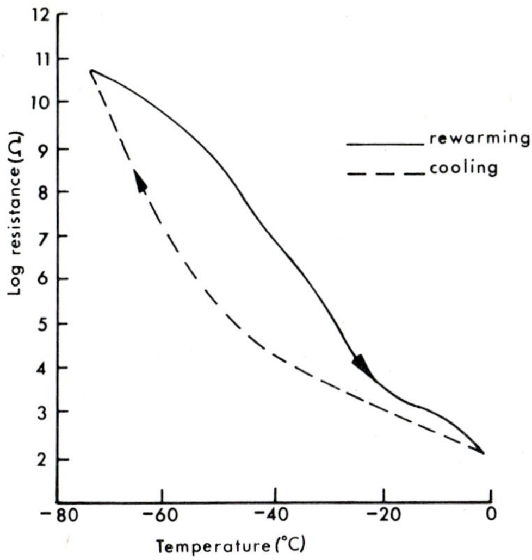

FIG. 13.5. Variation of electrical resistance of Earle's balanced salt solution during cooling and rewarming (after Rey, 1960a).

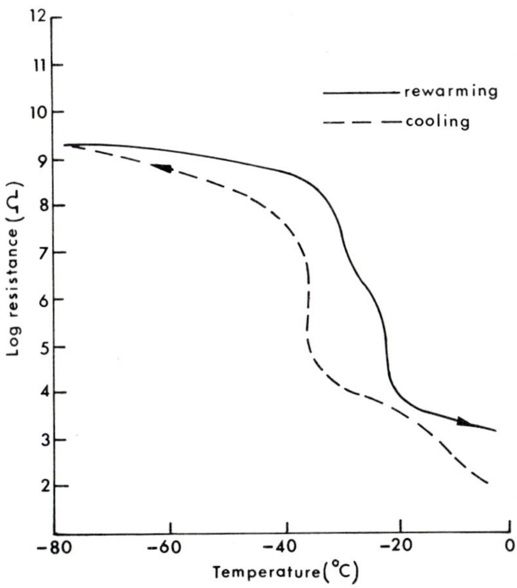

FIG. 13.6. Variation of electrical resistance of a 1 per cent NaCl solution during cooling and rewarming (after Rey, 1960a).

the structure depends on the way in which the substance is frozen and thermally treated afterwards. For example, a solution slowly frozen to −40°C differs widely from one which is first cooled to −80°C and later rewarmed to −40°C. The failure of fruit juices and certain other substances to remain in a frozen crystalline state during drying is well known and usually manifests itself as "puffing", foaming, or local melting (stickiness). For example, troubles of this sort with 28 per cent sucrose solution and also with 40 per cent coffee concentrate drying at −20°C have been reported by Quast and Karel (1968), and with apple juice drying in a fluidized bed at −34°C by Malecki *et al.* (1969). Another trouble is the tendency for the dry layer of a partially freeze-dried product to collapse at a certain temperature which, according to MacKenzie (1966), almost coincides with the crystallization temperature.

Earlier Rey (1960a) and Rey and Bastien (1962) controlled the sublimation temperature of frozen orange juice and of Earle's balanced salt solution by monitoring the reading of electrical resistance of the subliming frozen solids. However, the freeze-dried orange juice proved to be hygroscopic and developed poor quality in storage.

Satisfactory results were obtained by an alternative heat treatment (Rey and Bastien, 1962) in which the orange juice was frozen at −40°C in the ordinary way; later hardened at −196°C and then slowly rewarmed to −65°C before freeze-drying, as had been done previously. The product was less hygroscopic and had a faster drying rate. Some devitrification of orange juice must have taken place and hence the resulting crystallization provided a higher sublimation temperature zone for freeze-drying.

In a sense the frozen substance is subjected to a true "tempering" treatment. The method is clearly outlined in Section 1.4 and seems to be sufficiently convenient for most practical purposes, with the proviso that many fragile and labile substances, having a "eutectic zone" situated between −35° and −45°C, should be frozen to below −80°C and desiccated at about −40°C; even after this treatment they are still slow to freeze dry.

In summarizing the determinations to be made to satisfy the "eutectic requirements" of a substance to be freeze dried, Rey recommends:

1. Measuring the maximum temperature of complete solidification as a resistance, i.e. the highest temperature at which it is neces-

sary to cool to be sure supercooling effects are no longer present and all liquid phases are solid.
2. Measuring by DTA the minimum temperature of incipient melting by, i.e. the lowest temperature at which some endothermic phase begins that can be correlated to partial or total melting.

13.2 Residual moisture

In affording protection for microorganisms sugars have been used (Section 14.2). A point in their favour is that some retention of residual moisture is possible and necessary, especially for storage of freeze-dried specimens. Originally it was thought that dried organisms should be processed until they were as "dry as dust"; however, this was found to detract from viability (Fisher, 1950; Fry and Greaves, 1951). An optimal residual moisture content of 0·5–1·5 per cent which can be regulated by the presence of glucose in the suspending fluid is now generally accepted as satisfactory for microorganisms, and for viruses a little less than 1 per cent is recommended. The glucose acts by binding a certain small quantity of moisture thereby preventing overdrying.

The residual moisture may be determined by the methods of Flosdorf, Karl Fischer or Beckett—citing Szakmary and Vali (1963). Precisely, the methods are known as:

1. Gravimetric
2. Titrations
3. Vapour pressure

13.2.1 METHODS OF DETERMINATION

In the usual methods of moisture determination by heating in a vacuum oven or by chemical titrations it is impossible to differentiate between "free" and chemically "bound" water in biological materials and foodstuffs. The test sample is very often destroyed and is irrecoverable. As a standard, a moisture determination is made on a few grams of substance in a weighing bottle which is dried under a coarse vacuum at 50°C for about 24 hours. Alternatively, the drying can be accomplished in an evacuated desiccator at less than 13·3 Pa for several days. Both methods involve serial weighings on duplicate samples which are tedious and are subject to errors. Often the weight loss is to some extent due to decomposition because of overheating.

Similar decomposition and final destruction problems arise with chemical titrations for determining residual moisture. In the best known of these, due to Karl Fischer (1935), a weighed portion of the test sample is mixed with anhydrous pyridine-methanol and titrated with a solution of iodine and anhydrous sulphur dioxide until the colour changes from dark brown to yellow due to the pressure of water. It is claimed that with this titration as little as 0·5 mg water can be detected, or better than 0·1 mg using an electrometric end-point.

Measurements of the equilibrium vapour pressure over a substance may be employed to determine moisture content. Vincent and Simmons (1940) devised one such absolute apparatus, and Makower and Myers (1943) another. Rey (1964c) has described other types. Beckett (1954) has described fully a micro-moisture determination apparatus based on both these units.

Briefly, the procedure with Beckett's apparatus is to evacuate a glass system (Fig. 13.7) containing the test sample to below 13·3 Pa, with

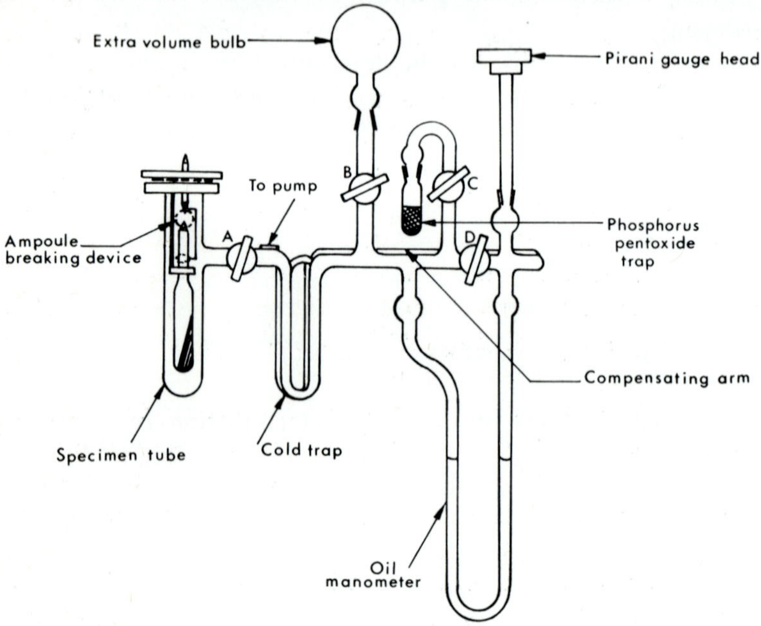

FIG. 13.7. Layout of micro-moisture determination apparatus (after Robson, 1954).

the sample held at a uniform temperature, and to collect the evaporated moisture in a cold trap at −78°C by opening taps A and D. The cold trap is then allowed to warm up with taps A and D closed but with tap C open, and the moisture is reabsorbed by the phosphorus pentoxide trap. When equilibrium was reached in 0·5 to 2 hours the pressure was read on an oil manometer which was filled with Octoil diffusion pump fluid. A continuously running vacuum pump connected to one arm of the manometer maintains a stable zero, and a compensating glass vacuum line, as shown in Fig. 13.7, follows the path taken by the cold trap before a connection is made to the pump.

The apparatus can be made independent of calibration simply by accurately determining the volume of the system between the taps A and D and including the head space in one arm of the oil manometer. Now, for a given vapour pressure rise in the system, the mass of water associated with this rise can be shown to be given by

$$m = p_{\text{oil}} V\left(\frac{1}{101\,308}\,\frac{273}{T}\,\frac{\rho_{\text{oil}}}{\rho_{\text{M}}}\,\frac{18}{22{\cdot}4}\right). \qquad (13.1)$$

where m = mass of water, V = volume of the system, T = absolute temperature, p_{oil} = oil pressure, ρ_{oil} = density of oil at T, ρ_{M} = density of mercury at T.

The manometer can be calibrated in scale divisions (10^{-5} g water) for direct measurement of the mass of moisture evaporated from the test sample. Refinements include a magnetically operated hammer for breaking open the necks of glass ampoules and vials which have been nicked at the stem and heated with a hot nichrome wire loop, and an extra volume bulb which is used to extend the upper range of the apparatus. Also a phosphorus pentoxide trap is fitted to absorb moisture on completing a reading. Both the bulb and the trap fittings are connected through isolating valves.

Beckett (1954) found that large corrections needed to be applied for the water vapour leaking in through the O-ring seal on the specimen tube. These were obtained from blank runs on the apparatus from time to time and are clearly indicated in Fig. 13.8, the theory of which is given in Section 14.3.

The apparatus has been employed by Robson (1954) for (1) studying the effect of the drying rate and adsorpton on freeze-dried substances, (2) examining water-vapour diffusion through rubber seals used on plasma bottles and others for holding freeze-dried products, and (3) determining the efficiencies of desiccants.

Comparisons with one vacuum-oven determination have shown that, whereas the vacuum oven gives no indication of sample decomposition, the micro-moisture apparatus indicates decomposition during a test by a plateau in the drying curve.

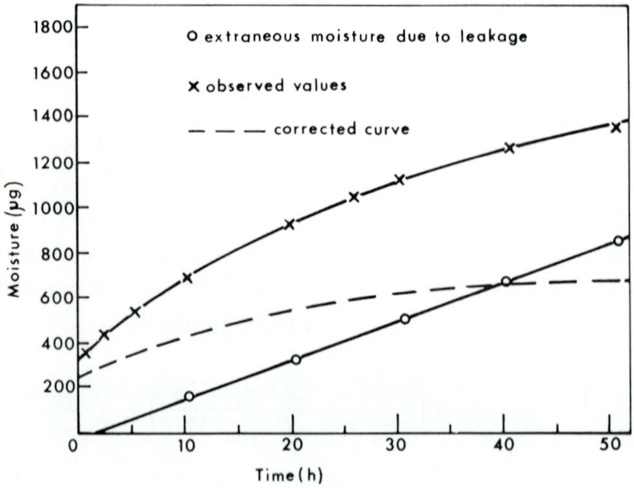

FIG. 13.8. Drying curves for an ampoule containing freeze-dried human blood plasma, extraneous moisture, and corrected result (after Beckett, 1954).

13.3 Measurement of end-point

A simple method of determining the end-point of freeze-drying (Mellor and Irving, 1969; Mellor, 1971), called by Mellor "end-point by trapped pressure ratio" (TPR), was developed in connection with the cyclic process; it suits equally well any type of freeze-drying process conducted *in vacuo*. The gas space above the product is sampled on several occasions during the final stages of secondary drying, and the partial pressure of water vapour is estimated in terms of a ratio, as an indication of the moisture content of the product. The measurements actually made are the total pressure in the chamber and the partial pressure of air when the water vapour is condensed out; the ratio of this partial pressure to the total pressure is then the TPR.

The product is unloaded when the TPR approaches a constant value in the range 0·6–0·8. The actual value in this range depends upon the type of food being freeze dried, and is defined on the basis of prior determinations of moisture contents in the product at several points in the range.

The apparatus (Fig. 13.9) consists of a copper U-tube; one arm is bent at right angles and attached horizontally to a manual vacuum

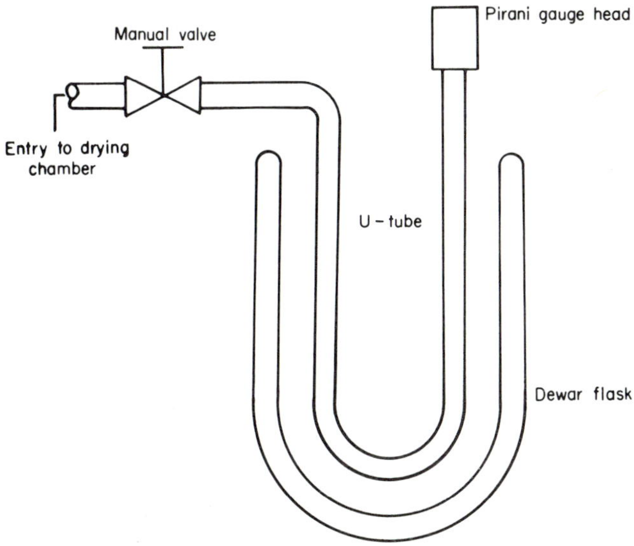

FIG. 13.9. Trapped pressure ratio apparatus.

valve, which in turn is connected by an elbow to the side of the drying chamber. A Pirani vacuum gauge head (range 0·13–1333 Pa) is connected to the top of the other arm. The whole assembly is mounted on the outside wall of the drying chamber so that the tube enters at a point in the middle of the drying end and has a short tube on the inside which can be directed to the vicinity of the product. A Dewar flask, containing a cold refrigerant, e.g. dry ice (CO_2) chips and alcohol, is used for condensing out the water vapour in the gas sample. After the valve has been closed the Dewar flask is slowly moved up from below the U-tube and clamped in position when the tube is fully immersed.

To operate the apparatus, the absolute pressure (water and air) is read on the Pirani gauge after closing the valve. The U-tube is then cooled by immersion in the Dewar flask and the gauge is again read when steady (the reading may be corrected for the dry ice temperature (−78·4°C) if high accuracy is required. The TPR is then the ratio of the second to the first reading). After the readings have been taken the valve is opened again to receive the next sample, and the Dewar flask removed to allow the tube to warm up. Normally about 30 minutes is allowed to elapse between readings, so that the gas sample in the tube may again come to equilibrium with the gas in the chamber. A typical response curve of TPR against drying time is shown in Fig. 13.10, which indicates that the approach to the end-point under cyclic pressures is about twice as fast as under constant pressures.

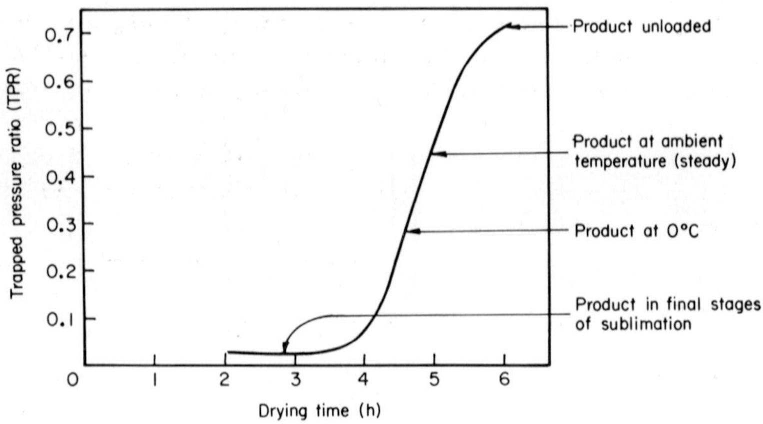

FIG. 13.10. Changes in the trapped pressure ratio during the freeze-drying of peas.

To compare the TPR against moisture content, the latter should be measured by taking three or four samples of the product after a drying run. In practice, the moisture content of a particular product need only be initially determined twice at two different values of TPR. Referring to Fig. 13.11, it should be possible to determine the moisture content in terms of TPR for any freeze-dried food by reading off the moisture content from the curve against the TPR that has been measured. It is seen that the data for six different foods are scattered about the regression line, as well as the data for a single product, e.g. mince meat, peas and yoghurt.

It should be possible to take TPR readings at regular intervals by operating automatically a small cryogenic cooler for condensing out the water vapour in a closed-off sample tube, and a heater to return the system to ambient temperature ready for the next sample. The device could be calibrated in terms of moisture content and it ought to be possible to determine TPR values in the range 0·5–5 per cent with an accuracy of 0·5 per cent.

Bouldoires and Bricout (1969) have described a similar end-point method in which the gas is sampled by suction instead of by diffusion down a tube. The method permits rapid sampling into a previously evacuated tube where the vapour can be condensed and discharged finally by an auxiliary vacuum pump. By allowing a feeble air injection continuously into the chamber, the strength of the vapour flux, in terms of the partial vapour pressure (difference between the two readings, as discussed above), can be followed all along as heating is applied during the course of sublimation to the point where the flux diminishes as it gives way finally to the presence of air from the injection. Sublimation is then complete. Bouldoires and Bricout have not attempted a calibration of the apparatus in terms of moisture content for partial vapour-pressure points beyond the end of the sublimation process.

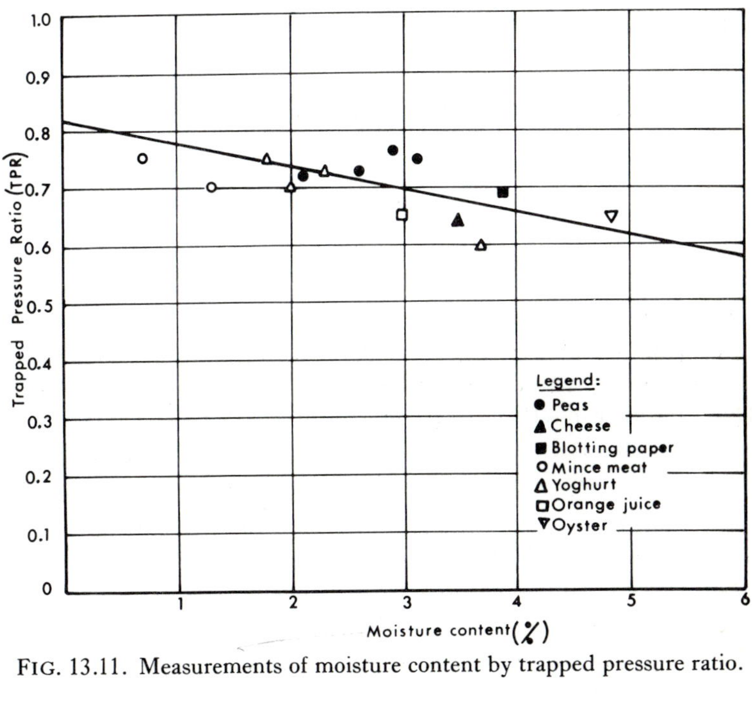

FIG. 13.11. Measurements of moisture content by trapped pressure ratio.

14 Biological Aspects

14.1 Living organisms

14.1.1 INTRODUCTION

Scientists have often contemplated the problem of the origin of life and of possible extra-terrestrial signs of it. At the end of the last century, Arrhenius put forward the idea of "Panspermia"—that life came to the Earth as bacterial spores after being wafted through space and that, if it must start somewhere, then it must simply be eternal.

One of the strongest arguments for believing in Panspermia is that organisms survive very long periods of time in the freeze-dried state at low moisture contents, without light and oxygen, and at temperatures close to absolute zero (Becquerel, 1950; Havelka, 1969). Another concerns the likelihood of long interplanetary journeys with the organism protected from radiation exposure buried in a meteorite or in a space probe which might be the work of an advanced extra-terrestrial civilization. It is believed that conditions for freeze-dried storage of biological materials exist on Mars, since temperatures are generally very low, low water and oxygen tensions exist, and the atmospheric pressure is about 7 mbar; however, these views could be outdated by most recent satellite work on the planet. Besides, recent indications that amino acids were probably synthesized on earth under conditions of heavy electrical storms have dampened a lot of enthusiasm for believing in unlimited life and its existence on other planets.

It is no secret that "water is the *mater* of life and also the matrix" in so far as the metabolism of living things is concerned, and this is equally true when much of the water is immobilized through freezing, macromolecular water binding, and direct removal through drying. Living things in which the water requirements are meagre lie dormant and survive; the simplest organisms surviving longer than the more complex ones. Becquerel (1951) estimated, assuming continuous chemical change according to van t'Hoff's law, that the reduction in metabolic rate is $8{\cdot}5\times10^4$ at -100°C and $7{\cdot}7\times10^{13}$ at -270°C compared with a normal life of one year at 10–20°C.

Knowledge of the behaviour of such organisms gives an insight into water structure of extra- and intra-cellular freezing and unfreezable water, since a residual moisture level is believed to be responsible for a continuous though imperceptible metabolism and, if some of the water is not required, it is thought to be responsible for maintaining a spatial configuration or protective barrier preventing oxidation. In freezing, at least to -100°C, the percentage water unfrozen in an organism is generally accepted as 5–10 per cent and intimately involved with the cells as unfreezable water which cannot be frozen out beyond this point, but is held by other forces such as van der Waals' forces. For example, freezing experiments on molluscs and human erythrocytes (Davies, 1968a) show that the viability is not really related to temperature but to the percentage water unfrozen.

The two principal uses to which freeze-dried microorganisms have been put are for the maintenance of laboratory strains and for vaccination. Storage banks, containing many different types of microorganisms, are in service at many laboratories throughout the world, so allowing the maintenance of standards and proof of identification of microorganisms during epidemics. Nowadays, the role of freeze-drying in vaccination is well recognized; it is widely used for BCG and *Brucella abortus* vaccines and for a number of viral suspensions which are important in medicine. Since the freeze-dried vaccine must be reconstituted when the vaccination takes place, the suspending fluid must have some special properties, such that it can be readily rehydrated and safely injected into man or animals without producing toxic or immunological effects. It must also be able to store well in the tropics.

14.1.2 PROTECTION AGAINST FREEZING PRIOR TO DRYING AND CHANGES IN THE STRUCTURE OF WATER

Freezing damage occurs in three ways: (1) as extra-cellular ice crystal growth causing mechanical disruption of cells, because of the large crystals formed on slow cooling, (2) as intra-cellular ice crystals which, although fine as a result of rapid freezing, nevertheless cause some disarray in the structure of water, and (3) the so-called protein denaturation, probably caused by the increasing concentration of salts in the cells as water is being removed as ice.

An important development in freeze preservation of cells is the protective effect afforded by glycerol (Lovelock, 1953) in reducing the amount of water available for freezing and therefore the concentration of electrolytes in the cell.

The glycerol concentration giving the freezing temperature at which irreversible damage first appears can be derived as follows:

The freezing-point depression ΔT for an aqueous solution of sodium chloride and glycerol with concentrations c_N and c_G respectively is given by

$$\Delta T = K^*(2c_N + c_G), \tag{14.1}$$

where K^* is the freezing-point depression of water per unit weight of solute. In terms of the critical salt concentration nc_N and the freezing-point depression at which irreversible damage first appears, equation (14.1) becomes

$$\Delta T = K^* n(2c_N + c_G). \tag{14.2}$$

Lovelock has shown (Fig. 14.1) that the equation is quite satisfactory for predicting the glycerol concentration required to prevent damage in rabbit spermatozoa, bull spermatozoa and human blood cells. Glycerol acts as a salt buffer and its low molecular weight, penetrability and lack of toxicity make it ideal for the protection of cells. However, it is being superseded by dimethyl sulphoxide (DMSO) because it is superior to most others, such as glucose, but nobody knows why. Unfortunately the vapour pressures of these compounds, being lower than for water, preclude them from being used in freeze-drying because of the increased toxicity as the concentration and temperature rises. However, in freeze-drying, sugars have sometimes replaced glycerol and DMSO, and this raises the question of whether the sugars act by protecting the organisms in freezing and not really in freeze-drying. Another difficulty is with thawing. Where small crystals occur through rapid freezing, Mazur (1960) has sugges-

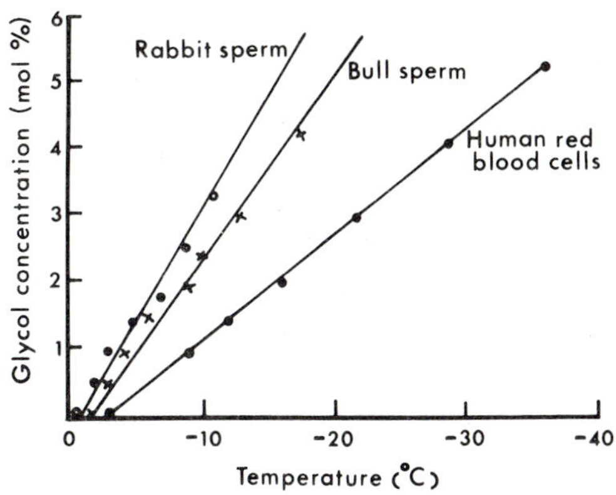

FIG. 14.1. The glycerol concentration required to protect rabbit spermatozoa, bull spermatozoa, and human red blood cells when frozen at various temperatures for 10 min (after Lovelock, 1953).

ted quick thawing to prevent intra-cellular crystals enlarging to lethal sizes before melting. If this is impossible then the recourse is to freeze-drying instead of thawing.

14.1.3 DRYING

To overcome the difficulties associated with freezing, thawing and drying the protection afforded by excipients—substances which in small amounts enhance or protect the properties of others—is of considerable importance. The outstanding example is peptone (Davies, 1968b), which has been shown to protect *Pseudomonas* sp. (10H) against damage in slow as well as fast freezing and in subsequent freeze-drying (Fig. 14.2). The same has been found to apply to *bacteriophage* T_4 (Fig. 14.3). Further, the marked difference in percentage survival between 20 per cent peptone and 10–12.5 per cent sucrose suspensions of these organisms at low rates of freezing (10–100°C min^{-1}) was in complete contrast to that at 900°C min^{-1} when they came together. It is therefore inferred that sucrose protects by stabilizing the low-energy levels of water binding whereas peptone protects by stabilizing both high- and low-energy levels in lattice and clathrate water structures. Thus, it is more difficult to remove water in the presence of peptone. It should therefore be possible to design

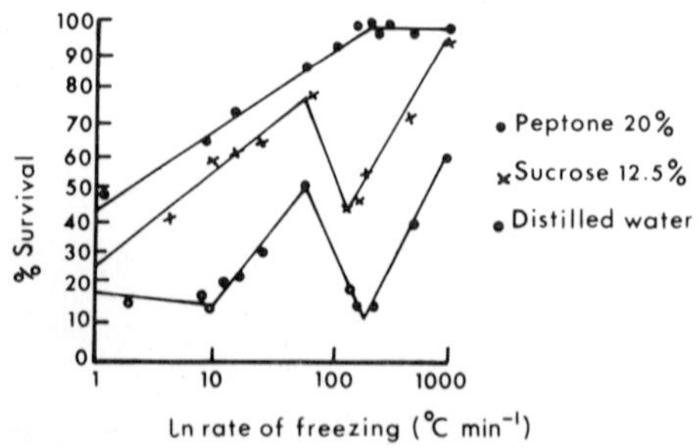

FIG. 14.2. The effect of freezing rate on the survival of *Pseudomonos sp.* (10H) suspended in various media.

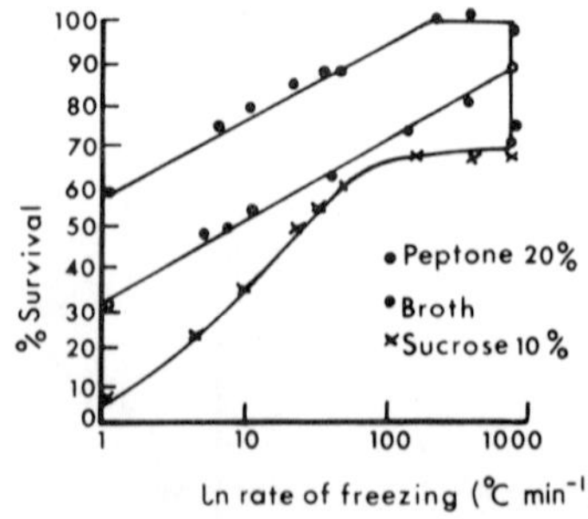

FIG. 14.3. The effect of freezing rate on the survival of *bacteriophage* T_4 suspended in various media.

freezing and drying experiments whereby the "binding" energies with structured water can be studied from viability counts on different protective suspensions and organisms.

14.2 Chemical degradation in storage

Loss of viability during slow removal of water and storage is due to chemical degradation, which is a function of the temperature, water activity and storage time. It will not be advantageous always to reduce chemical degradation by producing the right residual moisture–water activity levels if the sublimation and desorption processes are going to take an unduly long time at low temperatures or a shorter time at high temperatures. This is because the kinetics of degradation depend on temperature as well as time. This is immediately obvious from Arrhenius' plot of the rate constants for the reactions of different activation energies E_A (Fig. 14.4), which are for the latent heat of vaporization of water 40 kJ mol^{-1}, and for degradative reactions 130 kJ mol^{-1}. The higher activation energy process, therefore, has a rate which increases more rapidly with temperature.

14.2.1 CARBONYL COMPOUND REACTIONS

Reactions between carbonyl compounds and amino groups (Maillard reaction) and the oxidation of cell constitutents are the two most important ways in which the viability of freeze-dried organisms are lowered during storage.

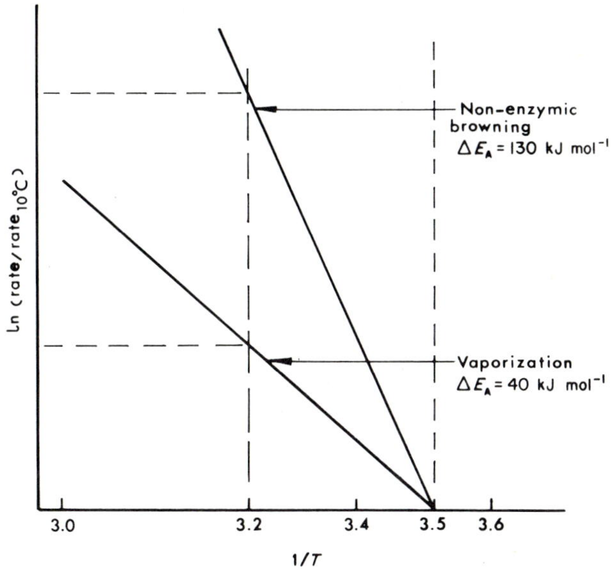

FIG. 14.4. Arrhenius' plot of activation energies for degradative and vaporization processes.

Thus a physical process which cannot accommodate temperature and time satisfactorily led Scott (1958, 1960) to claim that "a fundamental cause of death in dried organisms arises from alterations in cell protein brought about by reactions between carbonyl compounds present in the cells and amino groups of the cell protein resulting in the removal of amino side chains". The finding caused Greaves (1960a) to investigate accelerated storage trials using as his criterion, not long-term survival at 20°C but short-term survival at 100°C, and this led to sucrose being substituted for glucose, especially for long-term storage at relatively high temperatures.

14.2.2 EFFECT OF OXYGEN—THEORY

The storage life of microorganisms and vaccines is impaired by the presence of oxygen, even in the smallest quantities. The literature on the subject abounds with statements such as the sealing pressure of air should be as low as possible, about 8 Pa (Christian and Stockton, 1956); oxygen should be excluded (Scott, 1960); and free oxygen exposure to very dry microorganisms is detrimental (Stark and Herrington, 1931). Thus, in order to overcome the difficulty and to ensure ease and safety in opening, tubes containing an organism are filled with dry nitrogen at atmospheric pressure before being sealed (Rosenthal, 1948; Miller and Goodner, 1953; van Drimelen and Steyn, 1958; Muggleton, 1960). Commercial "dry" nitrogen is not entirely free of oxygen or water vapour and so other inert gases have often been tried—for example, argon (Gheorghiu and Sturdza, 1969).

Alternatively, storage *in vacuo* seems ideal but for the hazards on breaking open the neck of the ampoules (Section 14.3).

A theory for the effects of oxygen on the storage life of dried microorganisms has been proposed by Cox *et al.* (1973), who found a mathematical expression fitting the viability data in the literature as a function of storage time and oxygen concentration.

Firstly, it was found that the logarithm of viability was directly proportional to oxygen concentration (Fig. 14.5) and falls off rapidly at low concentration and, secondly, it is thereafter fixed at high concentration. However, as with all storage trials, viability falls off exponentially with time and so the relationship is more complex.

Cox *et al.* have chosen as their starting point the analogy with enzyme kinetics by assuming the reversibility of a complex formed between oxygen and a carrier compound X, that is

$$X + O_2 \underset{K_x}{\leftrightarrows} XO_2 \tag{14.3}$$

By postulating a substance A, being essential for viability and biologically inactive in its oxidizable state, they arrived at the relation:

$$XO_2 + A \overset{k^*}{\rightarrow} AO_2 + X \tag{14.4}$$

The above reaction equations lead to a solution provided that $d[XO_2]/dt = 0$ and $d(O_2]/dt = 0$, which have been deduced to be mostly true in practice; therefore

$$\ln \frac{[A]_0}{[A]} = \frac{k^*[X]_0[O_2]}{K_x + [O_2]} t, \qquad (14.5)$$

where $[A]_0$ = concentration of A at $t = 0$, $[A]$ = concentration of A at t; $[X]_0$ = concentration of X at $t = 0$, $[O_2]$ = concentration of oxygen (per cent), K_x = equilibrium constant (per cent), k^* = velocity constant for conversion: A to AO_2.

Turning again to the postulate for the concentration [A]; from equation (14.5), $-\ln [A] \propto t$, yet the gradient of the curve $\ln V_i$ against time decreases with time.

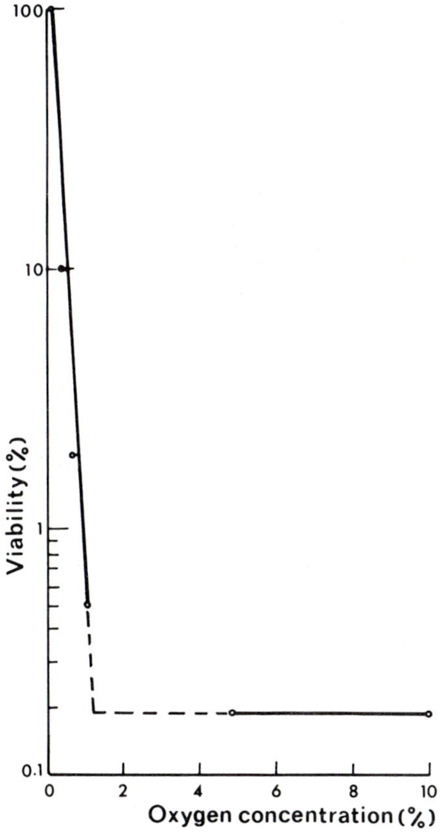

FIG. 14.5. Viability (per cent) versus oxygen concentration (per cent) for oxygen induced death of *Serratia marcescens* 8UK/2, stored under oxygen for 3 hours (after Cox and Heckly, 1973).

Given

$$P_D n_L = -\delta n_L$$

$$P_D = -k_1 \delta [A] \tag{14.6}$$

$$V_i = \frac{n_L}{n_L + n_D},$$

where P_D = probability that the organism dies, n_L = number of live organisms, n_D = number of dead organisms, V_i = viability.

Rearrangement of equations (14.6) and integration gives

$$\ln V_i = k_1[A] + k_2. \tag{14.7}$$

Substitution from equations (14.5) and (14.7) gives

$$\ln V_i = k_1[A]_0 \exp\left\{-\frac{k^*[X]_0[O_2]}{K_x + [O_2]}t\right\}, \tag{14.8}$$

where the terms $[A]_0$ and $[X]_0$ in the above equation can represent average values for the concentrations.

The experimental data of Cox and Heckly (1973) have been compared with theoretical curves obtained from equation (14.8) for the survival of *Serratia marcescens* 8UK (Fig. 14.6). It would appear from the curves that the theory predicts the reduction of viability with time fairly accurately at the higher concentrations, 5–10 per cent oxygen concentration being particularly good. Extensions to the theory have been dealt with by Cox (1973) for the kinetics of the formation of oxygen-induced free radicals in freeze-dried *Serratia marcescens*; however, severe viability loss occurs without any oxygen-induced free radicals being detected.

14.2.3 REHYDRATION

The ease with which freeze-dried microorganisms rehydrate is outstanding, considering the difficulty of water removal. Micro-organisms which survive the stages of freezing, drying and storage may yet die during rehydration—the last phase in the process cycle, which is not very well understood. Contributing factors to the death of microorganisms during rehydration appear to be excessive dilution, the effect produced by aids to reconstitution and temperature. Leach and Scott (1959) have examined these and have found that rehydration in small volumes of water and in solutions of 1–2 model sorbitol, mannitol, ribitol and erthritol helps reduce the mortality of *Vib metchnikovi.*

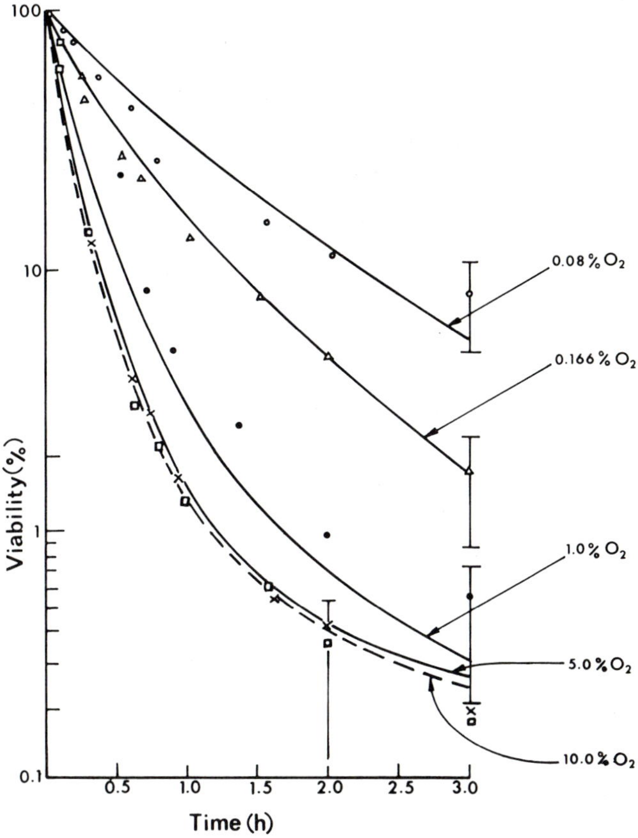

FIG. 14.6. Comparison of experimental and theoretical values of per cent viability of *Serratia marcescens* 8UK/2 as a function of oxygen concentration (per cent) and time (after Cox and Heckly, 1973).

14.3 Hazards during drying and storage

14.3.1 SEALING AND OPENING AMPOULES

The sealing of ampoules and vials in glass is almost universal for the storage of freeze-dried biological products. Under a vacuum seal it is dangerous to break open the neck of the tube because the sudden rush of air causes the contents to scatter. For safety, Cowan (1951) and Fry and Greaves (1951) recommended opening the neck of the tube by making a nick with a file and pressing a hot bead of glass against this, to cause a fine crack to develop. Then, after the air leaks gently in, the top of the tube may be lifted off. It is better to use soda glass than Pyrex, because of its higher expansion rate. Instructions on how to open tubes in this way are issued by the National Type Culture Collection, Colindale, England.

14.3.2 LEAKAGE OF PLASMA BOTTLE GASKETS—THEORY

Another way of sealing, for example, plasma bottles, is with rubber stoppers or gaskets. The products in such bottles may degrade from the effects of oxygen and moisture leaking through the rubber.

The leakage rate of rubber membranes (Henry, 1972) depends on two distinct phenomena: gas solubility and diffusivity. In the steady state the gas flow rate is given by

$$G_l = -D\frac{\mathrm{d}c}{\mathrm{d}x} = \frac{D(c_1 - c_2)}{l}, \tag{14.9}$$

where l is the thickness, D is the diffusivity, and c_1, c_2 are the surface concentrations. In practice, c_1 and c_2 may not be known but can be replaced by vapour pressures p_1, p_2; then

$$G_l = \frac{K(p_1 - p_2)}{l}, \tag{14.10}$$

where K = permeability constant—a less fundamental quantity than diffusivity.

If the sorption isotherm is linear and D a constant the relationship between the external vapour pressure p_a and the equilibrium concentration c_e in the membrane is linear; thus

$$c_e = Sp_a, \tag{14.11}$$

where S is the solubility.

This implies that equations (14.9) and (14.10) are equivalent and so

$$K = DS. \tag{14.12}$$

All the coefficients—K, D and S—vary with temperature in a like manner; accordingly,

$$K = K_{\infty}\, e^{\Delta H_m/RT} \tag{14.13}$$

where ΔH_m is the molar heat content and R the universal gas constant.

Crank (1956) analysed gaseous diffusion with zero concentration initially in a membrane and at the emergent face, obtaining the total quantity of diffusing gas;

$$G_l = \frac{Dc_2 t}{l} - \frac{lc_2}{6} - \frac{2lc_2}{\pi^2}\sum_1^{\infty}\frac{(-1)^{\nu}}{\nu^2}\exp\left(-\frac{\nu^2 D^2 t}{l^2}\right). \tag{14.14}$$

In the approach to a steady state $t \to \infty$ and the equation becomes

$$G_l = \frac{Dc_2}{l}\left(t - \frac{l^2}{6D}\right). \tag{14.15}$$

From which the steady state is established in time

$$t_{ss} = \frac{l^2}{6D}. \tag{14.16}$$

Reverting to pressures as shown above, equation (14.15) becomes

$$G_l = \frac{Kp_2}{l}\left(t - \frac{l^2}{6D}\right). \tag{14.17}$$

And, in conformity with Henry's law, the quantity of gas per unit volume equals the product of the solubility and the pressure; that is, $G_l/V = Sp_a$ (G_l is considered as the potential energy with the units Pa m^3, the units of pressure become Pa, whilst those of solubility become dimensionless), and this, by substitution from equations (14.11), (14.12) and (14.15), leads to

$$p_a = \frac{Dp_2}{Vl}\left(t - \frac{l^2}{6D}\right), \tag{14.18}$$

where the normal volume V of the condensed phase is assumed for the permeating gas filling a rubber-sealed bottle after sorption by the rubber and permeation through to the inside.

As an example, the leakage rate of oxygen into a plasma bottle will be given. Let the volume of the bottle be 16 cm^3; its gasket thickness 6 mm and surface area 0·68 cm^2. From van Amerongen's (1950) table the diffusivity and permeability values of butyl rubber, say, relative to oxygen are $8{\cdot}1 \times 10^{-12}\,\mathrm{m^2\,s^{-1}}$ and $0{\cdot}99 \times 10^{-12}\,\mathrm{m^2\,s^{-1}}$ respectively. By substituting these in equations (14.16) and (14.18) t_{ss} is established in 206 h, and it takes 720 h (1 month) for the oxygen partial pressure in the bottle to reach 27·5 Pa.

References

ABOUT, E. F. V. (1862). "L'homme à l'oreille cassée".

ALTMANN, R. (1890). Die Elementarorganismen und ihre Beziehungen zu den Zellen. Viet. Leipzig.

ALTY, T. (1933). Maximum rate of evaporation of water. *Phil. Mag.* **15**, 82.

VAN AMERONGEN, G. J. (1950). Influence of structure of elastomers on their permeability to gases. *J. Polymer Sci.* **5**, 307.

ANET, E. F. L. J. AND REYNOLDS, T. M. (1957). Chemistry of non-enzymic browning. I. Reactions between amino acids, organic acids, and sugars in freeze-dried apricots and peaches. *Aust. J. Chem.* **10**, 182.

ANON (1968). Freezing times of boneless meat. *Aust. Refrig. Air Cond. Heat.* **22**, no. 3, 41.

ANQUEZ, M. (1966). Presented at 5th International Course on Freeze-Drying, Lyon and Dijon.

VAN ARSDEL, W. B. (1963). "Food Dehydration. I-Principles". Avi Publishing Co., Westport, Connecticut.

BARRATT, R. W. AND TATUM, E. L. (1950). A simplified method of lyophilizing micro-organisms. *Science*, **112**, 122.

BARRER, R. M. (1948). Fluid flow in porous media. *Disc. Faraday Soc.* No. 3, 61.

BARRER, R. M. (1963). Diffusion in porous media. *Appl. Mat. Res.* **2**, 129.

BARTHOLAMI, G., BRENNAN, J. G. AND JOWITT, R. (1974). Mechanism of volatile retention in freeze-drying. The contribution of adsorption. *Proc. 4th Int. Congr. Fd. Sci. Technol.* **4**, 169.

BARTLETT, L. H. (1944). The specific heat of foodstuffs. II—A mathematical and thermodynamic determination. *Univ. Texas Publ.* No. 4432.

BECKETT, L. G. (1954). The effects of residual moisture in frozen-dried materials, and its measurement. *In* "Biological Applications of Freezing and Drying" (R. J. C. Harris, ed.), p. 285. Academic Press, London and New York.

BECQUEREL, P. (1936). La vie latente de quelques algues et animaux inférieurs aux base températures et la conservation de la vie dans l'univers. *C.r. hebd. Séanc. Acad. Sci., Paris*, **202,** 978.

BECQUEREL, P. (1950). La suspension de la vie des spores des bacteries et des moissessures desséchlées dans le vide vers le zéro absolu. Les conséquences pour la dissémination et la conservation de la vie dans l'univers. *C.r. hebd. Séanc. Acad. Sci., Paris*, **231**, 1392.

BECQUEREL, P. (1951). La suspension de la vie au confins du zéro absolu entre 0,0075°K et 0,047°K. Rôle de la synérèse reversible cytonucléoplasmique. *Proc. 8th Int. Congr. Refrig.*, p. 326.

BEKE, G. (1969). The effect of the sublimation temperature on the rate of the freeze-drying process and upon the volumetric change in meat muscle tissue. *Proc. 12th Int. Congr. Refrig.* **3**, 969.

BEKE, G. (1973). The effect of the sublimation temperature on the volumetric change in meat muscle tissue. *Proc. 13th Int. Congr. Refrig.* **3**, 955.

BELLOWS, R. J. AND KING, C. J. (1972). Freeze-drying of aqueous solutions: maximum allowable operating temperatures. *Cryobiology*, **9**, 559.

BENGTSSON, N. E. (1975). Freeze drying of sliced beef. *In* "Freeze Drying and Advanced Food Technology" (S. A. Goldblith, L. Rey and W. W. Rothmayr, eds), p. 253. Academic Press, London and New York.

BENGTSSON, O. AND BENGTSSON, N. E. (1968). Freeze-drying of raw beef. I–II. *J. Sci. Fd Agric.* **19**, 317 *and* 481.

BIRD, B. R., STEWART, W. E. AND LIGHTFOOT, E. N. (1960). "Transfer Phenomena". Wiley, New York.

BLOND, G. (1972). Etude de la texture des produits lyophilisés. *Nouveautés en Lyophilisation*. Centre de Perfectionnement Technique, Paris (Juin 1972).

DE BOER, J. H. (1953): "The Dynamical Character of Adsorption". Oxford University Press, London.

BOMBEN, J. L., BRUIN, S., THIJSSEN, H. A. C. AND MERSON, R. L. (1973). Aroma recovery and retention in concentration and drying of foods. *In* "Advances in Food Research" (E. M. Mrak and G. F. Stewart, eds), vol. 20, p. 1. Academic Press, New York and London.

BORDAS, F. AND D'ARSONVAL, A. (1906). Les basses températures et l'analyse chimique. *C.r. hebd. Séanc. Acad. Sci., Paris*, **142**, 1058; **143**, 567.

BOSANQUET, C. H. (1944). Unpublished results quoted by Hewitt, G. F. and Sharratt, E. W. (1963). Gaseous diffusion in porous media with particular reference to graphite. *Nature, Lond.* **198**, 952.

BOULDOIRES, J. P. AND BRICOUT, J. (1969). Contribution récente à l'étude des transferts de vapeur en cours de lyophilisation. *Rev. gén. Froid*, **60**, 247.

BRALSFORD, R. (1967). Freeze-drying of beef, I and II. *J. Fd Technol.* **2**, 339.

BRYNKO, C. AND SMITHIES, W. R. (1958). Rapid vacuum freeze-drying of meat. *J. Sci. Fd Agric.* **9**, 576.

DE BUHR, J. G. (1967). Apparatus and process for freeze-drying. U.S. Pat. No. 3, 262, 212.

BULL, H. B. (1944). Adsorption of water vapour by proteins, *J. Am. chem. Soc.* **66**, 1499.

BURKE, R. F. AND DECAREAU, R. V. (1961). Unpublished data. Thomas J. Lipton, Inc.

BURKE, R. F. AND DECAREAU, R. U. (1964). Recent advances in the freeze-drying of food products. *In* "Advances in Food Research" (E. M. Mrak and G. F. Stewart, eds), vol. 13, p. 1. Academic Press, New York and London.

BURROWS, G. (1957). Evaporation at low pressures. *J. appl. Chem.* **7**, 375.

BURROWS, G. (1965). Evaporation in an evacuated container. *Vacuum*, **15**, 389.

CAMBELL, W. L., PROCTOR, B. E. AND SLUDER, J. C. (1943–44). Development of dehydrated citrus powders. *Res. Rep. Quartermaster Contract Projects*, Mass. Instit. Technol. Cambridge.

CARLSON, R. A., RANDALL, J. M., GRAHAM, R. P. AND MORGAN, A. I. (1967). The rotary steam-coil vacuum evaporator. *Fd Technol.* **21**, no. 2, 90.

CARMAN, P. C. (1948). Molecular distillation and sublimation. *Trans. Faraday Soc.* **44**, 529.

CARREL, A. (1912). The preservation of tissue and its application in surgery. *J. Am. med. Ass.* **59**, 523.

CARSLAW, H. S. AND JAEGER. J. C. (1959). "Conduction of Heat in Solids" (2nd edition). Oxford University Press, London.

CHANDRASEKARAN, S. K. (1971). Volatile retention during drying of food liquids. Ph.D. Thesis, University of California, Berkeley.

CHANDRASEKARAN, S. K. AND KING, C. J. (1971a). *Chem. Engng Prog. Symp. Ser.* **67**, no. 107.

CHANDRASEKARAN, S. K. AND KING, C. J. (1971b). Retention of volatile flavor components during drying of fruit juices. *Chem. Engng Prog. Symp. Ser.* **67**, no. 108, 122.

CHAPMAN, S. AND COWLING, T. G. (1970). "The Mathematical Theory of Non-Uniform Gases" (3rd edition). Cambridge University Press, London.

CHARM, S. E. (1963). "Fundamentals of Food Engineering", p. 359. Avi Publishing Co., Westport, Connecticut.

CHIRIFE, J. AND KAREL, M. (1973). Volatile retention during freeze-drying of aqueous suspensions of cellulose and starch. *J. agric. Fd Chem.* **21**, 936.

CHRISTIAN, R. T. AND STOCKTON, T. T. (1956). The influence of sealing pressure on survival of *serratia marcescens* and *micrococcus pyogenes var. aureus* desiccated from the frozen state. *Appl. Microbiol.* **4**, 88.

CIBOROWSKI, J. AND SURGIEWIEZ, J. (1961). Examination of the yield of the condensation process of a sublimable material by mixing. *Przem. Chem.* **40**, 646.

CLARK, C. J. AND KING, J. P. (1968). An improved freeze-drying process using convective heat transfer. *Fd Technol.* **22**, 1235.

CLAUSING, P. (1929). Stationary flow of very dilute gases. *Physica, Eindhoven*, **9**, 65.

COBBIN, L. B. and THORP, R. H. (1957). Occurrence of a cardio-active principle in spleen. *Nature, Lond.* **180**, 242.

COOKE, N. E. AND SHERWOOD, T. K. (1955). The Effect of pressure on the rate of sublimation. *Proc. 9th Int. Congr. Refrig.* **2**, 133.

COPSON, D. A. (1962). "Microwave Heating". Avi Publishing Co., Westport, Connecticut.

COWAN, S. T. (1951). Infection and drying techniques. *In* "Freezing and Drying" (R. J. C. Harris, ed.) p. 127. The Institute of Biology, London.

COWELL, N. D. (1969). The calculation of food freezing times. *Proc. 12th Int. Congr. Refrig.* **2**, 667.

COX, C. S. (1973). Oxygen induced free radicals and viable decay in freeze-dried bacteria. Lyophilisation de matières biologiques. *Bull. int. Inst. Refrig.* Annexe 1973–5, 55.

COX, C. S., BAXTER, J. AND MAIDMENT, B. J. (1973). A mathematical expression for oxygen-induced death in dehydrated bacteria. *J. gen. Microbiol.* **75**, 179.

COX, C. S. AND HECKLY, R. J. (1973). Effects of oxygen upon freeze-dried and freeze-thawed bacteria: viability and free radical studies. *Canad. J. Microbiol.* **19**, 189.

CRANK, J. (1956). "The Mathematics of Diffusion". Oxford University Press, London.

CRANK, J. AND NICOLSON, P. (1947). A practical method for numerical evaluation of solutions of partial differential equations of the heat conduction type. *Proc. Camb. phil. Soc. biol. Sci.* **43**, 50.

DALGLEISH, J. M. (1962). Leybold continuous plant equipment. *In* "Freeze-drying of foodstuffs" (S. Cotson and D. B. Smith, eds), p. 201. Columbine Press, Manchester.

DALGLEISH, J. M. (1963). Brit. Pat. 928, 925.

DAUVOIS, PH. (1970). Contribution à l'étude de la congélation et de la lyophilisation de particules. Thèse de Docteur-Ingénieur, Faculté des Sciences, Université de Dijon.

DAUVOIS, PH. (1971). Envols en lyophilisation de particules. *Rev. gén Froid*, **62**, 827.

DAVIES, J. D. (1968a). Aspects of freezing important in freeze-drying of living cells. *In* "Low Temperature Biology of Foodstuffs" (J. Hawthorn, ed.), p. 177. Pergamon Press, Oxford.

Davies, J. D. (1968b). Freeze-drying biological materials. *Process Biochem.* **3**, no. 1, 11; no. 2, 48.

Decareau, R. V. (1962). Microwave freeze-drying. *In* "Freeze-Drying of Foodstuffs" (S. Cotson, and D. B. Smith, eds), p. 60. Columbine Press, Manchester.

Diels, K. and Jaeckel, R. (1962). "Leybold Vakuum-Taschenbuch", p. 61. Springer-Verlag, Berlin.

Döge, F. and Vêlebil, G. (1968) Vergleich der Productionskosten zwischen—Gefriertrocknung und Anlagekombination Gefrierkonzentration—Gefriertrocknung. *Informationsblatt.* **303**.

van Drimelen, G. C. and Steyn, H. S. (1958). The keeping quality of freeze-dried *brucella abortus* strain 19 vaccine. *J. gen. Microbiol.* **19**, 324.

Dumoré, J. M. (1964). Stability considerations in downward miscible displacements. *Soc. Petrol. Engng J.* 356.

Dunoyer. J. M. and Larousse, J. (1961). Expériences Nouvelles sur la Lyophilisation. *Trans 8th Nat. Vac. Symp. (Amer. Vac. Soc. and 2nd Int. Congr. Vac. Sci. Technol.)* p. 1059. Pergamon Press, Oxford.

Dyer, D. F. (1966). Transport phenomena in sublimation dehydration, Ph.D. Thesis, Georgia Institute of Technology, Atlanta, Georgia.

Dyer, D. F. and Sunderland, J. E. (1966). Bulk and diffusional transport in the region between the molecular and viscous conditions. *Int. J. Heat Mass Transf.* **9**, 519.

Dyer, D. F. and Sunderland, J. E. (1967). The transient temperature distribution during sublimation dehydration, *Trans. ASME J. Heat Transf.* **89**c, 109.

Dyer, D. F. and Sunderland, J. R. (1968a) Heat and mass transfer mechanism in sublimation dehydration. *Trans. ASME J. Heat Transf.* **90**C, 379.

Dyer, D. F. and Sunderland, J. E. (1968b). The role of convection in drying. *Chem. Engng Sci.* **23**, 965.

Dyer, D. F. and Sunderland, J. E. (1969). The influence of varying interface temperature on freeze-drying. *Bull. int. Inst. Refrig.* Annex 1969–4, 37.

Dyer, D. F. and Sunderland, J. E. (1971). Freeze-drying of bodies subject to radiation boundary conditions. *Trans. ASME J. Heat Transf.* **93**C, 427.

Earle, R. L. and Fleming, A. K. (1967). Cooling and freezing of lamb and mutton carcasses. I. Cooling and freezing rates in legs. *Fd Technol.* **21**, 79.

Ede, A. J. (1947). Freeze-drying chamber: Effect of shape. *Nature, Lond.* **159**, 610.

Ede, A. J. (1949a). The low-temperature vacuum drying process. *J. Soc. chem. Ind., London.* **68**, 336.

Ede, A. J. (1949b). A method for calculating freezing time of a rectangular package of food. *Mod. Refrig.* **52**, 52.

EHLERS, H., HACKENBERG, U. AND OETJEN, G. W. (1962). Wärmeübertragung und Temperaturkontrolle beim CQC-Gefriertrocknungs-prozess, *Trans. 8th Nat. Vac. Symp. (Amer. Vac. Soc. and 2nd Int. Congr. Vac. Sci. Technol.)* p. 1069. Pergamon Press, Oxford.

EIJKELBOOM, A. J. (1966). Experiences with the continuous concentration of fruit juices by freezing. *Bull. int. Inst. Refrig.* Annexe 1966–3, 141.

EVANS, G. W., ISAACSON, E. AND MACDONALD, J. K. L. (1950). Stefan-like problems. *Q. appl. Math.* **8**, 312.

FISCHER, K. (1935). A new method for the analytical determination of the water content of liquids and solids. *Angew. Chem.* **48**, 394.

FISHER, P. J. (1950). Viability of dried cultures. A note on the immediate death rate. *J. gen. Microbiol.* **4**, 455.

FLINK, J. (1969). Organic volatile retention in freeze-dried carbohydrate solutions. Ph.D. Thesis, Massachusetts Institute of Technology, Cambridge, Massachusetts.

FLINK, J. AND KAREL, M. (1969). Studies on the mechanism of retention of organic volatiles in freeze-dried systems. *Inst. Fd Technol. Nat. Conv.*, 1969, Paper No. 85.

FLINK, J. AND KAREL, M. (1970a). Retention of organic volatiles in freeze-dried solutions of carbohydrates. *J. agric. Fd Chem.* **18**, 295.

FLINK, J. AND KAREL, M. (1970b). Effect of process variables on retention of volatiles in freeze-drying. *J. Fd Sci.* **35**, 444.

FLINK, J. AND KAREL, M. (1972). Mechanism of retention of organic volatiles in freeze-dried systems. *J. Fd Technol.* **7**, 199.

FLOSDORF, E. W. (1949). "Freeze-drying". Reinhold, New York.

FLOSDORF, E. W. AND HYATT, G. W. (1952). The preservation of bone grafts by freeze-drying. *Surgery*, **31**, 716.

FOLSOM, T. R. (1946). Freezing and drying semi-solids. U.S. Pat. No. 2, 411, 152.

FORREST, J. C. (1959). Large-scale freeze-drying equipment for foodstuffs. *Br. Chem. Engr.* **7**, 390.

FORREST, J. C. (1960). The accelerated freeze-drying process for foodstuffs. Presented at 2ème Cours Internationaux de Lyophilisation. *Inst. int. Froid, Lyon*, **2**.

FORREST, J. C. (1961). Evacuating equipment for the accelerated freeze-drying of foodstuffs. Symposium on User Experience of Large-Scale Industrial Vacuum Plant. Institution of Mechanical Engineers, London.

FORREST, J. C., DALGLEISH, J. M. AND THOMPSON, H. P. (1962). Heat and mass transfer in sublimation drying of foods. *2nd Int. Heat Transf. Conf.* (Inst. Mech. Engrs), p. 926. Boulder, Colorado and London.

FRY, R. M. (1954). The preservation of bacteria. *In* "Biological Applications of Freezing and Drying" (R. J. C. Harris, ed.), p. 217. Academic Press, New York and London.

FRY, R. M. AND GREAVES, R. I. N. (1951). The survival of bacteria during and after drying. *J. Hyg. Camb.* **49**, 220.

GAFFNEY, J. J. AND STEPHENSON, K. W. (1968), *Trans. ASAE*, **11**, 874.

GERSH, I. (1932). The Altmann technique for fixation by drying while freezing. *Anat. Rec.* **53**, 309.

GERSH, I. (1948). Application in pathology of the method of fixation by freezing and drying tissues. *Bull. int. Ass. med. Museums*, **28**, 179.

GERSH, I. AND STEPHENSON, J. L. (1954). Freezing and drying of tissues for morphological and histochemical studies. *In* "Biological Applications of Freezing and Drying" (R. J. C. Harris, ed.), p. 329. Academic Press, New York and London.

GHEORGHIU, M. AND STURDZA, S. A. (1969). L'argon, tampon protecteur dans le conditionnement du BCG lyophilisé. *Bull. Inst. int. Froid*, Annexe 1969–9, 211.

GINNETTE, L. F., GRAHAM, R. P. AND MORGAN, A. I. (1958). Freeze-drying rates. *Trans. 5th Nat. Vac. Symp.* (*Amer. Vac. Soc.*) p. 268. Pergamon Press, Oxford.

GINSBURG, A. S., SYROYEDOV,, V. I. AND LYAKHOVITSKY, B. M. (1969). Freeze-drying effect on quality of foodstuffs and biological materials. *Bull. int. Inst. Refrig.* Annex 1969-9, 111.

GLASSTONE, S. (1947). "Thermodynamics for Chemists". Van Nostrand, New York.

GLICK, D. AND MALMSTROM, B. G. (1952). Simple and efficient freeze-drying apparatus for the preparation of embedded tissue. *Expl Cell Res.* **3**, 125.

GOLDBLITH, S. A. (1964). Freeze-dehydration of foods. *In* "Researches and Developments in Freeze-Drying" (L. Rey, ed.), p. 555. Hermann, Paris.

GOLDBLITH, S. A., KAREL, M. AND LUSK, G. (1964). The role of food science and technology in the freeze dehydration of foods. *In* "Aspects Théoriques et Industriels de la Lyophilisation" (L. Rey, ed.), p. 527. And *Fd Technol.* **17**, 139; **17**, 258.

GOODING, E. G. B. (1962). The storage behaviour of dehydrated foods. *In* "Recent Advances in Food Science" (J. Hawthorn and J. Muil Leitch, eds), p. 31. Butterworth, London.

GOODING, E. G. B. AND ROLFE, E. J. (1957). Some recent work on dehydration in the United Kingdom. *Fd Technol.* **11**, 302.

GÖRLING, P. (1956). Untersuchungen zur Aufklarung des Trocknungsverhaltens pflanzlicher Stoffe. *VDI-Forschungsheft*, 458.

GÖRLING, P. (1958). Physical phenomena during the drying of foodstuffs. *In* "Fundamental Aspects of the Dehydration of Foodstuffs", p. 42. Macmillan, New York.

GREAVES, R. I. N. (1942). The freezing of human serum and plasma in Medical Research Council transfusion bottles, before drying by sublimation from the frozen state. *J. Hyg. Camb.* **41**, 489.

GREAVES, R. I. N. (1944). Centrifugal vacuum freezing. Its application to drying of biological materials from the frozen state. *Nature, Lond.* **153**, 485.

GREAVES, R. I. N. (1954). Theoretical aspects of drying by vacuum sublimation. *In* "Biological Applications of Freezing and Drying". (R. J. Harris, ed.), p. 87. Academic Press, London and New York.

GREAVES, R. I. N. (1960a). Some factors which influence the stability of freeze-dried cultures. *In* "Recent Research in Freezing and Drying" (A. S. Parkes and A. U. Smith, eds), p. 203. Blackwell, Oxford.

GREAVES, R. I. N. (1960b). The application of heat in freeze-drying systems, *Ann. N.Y. Acad. Sci.* **85**, 682.

GREAVES, R. I. N. (1962). The application of heat in freeze-drying systems. *In* "Progrès Récents en Lyophilisation" (L. Rey, ed.), p. 71. Hermann, Paris.

GREAVES, R. I. N. (1964). Fundamental aspects of freeze-drying bacteria and living cells. *In* "Aspects Théoriques et Industriels de la Lyophilisation" (L. Rey, ed.), p. 407. Hermann, Paris.

GREENFIELD. P. F. (1972). Mechanistic studies in cyclic-pressure freeze drying. Ph.D. Thesis, University of New South Wales.

GREENFIELD, P. F. AND MELLOR, J. D. (1974). The effect of cycled-pressure on drying conditions during freeze drying. *J. Fd Technol.* **9**, 405.

GREIFF, D. AND PINKERTON, H. (1954). The Effects on biological materials of freezing and drying by vacuum sublimation. I. Development and testing of apparatus. *J. exp. Med.* **100**, 81.

VON GUERICKE, O. (1672). Experimenta nova (ut vocantur) Magdeburgica, Amsterdam.

GUEX, M. (1969). Etude de la surface spécifique de substances lyophilisées en couche. *Bull. Inst. int. Froid,* Annexe 1969-9, 71.

HAAS, G. J., PRESCOTT, H. E. AND D'INTINO, J. (1972). Pressure freezing air-drying. A new technique to reduce deterioration in drying tissue. *J. Fd Sci.* **37**, 430.

HACKENBERG, U. (1965). Freeze-drying and packaging under protective gas. *Proc. 11th Int. Congr. Refrign.* **2**, 1507.

HACKENBERG, U. (1966). U.S. Patent No. 3,234,658.

HAMILTON, W. H. (1966). Freeze-drying apparatus method. U.S. Pat. No. 3,230,633.

HANSON, S. W. F. (ed.) (1961). "The Accelerated Freeze-Drying Method of Food Preservation". HMSO, London.

HARDIN, T. C. (1965). Heat and mass transfer mechanics in freeze-drying. Ph.D. Thesis, Georgia Institute of Technology, Atlanta.

HARPER, J. C. (1962). Transport properties of gases in porous media at reduced pressures with reference to freeze drying. *A.I.Ch.E.J.* **8**, 298.

HARPER, J. C. AND CHICHESTER, C. O. (1963). Improvements in rates of freeze-drying. *Trans. 10th Nat. Vac. Symp.* (*Am. Vac. Soc.*), p. 47.

HARPER, J. C., CHICHESTER, C. O. AND ROBERTS, T. E. (1962). Freeze-drying of foods: dielectric heating applied to dehydrated food production. *Agric. Engng,* **43**, 78.

HARPER, J. C. AND EL SAHRIGI, A. F. (1964). Thermal conductivities of gas-filled porous solids. *I and E.C. Fundamentals,* **3**, 318.

HARPER, J. C. AND TAPPEL, A. L. (1957). Freeze-drying of food products. *In* "Advances in Food Research" (E. M. Mrak and G. F. Stewart, eds), vol. 7, p. 210. Academic Press, New York and London.

HAVELKA, J. (1969). La lyophilisation, l'eau et la vie dans l'espace cosmique *C. r. 13ᵉ Congrès Int. Froid,* **3**, 925.

HEISS, R. (1933). The quantities of water frozen out in the freezing of foodstuffs. *Biochem. Z.* **267**, 438.

HEISS, R. AND SCHACHINGER, L. (1951). Fundamentals of freeze-concentration of liquids. *Fd Technol.* **5**, 211.

HELDMAN, D. R. AND HOHNER, G. A. (1972), Atmospheric freeze-drying processes for food, *In Proc. Int. Symp. Heat and Mass Transf. Prob. in Fd Engng,* **2**, E6, Wageningen.

HENDRICKSON, H. M. (1958). Saline water conversion by freezing. *Refrig. Engng,* **66**, no. 8, 31.

HENRY, R. P. (1972). La permeation des gaz à travers les élastomeres. Applications à la conservation de produits lyophilisés. *C. r. Nouveautés en Lyophilisation.* Centre de Perfectionnement Technique, Paris (Juin 1972).

HEWITT, G. F. AND SHARRATT, E. W. (1964). The effect of back-pressure on the counter diffusion of gases in graphite. *J. nucl. Mater.* **13**, 197.

HILL, J. E. (1967). Sublimation dehydration in the continuum, transition and free molecule flow regimes. Ph.D. Thesis, Georgia Institute of Technology, Atlanta.

HIRSCH, E. H. (1961). Flow of gas through porous media in the region between molecular and viscous condition. *J. appl. Phys.* **32**, 977.

HIRSCH, M. (1946). Drying and Freezing. *Refrig. Engng,* **51**, 331.

HOLDEN, J. F. AND SCOTT, J. (1951). A simple apparatus for drying small specimens of sera. *Aust. J. exp. Biol. med. Sci.* **29**, 47.

HOLDSWORTH, S. D. (1968). Heat and mass transfer in freeze drying. *Proc. Biochem.* **3**, 59.

HUIGE, N. J. J, SENDEN, M. M. G. AND THIJSSEN, H. A. C. (1972). Nucleation and growth kinetics for the crystallization of ice from dextrose solutions in a continuous stirred tank crystallizer with supercooled feed. Chisa Symp., Prague. (11–15 September).

ITO, K. (1971). Freeze-drying of pharmaceuticals. Eutectic temperatures and collapse of solute matrix upon freeze-drying of three-component systems. *Chem. Pharm. Bull.* **19**, 1095.

JASON, A. C. (1958). A study of evaporation and diffusion processes in the drying of fish muscle. *In* "Fundamental Aspects of the Dehydration of Foodstuffs", p. 103. Macmillan, New York.

JENSEN, W. A. (1954). A new approach for freeze-drying tissue. *Expl. Cell Res.* **7**, 572.

JENSEN, W. A. AND KAVALJIAN, L. J. (1957). Notes on the freeze-drying of plant tissue. *Stain Technol.* **32**, 33.

JEPPSON, M. R. (1962). U.S. Pat. No. 3,222,796.

KALLISTRATOS, G. AND VON SENGBUSCH, R. (1964). Vergleichende Untersuchungen über die Verluste verschiedener Nahrungsmittelbestandteile durch Gefriertrocknung und durch andere Trocknungsarten. *Nutr. Dieta*, **6**, 193.

KAN, B. AND DE WINTER, F. (1968). Accelerating freeze drying through improved heat transfer. *Fd Technol.* **22**, 67.

KAREL, M. (1963). Physical and chemical considerations in freeze-dehydrated foods. *In* "Exploration in Future Food-Processing Techniques" (S. A. Goldblith ed.), p. 54. Massachussetts Institute of Technology Press.

KAREL, M. (1972–73). Recent research and development in the field of low-moisture and intermediate-moisture foods. *CRC Crit. Revs in Fd Technol.* (T. E. Furia, ed.) **3**, 329.

KAREL, M. (1974). Fundamentals of dehydration processes *In* "Advances in Preconcentration and Dehydration of Foods" (A. Spicer, ed.), p. 45, Applied Science Publishers, London.

KAREL, M. AND FLINK, J. (1969). Mechanism of volatile retention in freeze-dried carbohydrate systems, paper presented at *A.I.Ch.E. Meeting*, Washington, D.C.

KATAYAMA, K., HAYASKI, Y. AND KIMURA, T. (1969). Researches on heat and mass transfer by sublimation of ice. *Bull. J.S.M.E.* **12**, 257.

KEEY, R. B. (1972). "Drying: Principles and Practice", p. 334. Pergamon Press, Oxford.

KEKWICK, R. A. (1953–54). Some considerations relating to the design of freeze-drying plant. *Proc. Inst. Refrig.* **50**, 65.

KESSLER, H. G. (1962). Wärme-und Stoffaustausch beider Gefriertrocknung poriger Güter. *Chemie-Ing-Techn.* **34**, 163.

KESSLER, H. G. (1975). Heat and mass transfer in freeze-drying of mixed granular particles. *In* "Freeze Drying and Advanced Food Technology (S. A. Goldblith, L. Rey and W. W. Rothmayr, eds), p. 223. Academic Press, London and New York.

KIDD, F. (1941). Improvements in and relating to the low temperature treatment for the preservation of perishable foods and other materials. Brit. Pat. No. 539, 477.

KING, C. J. (1968). Rates of moisture sorption and desorption in porous dried foodstuffs, *Fd Technol.* **22**, 509.

KING, C. J. (1970). Freeze-drying of foodstuffs. *CRC Crit. Revs in Fd Technol.*, p. 379.

KING, C. J. (1973). Applications of freeze-drying to food products. *In* "Freeze Drying and Advanced Food Technology" (S. A. Goldblith, L. Rey and W. W. Rothmayr, eds), p. 333, Academic Press, London and New York.

KING, C. J. (1974). Pressurized freezing for retarding shrinkage after drying: A quantitative interpretation, *Cryobiology*, **11**, 121.

KING, C. J. AND CHANDRASEKARAN, S. K. (1973). Analysis of volatile loss from liquid foods during freeze-drying and evaporative drying as a ternary diffusion process. *Proc. 13th Int. Congr. Refrig.* p. 649, Washington.

KING, C. J., LAM, W. K. AND SANDALL, O. C. (1968). Physical properties important for freeze-drying poultry meat. *Fd Technol.* **22**, 1302.

KNIEPEN, P. (1929). Verfahren zum Entfernen adhärierender Gase and zum Trocknen von Gegenständen unter Anwendung eines periodischen Vakuums. German Pat. No. 487,769.

KRAMERS, H. (1958). Rate-controlling factors in freeze-drying. *In* "Fundamental Aspects of the Dehydration of Foodstuffs". *Soc. Chem. Ind.* Macmillan, New York.

KRAMERS, H. AND STEMERDING, S. (1951). Sublimation of ice in vacuum. *Appl. Sci. Res.* **A3**, 73.

KRISCHER, O. (1956). "Die wissenschaftlichen Grundlagen der Trocknungstechnik". Springer-Verlag, Berlin.

KRYLOV, V. V., KAMOVNIKOV, B., IAOUCHEVA, E. AND KAUKHCHESHVILI, E. (1967). Control of heat supply in sublimation drying based on extreme temperature zones. *Pischevaya Tekhuologiya*, no. 1, 120.

KUIVENHOVEN, A. C. J. (1966). Wash columns for the continuous removal of ice crystals from aqueous solutions. *Bull. int. Inst. Refrig.* Annex 1966-3, 85.

KUPRIANOFF, J. (1964). Fundamental and practical aspects of the freezing of foodstuffs. *In* "Aspects Théoriques et Industriels de la Lyophilisation" (R. Rey, ed.), p. 497. Hermann, Paris.

LAMBERT, J. B. AND MARSHALL, W. R. (1962). Heat and mass transfer in freeze drying. *Nat. Acad. Sci., Nat. Res. Council*, p. 105, Washington.

LEACH, R. H. AND SCOTT, W. J. (1959). The influence of rehydration on the viability of dried microorganisms. *J. gen. Microbiol.* **21**, 295.

LESLIE, J. (1811). Méthode nouvelle de produire et d'entretenir la congélation. *Ann. Chim. Phys.* **78**, 177.

LEVENSON, L. L., MILLERON, N. AND DAVIS, D. H. (1961). Optimization of molecular flow conductance. *Trans. 7th Nat. Vac. Congr.* (*Am. Vac. Soc.*), p. 372. Pergamon Press, Oxford.

LEVINSON, S. O. AND OPPENHEIMER, F. (1944). Drying by ice sublimation. The principle of infra-red radiation for rapid desiccation of plasma and other biologics. Samuel Deutsch Serum Center, Chicago, and U.S. Pat. No. 2,445,120.

LEWIN, L. M. AND MATELES, R. F. (1962). Freeze-drying without vacuum, a preliminary investigation. *Fd Technol.* **16**, 94.

LINDE, C. (1895). Brit. Pat. No. 12,528.

LINDE, C. (1873). Bayer. Ind. u. Gew. Bl.

LOVELOCK, J. E. (1953). The haemolysis of human red blood-cells by freezing and thawing. *Biochim. Biophys. Acta*, **10**, 414.

LOVELOCK, J. E. AND BISHOP, M. W. H. (1959). Prevention of freezing damage to living cells by dimethyl sulphoxide. *Nature, Lond.* **183**, 1394.

LUIKOV, A. V. (1950). *Teoriya Sushki*, G.E.N., Moscow/Leningrad.

LUIKOV, A. V. (1963). Heat and mass transfer in evaporation. *Int. Chem. Engng.* **3**, 195.

LUIKOV, A. V. (1969). Heat and mass transfer in freeze-drying at high vacuum. *Bull. int. Inst. Refrig.* Annexe 1969-4, 143.

LUIKOV, A. V., PERELMAN, T. L. AND ANISIMOV, S. I. (1971). Evaporation of a solid into vacuum. *Int. J. Heat Mass Transf.* **14**, 177.

LUSK, G., KAREL, M. AND GOLDBLITH, S. A. (1964). Thermal conductivity of some freeze-dried fish. *Fd Technol.* **18**, 121.

LUYET, B. J. (1939). Vitrification of water. *Physiol. Rev.* **56**, 1244.

LUYET, B. J. (1960). On various phase transitions occurring in aqueous solutions at low temperatures. *Ann. N. Y. Acad. Sci.* **85**, 549.

MACKENZIE, A. P. (1966). Basic principles on freeze-drying for pharmaceuticals. *Bull. Parenter. Drug Ass.* **20**, 101.

MCCULLOCH, J. W. AND SUNDERLAND, J. E. (1970). Integral techniques applied to sublimation dehydration with radiation boundary conditions. *J. Fd Sci.* **35**, 834.

MAGNUSSEN, O. M. (1969). Measurements of heat and mass transfer coefficients during freeze drying. *Bull. int. Inst. Refrig.* Annex 1969-4, 65.

MAKOWER, J. AND MYERS, S. (1943). A new method for the determination of moisture in dehydrated vegetables. *Proc. Inst. Fd Technol.* p. 156.

MALECKI, G. J., SHINDE, P., MORGAN, A. I. AND FARKAS, D. F. (1969). Atmospheric fluidized-bed freeze-drying of apple juice and other liquid foods. *Inst. Fd Technol. Nat. Conv.*, 1969.

MALTINI, E. (1975). Thermal phenomena and structural behaviour of fruit juices in the pre-freezing stage of the freeze-drying process. *In* "Freeze Drying and Advanced Food Technology" (S. A. Goldblith, L. Rey and W. W. Rothmayr, eds), p. 121. Academic Press, London and New York.

MASON, P. B. (1963). Freeze-drying. Brit. Pat. No. 1,052,280.

MASSEY, W. M. AND SUNDERLAND, J. E. (1967). Measurement of thermal conductivity during freeze-drying of beef. *Fd Technol.* **21**, 408.

MAZUR, P. (1960). Physical factors implicated in the death of micro-organisms at sub-zero temperatures. *Ann. N.Y. Acad. Sci.* **85**, 610.

MEFFERT, H. F. Th. (1969). Heat and mass transfer problems in freeze drying. Dechema-Monographien: Lebensmittel-Konservierung 19 Vortäge: (Bristol Symp. 1968), 63, 145 (Verlag Chemie).

MEFFERT, H. F. Th. (1965). Heat transmission in freeze-drying materials. *Proc. 11th. Int. Congr. Refrig.* **2**, 1527.

MEHRLICH, F. P. AND HAUGH, R. R. (1965). U.S. Pat. No. 3,169,070.

MELLON, E. F., KORN, A. H. AND HOOVER, S. R. (1949). Water absorption of proteins. IV. Effect of physical structure. *J. Am. chem. Soc.* **71**, 2761.

MELLOR, J. D. (1954). A method of protecting the surfaces of some materials in freeze-drying. *Vacuum*, **4**, 341.

MELLOR, J. D. (1961). Control of final water content in freeze-dried substances. *Trans. 7th Nat. Symp. Vac. Tech.* (*Amer. Vac. Soc.*), p. 224.

MELLOR, J. D. (1962a). Accelerated freeze-drying of foodstuffs. *CSIRO Food Pres. Quart.* **22**, 41.

MELLOR, J. D. (1962b). Vapour phase conditions in freeze-drying. *Trans. 2nd Int. Congr. Vac. Sci. Tech.* (*Am. Vac. Soc.*), **2**, 1064.

MELLOR, J. D. (1964). Vacuum cooling of mushrooms. *Mushroom Growers' Bulletin*, no. 177, 396.

MELLOR, J. D. (1966a). A small freeze dryer for industry. *CSIRO Food Pres. Quart.* **26**, 37.

MELLOR, J. D. (1966b). Vapour transfer in the course of freeze-drying. "Lyophilisation—Recherches et Applications Nouvelles" (L. Rey, ed.), p. 75. Hermann, Paris.

MELLOR, J. D. (1967a). Freeze-drying process with cyclic vacuum pressure. Brit. Pat. No. 1,083,244.

MELLOR, J. D. (1967b). Freeze-drying process. U.S. Patent No. 3,352,024.

MELLOR, J. D. (1971). Gas pressure and flow. Instrumental Techniques. Specialist Courses for the Food Industry No. 1, CSIRO, p. 24.

MELLOR, J. D. (1974). Advances in the theory of ice sublimation. *Bull. int. Inst. Refrig.* Annexe 1974-3, 121.

MELLOR, J. D. AND GREENFIELD, P. F. (1972). Heat and vapour transfer problems in cyclic-pressure freeze drying. *Proc. Int. Symp. Heat and Mass Transf. Prob. in Food Engng.* **2**, p. E8, Wageningen.

MELLOR, J. D. AND IRVING, A. R. (1969). Role of an excipient in freeze-drying for improving the quality of the product. *Bull. int. Inst. Refrig.* Annexe 1969-9, 225.

MELLOR, J. D. AND LOVETT, D. A. (1968). Flow of gases through channels with reference to porous materials. *Vacuum*, **18**, 625.

MELLOR, J. D. AND MIDDLEHURST, J. (1971). The effect of cycling the vacuum pressure on the freeze-drying rate. *Proc. 13th Int. Congr. Refrig.* **3**, 717, Washington.

MELLOR, J. D. AND MUNNS, A. S. (1968). Freeze drying with cyclic vacuum pressure. *Aust. Refrig. Air Cond. Heat*, **22**, no. 8, 20.

MELLOR, J. D. AND OHYE, D. F. (1960). An ampoule freeze-dryer for microbiological research, *Vacuum*, **10**, 245.

MENTING, L. C. AND HOOGSTAD, B. (1967a). Volatile retention during the drying of aqueous carbohydrate solutions. *J. Fd Sci.* **32**, 87.

MENTING, L. C. AND HOOGSTAD, B. (1967b). Selectivity of carbohydrate films as influenced by their moisture content. *Experientia*, **23**, 738.

MERYMAN, H. T. (1959). Sublimation freeze-drying without vacuum. *Science*, **130**, 628.

MERYMAN, H. T. (1962). The biophysics of freeze-drying. *In* "Progrès Récents en Lyophilisation" (L. Rey, ed.). Hermann, Paris.

MERYMAN, H. T. (1964). Induction and dielectric heating for freeze-drying. *In* "Aspects Théoriques et Industriels de la Lyophilisation" (L. Rey, ed.), p. 183. Hermann, Paris.

MICHAELS, A. S. (1974). Tailored membranes. *In* "Advances in Pre-concentration and Dehydration of Foods" (A. Spicer, ed.), p. 213. Applied Science Publications, London.

MIDDLEHURST, J. AND PARKER, N. S. (1972). Analogue solutions to some heat and mass transfer problems. *In Proc. Int. Symp. on Heat and Mass Transf. Prob. in Fd Engng*, **2**, p. E5, Wageningen.

MILLER, R. AND GOODNER, K. (1953). Studies on the stability of lyophilized BCG vaccine. *Yale J. Biol. Med.* **25**, 262.

MINK, J. H. AND SACHSEL, G. F. (1961). Evaluation of freeze-drying mechanisms using mathematical models. *In* "Freeze-Drying of Foods", p. 84. Nat. Acad. Sci., Nat. Res. Council, Washington.

MONZINI, A. AND MALTINI, E. (1973). Studies on the freeze-drying of frozen concentrated orange juices. *Bull. int. Inst. Refrig.* Annexe 1973-4, 123.

MORGAN, A. I. (1967a). Liquid food evaporation and aroma recovery. Aroma und Geschmackstoffe in Lebensmitteln, *Fortbildungskurs*, 225. Forster Verlag, Zürich.

MORGAN, A. I. (1967b). Evaporator concepts and evaporator design for minimum fouling. *Fd Technol.* **21**, no. 10, 63.

MUGGLETON, P. W. (1960). Freeze-drying of bacteria with special reference to BCG. *In* "Recent Research in Freezing and Drying", p. 229. Blackwell, Oxford.

NEMITZ, G. (1964). Physico-chemical processes during the freeze-drying of proteins. *Kältetechnik.* **16**, 368.

NEUMANN, K. H. (1957). Fortschritte in Methodik und Anwendung der Gefriertrocknung. *Chem. Ing. Tech.* **29**, 267.

NEUMANN, K. H. (1960). Les problèmes de mesure et de rélage en lyophilisation. *In* "Traité de Lyophilisation" (L. Rey, éd.), p. 185. Hermann, Paris.

NEUMANN, K. H. (1963). Temperature responsive freeze-drying method and apparatus. U.S. Pat. No. 3,077,036.

NOMOTO, M. AND NARAHASHI, Y. (1959). A proteolytic enzyme of *Streptomyces griseus.* I. Purification of a protease of *Streptomyces griseus, J. Biochem,* **46**, 653.

OETJEN, G. W. AND EILENBERG, H. J. (1969). Heat transfer during freeze-drying with moved particles. *Bull. int. Inst. Refrig.* Annex 1969-4, 19.

OETJEN, G. W., EHLERS, H., HACKENBERG, U., MOLL, J. AND NEUMANN, K. H. (1963). Temperature measurement and control of freeze-drying process. *In* "Freeze-Drying of Foods", p. 25. *Nat. Acad. Sci., Nat. Res. Council,* Washington.

OLDENCAMP, H. A. AND SMALL, R. F. (1965). U.S. Pat. No. 3,199,217.

PACKER, D. M. AND SCOTT, G. H. (1942). A cryostat of new design for low temperature tissue dehydration. *J. tech. Meth.* **22**, 85.

PARKES, A. S. (1962). Biological aspects of freezing and drying. *In* "Progrès Récents en Lyophilisation" (L. Rey, ed.), p. 1. Hermann, Paris.

PEASE, D. C. (1947). The disappearance of salt from glass ice during low-temperature dehydration, and its implications in electron microscopy. *Science,* **106**, 543.

PENNY, I. F., VOYLE, C. A. AND LAWRIE, R. A. (1964). Some properties of freeze-dried pork muscles of high or low ultimate pH. *J. Sci. Fd Agric.* **15**, 559.

PLANK, R. (1913). *Z. ges. Kälte-Ind.* **20**, 109.

POLGE, C. AND ROWSON, L. E. A. (1952). Fertilizing capacity of bull spermatozoa after freezing at −79°C. *Nature,* **169**, 626.

POLLARD, W. G. AND PRESENT, R. D. (1948). On gaseous self-diffusion in long capillary tubes. *Physiol. Rev.* **73**, 762.

Popovsky, V. G., Timofeieva, O. A. and Soboleva, I. M. (1965). Gas chromographic analysis of the volatile aromatic compounds in fresh and dried fruits. *Work*, **6**, 126.

della Porta, P. M. (1967). Improvements in or relating to freeze-drying. Brit. Pat. No. 1,062,159.

Powell, R. W. (1939–40). On the rate of sublimation of ice. *Proc. Br. Ass. Refrig.* **36**, 61.

Quast, D. G. and Karel, M. (1968). Dry layer permeability and freeze-drying rates in concentrated fluid systems. *J. Fd Sci.* **33**, 170.

Ranz, W. E. and Marshall, W. R. (1952). Evaporation from drops. *Chem. Engng Prog.* **48**, 141; 173.

Rey, L. R. (1959). Procédé et dispositif pour le contrôle et la régulation de la congélation et du dégal de différentes substances et notamment pour le contrôle et la régulation d'opérations de congélation—dessiccation, Brevets d'invention du Centre National de la Recherche Scientifique, Paris.

Rey, L. R. (1960a). Thermal analysis of eutectics in freezing solutions. *Ann. N.Y. Acad. Sci.* **85**, 510.

Rey, L. R. (1960b). "Conservation de la Vie par le Froid". Hermann, Paris.

Rey, L. R. (1964a). Fundamental aspects of lyophilization. *In* "Aspects Théoretiques et Industriels de la Lyophilisation" (L. Rey, éd.), p. 23. Hermann, Paris.

Rey, L. R. (1964b). Orientations nouvelles de la lyophilisation. *In* "Aspects Théoriques et Industriels de la Lyophilisation" (L. Rey, éd.), p. 621. Hermann, Paris.

Rey, L. R. (1964c). L'humidité résiduelle des produits lyophilisés: Nature-Origine et méthodes d'étude. *In* "Aspects Théoretiques et Industriels de la Lyophilisation" (L. Rey, éd.), p. 199. Hermann, Paris.

Rey, L. R. and Bastien, M. C. (1962). Biophysical aspects of freeze-drying: Importance of the preliminary freezing and sublimation periods. *In* "Freeze-Drying of Foods" (F. R. Fisher, ed.), p. 25. Nat. Acad. Sci. Nat. Res. Council.

Riedel, L. (1950). Der Kältebedarf beim Gefrieren von Obst und Gemüse. *Kältetechnik*, **2**, 195; 501.

Rieutord, L. (1960). Caractéristiques générales des appareils industriels à lyophilisation. *In* "Traité de la Lyophilisation" (L. Rey, éd.), p. 141. Hermann, Paris.

Rieutord, L. (1965). Process of regulating freeze-drying operations and apparatus for carrying out said process. U.S. Pat. 3,192,643, and Brit. Pat. 925,343.

Robson, E. M. (1954). Some aspects of micro-moisture determination. *Vacuum*, **4**, 60.

DE ROISSART, H. AND LAEDERICH, T. (1966). Un développement de l'equation de conduction de chaleur appliqué à la condensation et à la congélation en lyophilisation. *In* "Lyophilisation—Recherches et Applications Nouvelles" (L. Rey, éd.), p. 55. Hermann, Paris.

ROLFE, E. J. (1958). The influence of the conditions of dehydration on the quality of vacuum-dried meat. *In* "Fundamental Aspects of Dehydration of Foodstuffs", p. 211. Society of Chemical Industry, London.

ROLFE, E. J. (1963). The freeze-drying of fish meat. *In* "Freeze-Drying of Foodstuffs" (S. Cotson and D. B. Smith, eds), p. 119, Columbine Press, Manchester.

ROSENTHAL, S. R. (1948). Conservation du bacille de Calmette et Guerin (BCG) par dessiccation après congélation. *Annls. Inst. Pasteur, Paris*, **75**, 209.

ROSTAND, J. (1946). Glycérine et la résistance du sperme aux basses températures. *C.r. hebd. Séanc. Acad. Sci., Paris*, **222**, 1524.

ROWE, T. W. G. (1964a). Energy, mass transfer and economy in large scale freeze-drying. *In* "Aspects Théoriques et Industriels de la Lyophilisation" (L. Rey, ed.), p. 141. Hermann, Paris.

ROWE, T. W. G. (1964b). Vacuum techniques. *Fd Trade Rev.* **35**, 42.

ROWE, T. W. G. (1970). Freeze-drying of biological materials: Some physical and engineering aspects. *In* "Current Trends in Cryobiology" (A. U. Smith, ed.), p. 61. Plenum Press, New York.

ROWE, T. W. G., ACHUCARRO, J. L. AND DRUMMOND SMITH, B. (1963). Factors affecting the economy of the freeze-drying process. *In* "Freeze-drying of Foodstuffs" (S. Cotson, and D. B. Smith, eds), p. 71. Columbine Press, Manchester.

RULKENS, W. H. AND THIJSSEN, H. A. C. (1972). Retention of volatile compounds in freeze-drying slabs of malto-dextrin. *J. Fd Technol.* **7**, 79.

SANDALL, O. C., KING, C. J. AND WILKE, C. R. (1967). The relationship between transport properties and rates of freeze-drying of poultry meat. *A.I.Ch.E. Journal*, **13**, 428.

SARAVACOS, G. D. (1965). Freeze-drying rates and water sorption in model food gels. *Fd Technol.* **19**, 625.

SARAVACOS, G. D. (1967). Effect of the drying method on the water sorption of dehydrated apple and potato. *J. Fd Sci.* **32**, 81.

SARAVACOS, G. D. AND CHARM, S. E. (1962). A study of the mechanism of fruit and vegetable dehydration. *Fd Technol.* **16**, 78.

SARAVACOS, G. D. AND MOYER, J. C. (1968). Volatility of some flavor compounds during freeze-drying of foods. *Chem. Engng Progr. Symp. Ser.* **64**, no. 86, 37.

SARAVACOS, G. D. AND PILSWORTH, M. N. (1965). Thermal conductivity of freeze-dried model food gels. *J. Fd. Sci.* **30**, 773.

Saravacos, G. D. and Stinchfield, R. M. (1965). Effect of temperature and pressure on the sorption of water vapour by freeze-dried food materials. *J. Fd Sci.* **30**, 779.

Sauvageot, F., Beley, P., Marchand, A. and Simatos, D. (1969). Quelques données expérimentales sur le comportement au cours de la cryodessiccation des composés volatils de jus de fruits. *Bull. Inst. int. Froid*, Annexe 1969-9, 133.

Schrage, R. W. (1953). "Theoretical Study of Interphase Mass Transfer". Columbia University Press, New York.

Schultz, T. H., Dimick, K. P. and Makower, B. (1956). Incorporation of natural fruit flavours into fruit juice powders. I. Locking of citrus oils into sucrose and dextrose. *Fd Technol.* **10**, 57.

Scott, W. J. (1958). Effect of residual water on the survival of dried bacteria during storage. *J. gen. Microbiol.* **19**, 624.

Scott, W. J. (1960). A mechanism causing death during storage of dried microorganisms. *In* "Recent Researches in Freezing and Drying" (A. S. Parkes and A. U. Smith, eds), p. 188. Blackwell Scientific Publications, Oxford.

Scott D. S. and Dullien, F. A. (1962). The flow of rarefied gases. *A.I.Ch.E. Journal*, **8**, 293.

Seffinga Engng. Co (1964). Process and apparatus for freezing and freeze-drying liquid substances. Brit. Pat. No. 1,050,032.

Shackell, L. F. (1909). An improved method of desiccation with some applications to biological problems. *Am. J. Physiol.* **24**, 325.

Sharp, J. G. (1953). Dehydrated meat. DSIR Fd Investigation Spec. Rep. No. 57, p. 43.

Short, B. E. (1944). The specific heat of foodstuffs. I. An experimental determination. University of Texas Publication, No. 4432.

Shumsky, K. (1958). Basic principles in the calculation of "sublimation" condensers. *Holodil'naja Technika.*

Slatter, M. A. and Jones, M. C. (1972). Development of an accurate method of predicting freezing rates in watery solid. *In Proc. Int. Symp. on Heat and Mass Transf. Prob. in Fd Engng*, **2**, D2-1, Wageningen.

Smithies, W. R. and Blakley, T. S. (1959). Design of freeze-drying equipment for the dehydration of foodstuffs. *Fd Technol.* **13**, 610.

Smithies, W. R. (1966). Freeze-drying of meat and meat products. *In* "Advances in Freeze Drying" (L. Rey, ed.), p. 177. Hermann, Paris.

Soboleva, I. M. and Popovsky, V. G. (1971). High boiling compounds of plant freeze-dried products. Transactions Scientific Research Work, Moldavian Food Research Institute, Vol. 11, p. 78.

Spanyār, P., Kevei, E. and Szarföldi, I. (1966). The possibilities of preserving coffee aroma during freeze-drying (A kavē aromaanyagainak megörzēsi lehetösēgei a liofilesēs folyaman). *Hütoipar*, **13**, 106; *Kälte*, **19**, 549.

SPIESS, W. E. L. (1964). Quality changes of vegetables and fruit during the freeze-drying process. *Kältetechnik*, **16**, 349.

SPIESS, W. E. L., WOLF, W., TIRTOHUSODO, H. AND SOLÉ, C. P. (1969). The influence of the structure on the mass transfer in freeze-drying in thermodynamic aspects of freeze-drying, *Bull. int. Inst. Refrig.* Annexe 1969-4, 155.

STARK, C. N. AND HERRINGTON, B. L. (1931). The drying of bacteria and the viability of dry bacterial cells. *J. Bact.* **21**, 13.

STEIN, C. D. AND ROGERS, H. (1950). Recovery of viable microorganisms and viruses from vapors removed from frozen suspensions of biologic material during lyophilization. *Am. J. vet. Res.* **11**, 339.

STEINBACH, G. AND ROTHLEY, G. (1973). Experimental determination of the coefficients for a dimensionless equation for the freeze-drying of single bodies and of vessels filled with frozen product. *Proc. 13th Int. Congr. Refrig.* **3**, 617.

STEPHENSON, J. L. (1953). Theory of the vacuum drying of frozen tissue. *Bull. math. Biophys.* **15**, 411.

STEPHENSON, J. L. (1954). Theory for the design of apparatus for drying frozen tissues. *Bull. math. Biophys.* **16**, 23.

STRASSER, J., HEISS, R. AND GÖRLING, P. (1966). The water vapour and heat transport during vacuum freeze-drying with reference to optimal process conditions. *Kältetechnik*, **18**, 286.

STRICKLAND-CONSTABLE, R. F. AND BRUCE, E. W. (1954). The mechanism of drying of solids. Part IV: The sublimation of pure ice. *Trans. Inst. Chem. Engrs.* **32**, 192.

SZAKMARY, G. AND VALI, A. (1963). Studies on the residual moisture content of freeze-dried vaccines. *Acta vet. Beogr.* **13**, 411.

TAKESAKI, M. (1921). Shimase type desiccator. Jap. Pat. No. 38,535.

TAMMAN, G. (1898). Über die Abhängigkeit der Zahl der Kerne, welche sich in verschiedenen unterkühlten Flüssigkeiten bilden, von der Temperatur. *Z. phys. Chem.* **25**, 441.

THILORIER, M. (1835). *Ann. Chim. Phys.* **60**, no. 2, 427; 432.

THIJSSEN, H. A. C. (1969). Recent developments in freeze concentration of food liquids. (Bristol Symp. 1968). *Dechema-Monografien*, **63**, 153.

THIJSSEN, H. A. C. (1970). Concentration processes for liquid foods containing volatile flavours and aromas. *J. Fd Technol.* **5**, 211.

THIJSSEN, H. A. C. (1974). Freeze-concentration. *In* "Advances in Preconcentration and Dehydration of Foods" (A. Spicer, ed.), p. 115. Applied Science Publications, London.

THIJSSEN, H. A. C. AND MIDDELBERG, L. R. W. A. (1966). Fundamentals in fruit juice concentration. *Bull. int. Inst. Refrig.*, Annexe 1966-3, 113.

THIJSSEN, H. A. C. AND RULKENS, W. H. (1968). Retention of aromas in drying food liquids. *De ingenieur—Chemische Technik*, **80**, 45.

THIJSSEN, H. A. C. AND RULKENS, W. H. (1969). Effect of freezing rate on rate of sublimation and flavour retention in freeze-drying. *Bull. int. Inst. Refrig.* Annexe 1969-4, 99.

THOMPSON, J. S., FOX, J. B. AND LANDMANN, W. A. (1962). The effect of water and temperature on the deterioration of freeze-dried beef during storage. *Fd Technol.* **16**, 131.

THORNE, A. L. C. (1953). Recovery of caprinized and lapinized rinderpest viruses from condensed water vapour removed during desiccation. *Nature, Lond.* **171**, 609.

THUSE, E., GINNETTE, L. F. AND DERBY, R. (1968). U.S. Pat. No. 3,362,835.

TISSOT, A. S. (1792). Avis au peuple sur sa santé, Vol. **2**, p. 253. Grasset, Lausanne.

TREFFENBERG, L. (1953). A method of freeze-drying of histological preparations. *Ark. Zool.* **4**, 295.

TRIEBES, T. A. AND KING, C. J. (1966). Factors influencing the rate of heat conduction in freeze drying. *I. and E.C. Process Design and Development*, **5**, 430.

TRIGGS, W. W. (1944). Brit. Pat. No. 586,693.

TSCHUDIN, K. (1946). Rate of vaporization of ice. *Helv. Phys. Acta*, **19**, 91.

VANSTEENBERGHE, P. (1903) Procédé de conservation du virus rabique à l'état sec. *C.r. Séanc. Soc. Biol.* **55**, 1646.

VINCENT, R. S. AND SIMMONS, A. (1940). Sensitive method for the determination of water applied to electrical insulating materials. *Proc. phys. Soc.* **52**, 489.

VUILLARD, G. (1957). Contribution à l'étude vitreux et de la cristallisation des solutions aqueuses. Thèse, Masson et Cie, éd., Paris.

WISMER-PEDERSEN, J. (1965). Effect of EDTA and pH on properties of freeze-dried pork muscle. I. Effect of pH and magnesium and calcium ions on freeze-dried myofibrils. *J. Fd Sci.* **30**, 85. II. Effect of injection of EDTA and NaOH before drying. *J. Fd Sci.* **30**, 91.

WOLLASTON, W. H. (1813). On a method of freezing at a distance. *Trans. R. Soc.* **103**, 71.

WOODWARD, H. T. (1961). Study of vapour removal systems in dehydration of food products having piece or block conformation. *Quartermaster Contract Rep.* DA 19-129-QM-1597.

WOODWARD, H. T. (1963). Freeze-drying without vacuum. *Fd Engng*, **35**, 96.

WYCHOFF, R. W. G. (1946). Frozen-dried preparations for the electron microscope. *Science*, **104**, 36.
YOUNG, F. E. AND JONES, F. T. (1949). The sucrose–water phase diagram. *J. Phys. Coll. Chem.* **53**, 1334.
ZAMZOW, W. H. AND MARSHALL, W. R. (1952). Freeze drying with radiant energy. *Chem. Engng Progr.* **48**, 21.

Additional Research References

AHLRICH, J. L. AND WHITE, J. L. (1961). Freezing and lyophilizing alters the structure of bentonite gels. *Science*, **136**, 1116.

AMOIGNON, J. (1972). Contribution à l'etude de la contamination physico-chimique des produits lyophilisés. *Nouveautés en Lyophilisation.* Centre de Perfectionnement Technique, Paris (Juin 1972).

ANG, T. K., FORD, J. D. AND PEI, D. C. T. (1977). Microwave freeze-drying of food: a theoretical investigation. *Int. J. Heat Mass Transf.* **20**, 517.

AUERBACH, E., WANG, H., MAYNARD, N., DOTY. D. M. AND KRAYBILL, H. R. (1954). A histological and histochemical study of beef dehydration. V.—Some of the factors influencing the rehydration level of frozen-dried muscle tissue. *Fd Res.* **19**, 557.

BARRETT, A. S. D. AND BECKETT, L. G. (1951). Aspects of the design of freeze-drying apparatus. Freezing and Drying. Symposium Report, p. 41. Institute of Biology, London.

BARRETT, J. R., LAXON, R. AND WEBSTER, P. H. N. (1964). Freeze-drying vacuum equipment—Economics of steam ejector and refrigerated systems. *Fd Technol.* **18**, 38.

BERLIN, E., KLIMAN, P. G. AND PALLANSCH, M. J. (1966). Surface areas and densities of freeze-dried foods. *J. agric. Fd Chem.* **14**, 15.

BRADISH, C. J. (1947). Freeze-drying. *Chem. Prod.* **10**, 60.

BRUCE, E. W. (1950–52). Sublimation of ice *in vacuo*. *J. Imp. Coll. Chem. Engng Soc.* **6**, 39.

BRYNKO, C., AND SMITHIES, W. R. (1956). Meat dehydration by rapid vacuum freeze-drying. *Food in Canada*, **16**, 25.

COPSON, D. A. (1958). Microwave sublimation of foods. *Fd Technol.* **12**, 270.

DAUVOIS, PH. AND BOUTEIL, R. (1973). Transferts de chaleur, de masse et de quantités de mouvement en congélation et lyophilisation en continu de particules sphériques de café en lit agité. *Proc. 13th Int. Congr. Refrig.* **2**, 337.

DELONCA, H. (1974). Apports de la lyophilisation dans l'industrie pharmaceutique *Journées d'Etude sur la Lyophilisation*. Centre Perfectionnement Technique, Paris.

DUNOYER, L. (1951). The theoretical factors of vacuum drying. *Vide*, **6**, 1025; **6**. 1077.

EDE, A. J. (1947). The use of the McLeod gauge with mixture of gas and vapour. *J. scient. Instrum.* **24**, 198.

ERMAKOVA, E. A. (1964). Some results of the experimental study of ice evaporation in vacuum. *Inzh. Fiz. Zh.* **7**, 37.

GAC, A., LETANG, G. AND POTTIER, M. (1974). Etude du piégeage de la vapeur d'eau en lyophilisation. *Bull. Inst. int. Froid*, Annexe 1974-3, 149.

GHEORGHIU, M. (1968). Theoretical principles of freeze-drying. *Microbiologia Parazit. Epidem.* **13**, 27.

GOLDBLITH, S. A., REY, L. AND ROTHMAYR, W. W. (eds) (1975). "Freeze Drying and Advanced Food Technolology", p. 730. Academic Press, London and New York.

GREENFIELD, P. F. (1973). An example of simulation applied to food engineering. *Food Tech. in Aust.* **25**, 502.

GREENFIELD, P.. F. (1974). Cyclic-pressure freeze-drying. *Chem. Engng Sci.* **29**, 2115.

GREIFF, D. AND ROWE, T. W. G. (1976). Recent advances and applications of freeze-drying technology. *Adv. cryogen. Engng*, **21**, 418.

GUIGO, E. (1952). Drying by sublimation. *Molochnaya Pnom.* **13**, no. 4, 32.

GOUIGO, E. I., KAUKHCHESHVILI, E. I., SAVCHENKO, A. F. AND SIDOROVA, N. D. (1974). Lyophilisation sous pression atmospherique des matières premières d'origine animale. *Bull. Inst. int. Froid,* Annexe 1974-3, 137.

GOUIGO, E. I. AND TSVETKOV, TS. (1974). Mécanisme du passage des gaz raréfiés dans les matériaux au cours de la lyophilisation. *Bull. Inst. int. Froid,* Annexe 1974-3, 141.

HAMM, R. (1963). Freeze drying of meat. Chemical and physical changes during freeze drying. *Fleischwirtschaft,* **15**, 201.

HAUSS, R. (1964). Chemical and physical changes during freeze-drying of meat. *Fleischwirtschaft,* **15**, 124.

HUMUR, R. (1963). Chemische und physikalische Veränderungen bei der Gefriertrocknung von Fleisch. *Fleischwirtschaft,* **15**, 201.

KAUKHCHESHVILI, E. I., JOURAVSKA, N. K., PALMINE, I. V. AND POTIPAEVA, N. N. (1974). Dessiccation de viande congelée sous vide dans un liquide caloporteur. *Bull. Inst. int. Froid*, Annexe 1974-3, 131.

KEARSEY, H. S., STRICKLAND-CONSTABLE, R. F. AND BRUCE, E. W. (1953–54). Freeze-drying. 2—A study of the factors governing the rate of freeze-drying of blood plasma. *Proc. Inst. Refrig.* **50**, 69.

KEPPELER, R. A. AND COWART. D. G. (1972). Theoretical properties of freeze-dried mushrooms. *J. Fd Sci.* **37**, 205.

KING, C. T. (1971). "Freeze-Drying of Foodstuffs". Chemical Rubber Co., Cleveland.

KOSTAROPOULOS, A. E., WOLF, W. AND SPIESS, W. E. L. (1975). Modelle für die theoretische Bestimmung der Wärmeleitzahl inhomogener poröser Stoffe. *Lebensm.-Wiss u-Technol.* **8**, 177.

LAMBERT, J. B. (1956). Heat and mass transfer in freeze-drying. *Diss. Abstr.* **16**, 2116.

LARSON, R. W., STEINBERG, M. P. AND NELSON, A. I. (1967). Effect of gas flushing at several pressures on rate of freeze-drying of beef. *Fd Technol.* **21**, 401.

LUIKOV, A. V. AND LEBEDEV, D. P. (1973). Study of the ice sublimation process. *Int. J. Heat Mass Transf.* **16**, 1087.

LUIKOV, A. V. AND VASILIEV, L. L. (1969): Heat and mass transfer of capillary porous bodies in a rarefied gas flow. *Bull int. Inst. Refrig.* Annexe 1969-4, 47.

MACKENZIE, A. P. (1965). Factors affecting the mechanism of transformation of ice into water vapour in the freeze-drying process. *Ann. N.Y. Acad. Sci.* **125**, 522.

MEHRLICH, F. P. (1960). Novel development can lower freeze-drying costs. *In* "Freeze-Dehydration of Foods" (F. R. Fisher, ed.), p. 43. Research and Development Associates Food and Container Institute.

MERYMAN, H. T. (1963). U.S. Pat. No. 3,096,163.

MERYMAN, H. T. (ed.) (1966). "Cryobiology", p. 775. Academic Press, London and New York.

METTIVIER MEYER, J. C. (1962). Raumbedarf und Wasseraufnahme von gefriergetrockneten Gemüsen. 5. Gefriertrocknungstagung in Köln, Leybold-Hochvakuum—Anlagen GmbH, Köln-Bayenthal.

NERGE, W. (1966): Controlled multi-pressure drying apparatus. U.S. Pat. No. 3,234,661.

NEI, T., SOUZU, H. AND HANAFUSA, N. (1963). Mechanism of drying during freeze-drying. X-Effect of the rate of cooling (freezing) upon the rate of drying and the residual moisture content of specimens. *Low Temp. Sci.* **B21**, 71.

OBAYASHI, Y. AND CHO, C. (1957). Further studies on the adjuvant for dried BCG vaccine. *Bull. Wld Hlth Org.* **17**, 255.

PASSEY, A. D. AND MOREAU, J. R. (1973). Quasi-steady state mathematical model for a spherical freeze-drying particle. *J. Can. Inst. Fd Sci. Technol.* **6**, no. 4, 270.

PASSEY, A. D., RIELAND, R. R. AND BOULET, M. (1973). Mathematical model for microwave freeze-drying of spherical particles. *Proc. 13th Int. Congr. Refrig.* **2**, 347.

PENNY, I. F., VOYLE, C. A. AND LAWRIE, R. A. (1963). A comparison of freeze-dried beef muscle of high or low ultimate pH. *J. Sci. Fd Agric.* **14**, 535.

REY, L. (éd.) (1960). "Traité de la Lyophilisation". Hermann, Paris.

REY, L. (éd.) (1962). "Progrès Récents en Lyophilisation". Hermann, Paris.

REY, L. (ed.) (1966). "Advances in Freeze-Drying". Hermann, Paris.

ROBSON, E. M. AND ROWE, T. W. G. (1960). The physics of secondary drying. *In* "Recent Research in Freezing and Drying" (A. S. Parkes and A. U. Smith, eds), p. 146. Blackwell Scientific Publications, Oxford.

ROSS, T. K. AND RUSHTON, A. (1961). The sublimation of granular solids. Internat. Dev. Heat Transf. *Proc. 1961 Internat. Heat Transf. Conf.* Pat IV, p. 707.

ROWE, T. W. G. (1960). The theory and practice of freeze-drying. *Ann. N.Y. Acad. Sci.* **85**, 641.

ROWE, T. W. G. (1962). Water vapour removal in food freeze-drying. *Vide*, **17**, 516.

SIMATOS, D. (1964). L'eau et les formes de liaison de l'eau dans les produits lyophilisés. *In* "Aspects Théoriques et Industriels de la Lyophilisation" (L. Rey, éd.), p. 235. Hermann, Paris.

SIMATOS, D., BLOND, G., DAUVOIS, PH. AND SAUVAGEOT, F. (1974). "La Lyophilisation Principes et Applications". Collection de l'A.N.R.T. Association Nationale de la Recherche Technique, Paris.

SMITH, A. U. (ed.) (1970). "Current Trends in Cryobiology". Plenum Press, New York.

SOLÉ, P. AND SPIESS, W. E. L. (1969). Heat transfer in the freeze-drying of model substances. *Verfahrenstechnik*, **3**, 340.

SOLÉ, P. AND SPIESS, W. E. L. (1971). Computer prediction of heat transport controlled freeze-drying. *Revista del Colegio de Quimicos e Ingenieros Quimicos de Costa Rica*, **1**, 79.

SPIESS, W. E. L. AND WIEN, K. J. (1975). Possibilities of direct surface sublimation in freeze drying. *In* "Freeze Drying and Advanced Food Technology" (S. A. Goldblith, L. Rey and W. W. Rothmayr, eds), p. 421. Academic Press, London and New York.

SPIESS, W. E. L., WOLF, W., BUTTMI, W. AND JUNG, G. (1974). A new concept for the freeze-concentration process. *In* "Advances in Preconcentration and Dehydration of Foods" (A. Spicer, ed.), p. 195. Applied Science Publishers, London.

STEPHENSON, J. L. (1960). Ice crystal formation in biological materials during rapid freezing. *Ann. N.Y. Acad. Sci.* **85**, 535

TODOROVIĈ, M. (1977). The temperature and pressure field in freeze-drying of one-dimensional body. *Bull. int. Inst. Froid*, Annexe 1977-4.

Tuma, V. (1974). Drying time calculation for ice granules in a plate freeze-dryer. *Bull. int. Inst. Froid*, Annexe 1974-3, 159.

Wang, H., Auerbach, E., Bates, V., Doty, D. M. and Kraybill, H. R. (1954). A histological and histochemical study of beef dehydration. IV.—Characteristics of muscle tissues dehydrated by freeze-drying techniques. *Fd Res.* **19**, 543.

Ward, K. (1961). Accelerated freeze-drying. 2—Fundamental design problems *Fd Manuf.* **36**, 60.

Author Index

Numbers in italics indicate the pages on which the references are listed.

Subject Index